Swati Sharma

Carbon for Micro and Nano Devices

Also of Interest

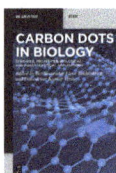

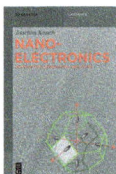

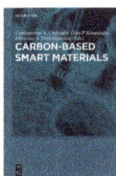

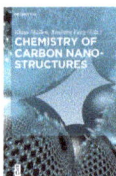

Swati Sharma

Carbon for Micro and Nano Devices

—

DE GRUYTER

Author
Prof. Swati Sharma
Indian Institute of Technology Mandi
School of Mechanical and Materials Engineering
A4-207
175005 Kamand
Himachal Pradesh India
swati@iitmandi.ac.in

ISBN 978-3-11-062062-7
e-ISBN (PDF) 978-3-11-062063-4
e-ISBN (EPUB) 978-3-11-062070-2

Library of Congress Control Number: 2023948762

Bibliographic information published by the Deutsche Nationalbibliothek
The Deutsche Nationalbibliothek lists this publication in the Deutsche Nationalbibliografie;
detailed bibliographic data are available on the Internet at http://dnb.dnb.de.

© 2024 Walter de Gruyter GmbH, Berlin/Boston
Cover image: Micro-scale Carbon Patterns, by Dr. Swati Sharma
Typesetting: VTeX UAB, Lithuania
Printing and binding: CPI books GmbH, Leck

www.degruyter.com

Preface

We know so much about carbon and yet so little. Most crystalline solids can only exist in their three-dimensional forms. But carbon has proven its stability in both three and two dimensions as graphite and graphene. There are reports on even one-dimensional carbon allotropes, known as carbynes, that are yet to find technological applications. Additionally, carbon manifests itself in complicated curved geometries that can co-exist with the layer-plane systems and results in physically amorphous yet electrically conductive disordered solids. Despite decades of investigations and widespread applications, the exact microstructure of many such carbon materials remains unfathomed. In parallel to carbon science, carbon technology continues to grow. The processing techniques of carbon materials keep evolving. Consequently, new and revived literature on this fascinating element is constantly needed.

The most significant technological advancement of the last few decades has been the realization of micro- and nano-scale devices. These miniaturized but high-performance devices demand much less material compared to their larger counterparts. A number of energy generation and storage devices, wearable sensors, bioelectronic medicine, human-machine interfaces and smart biomedical tools have entirely changed the way technology is perceived in today's scientific and commercial scenario. Initially micro- and nanofabrication was limited to silicon and noble metals. But carbon's utility was soon recognized when miniaturized electrodes became exceedingly popular. These electrodes were expected to operate in a wide range of electrochemical environments, where the optimum combination of electrical, electrochemical, thermal and mechanical properties of carbon proved to be of tremendous advantage. Over the years carbon has not only replaced the traditional materials in several devices, certain applications are only made possible due to carbon.

Carbon nanomaterials as well as disordered solids such as glass-like and porous carbon are used in the manufacturing of micro- and nano-scale devices. In order to select the most suitable carbon material for a given application, one needs to compare different forms of carbon, evaluate their pros and cons, and be aware of the trade-offs. This is only possible by understanding carbon's hybridization, elemental network formation, electron transport and molecular energy distribution. In addition, one needs to be well acquainted with the micro- and nanofabrication technique compatible with a specific carbon form.

Keeping these aspects in mind, this book project was initiated. The main features include (i) an explanation of carbon at atomic and molecular level that leads to the formation of different allotropes, (ii) hybridization in curved carbon materials, (iii) manufacturing processes specific to carbon-based micro/ nanofabrication, (iv) synthesis and characterization of various carbon nanomaterials, and (v) a comparison of different carbon materials for device applications. In addition, some latest carbon forms such as laser-patterned carbon, developed for flexible carbon electronics are also included.

https://doi.org/10.1515/9783110620634-201

One of the most interesting aspects of this book is the hybridization based structure-property correlation in curved and non-graphitizing carbons. Microstructural models of non-graphitizing carbons are provided and discussed in detail.

This idea of this book was conceived when I was working as a Scientist at the Karlsruhe Institute of Technology, Germany. In 2019, I moved to the Indian Institute of Technology Mandi. Students and scholars who worked (and are working) with me in both of these institutions have contributed indirectly to this book, as the results from their research work have enhanced my knowledge of the subject. There are also some direct contributions, in terms of drawing figures and tables from Mamta Devi, Jyoti Shikhar, Ashish Jaswal, Bhavika Chouhan and Aditi Bodhankar. With my parents Sushma and Shivkumar Upadhyay, and sister Shailee Upadhyay, I often talked about the broader impact of this book and carbon technology in general. I am sincerely thankful to all. Special thanks to my life-partner Erwin Fuhrer with whom I discussed the fundamentals of sensors, suitability of carbon for biomedical devices, and various technical aspects of LaTeX.

For the interested audience, some of my video lectures related to this book are available freely via the National Programme on Technology Enhanced Learning (NPTEL), an initiative of the Government of India. I hope this book will serve its purpose and the readers will enjoy the contents.

Contents

Preface —— V

1   **Micro- and nanodevice technology** —— **1**
1.1   Background —— **1**
1.1.1   Need for micro/nano devices —— 2
1.2   Micro- and nanofabrication: basic concepts —— 3
1.2.1   The idea of scale —— 3
1.2.2   Important definitions —— 5
1.2.3   Top-down and bottom-up manufacturing —— 9
1.3   Common micro- and nanoscale devices —— 10
1.3.1   Miniaturized battery —— 12
1.3.2   Supercapacitors —— 14
1.3.3   Fuel cells —— 17
1.3.4   Sensor —— 20
1.3.5   MEMS and NEMS —— 24
1.3.6   Biomedical devices —— 25
1.3.7   Function-specific performance evaluation —— 27

2   **Carbon materials for devices** —— **29**
2.1   Carbon and silicon —— **29**
2.2   Carbon materials in miniaturized devices —— **31**
2.3   Materials and their properties —— **33**
2.3.1   Isotropic and anisotropic materials —— **35**
2.4   Atomic structure of carbon and hybridization —— **36**
2.5   Allotropes of carbon —— **43**
2.5.1   Classification of carbon allotropes —— **45**
2.5.2   Composition diagram —— **46**
2.5.3   Allotrope conversion —— **47**
2.6   Phase diagram —— **49**
2.7   Comparison of different forms of carbon —— **52**

3   **Chip-based carbon devices** —— **56**
3.1   Introduction —— **56**
3.2   Fabrication techniques for chip-based device —— **57**
3.2.1   Photolithography —— **58**
3.2.2   Nano-imprint lithography —— **62**
3.2.3   X-ray lithography —— **63**
3.2.4   Two-photon lithography —— **66**
3.2.5   Support techniques —— **67**
3.3   Conversion of polymer patterns into carbon —— **70**

3.3.1   Pyrolysis —— 71
3.3.2   Carbonization and graphitization —— 78
3.3.3   Material characterization during heat-treatment —— 78
3.4   Polymer-derived carbon: Physicochemical aspects —— 79
3.4.1   Polymer-to-carbon conversion in nature —— 80
3.4.2   Early technological applications of synthetic carbon —— 82
3.4.3   Classification of polymer-derived carbon —— 86
3.5   Applications and example devices —— 98
3.5.1   Energy storage —— 99
3.5.2   Biosensing and cell culture —— 100
3.5.3   Topography sensing —— 101
3.5.4   Other devices —— 102
3.6   Current limitations and future trends —— 103

4   Carbon nanomaterial-based devices —— 107
4.1   Introduction to carbon nanomaterials —— 108
4.1.1   Formation and stability —— 108
4.1.2   Classification —— 110
4.1.3   Graphene —— 112
4.1.4   Carbon nanotube (CNT) —— 117
4.1.5   Carbon nanofiber —— 120
4.1.6   Buckminsterfullerene —— 121
4.1.7   Carbon black —— 122
4.2   Synthesis of carbon nanomaterials —— 125
4.2.1   Chemical vapor deposition (CVD) —— 125
4.2.2   Exfoliation of graphite —— 134
4.2.3   Epitaxial growth —— 136
4.2.4   Arc evaporation and laser irradiation of graphite electrodes —— 137
4.2.5   Manufacturing process of carbon black —— 139
4.3   Patterning of carbon nanomaterials for devices —— 140
4.3.1   Printing —— 140
4.3.2   Spray coating —— 151
4.3.3   Direct growth —— 151
4.3.4   Pattern transfer —— 152
4.4   Applications of carbon nanomaterials —— 152
4.5   Future prospects —— 154
4.5.1   Health and environmental safety of carbon nanomaterials —— 155

5   Carbon fiber based devices —— 157
5.1   Introduction —— 157
5.2   Spinning techniques —— 160
5.2.1   Electrospinning —— 161

5.2.2	Melt-spinning —— **166**	
5.2.3	Other spinning techniques —— **167**	
5.3	Carbon fiber precursors —— **167**	
5.3.1	Polyacrylonitrile (PAN) —— **168**	
5.3.2	Pitch —— **171**	
5.3.3	Rayon and natural fibers —— **173**	
5.3.4	Lignin —— **174**	
5.4	Precursor requirements —— **175**	
5.4.1	Viscoelasticity —— **175**	
5.4.2	Carbonizability —— **177**	
5.5	Properties of carbon fibers —— **177**	
5.5.1	Electrical conductivity —— **177**	
5.5.2	Mechanical properties —— **178**	
5.5.3	Thermal conductivity —— **181**	
5.6	Classification of carbon fiber —— **181**	
5.6.1	Classification based on manufacturing process —— **181**	
5.6.2	Classification based on microstructure —— **182**	
5.6.3	Classification based on physicochemical properties —— **183**	
5.7	Carbon fibers in devices —— **185**	
5.7.1	Tip-sensitive carbon fiber microelectrodes —— **186**	
5.7.2	Surface-sensitive carbon fiber electrodes —— **188**	
5.7.3	Direct use of carbon fiber mats —— **191**	
5.8	Advances in carbon fibers —— **191**	

6	**Flexible and futuristic carbon devices —— 195**	
6.1	Flexible electronics —— **195**	
6.1.1	Flexible substrates —— **196**	
6.2	Manufacturing of carbon-based flexible devices —— **197**	
6.2.1	Pattern transfer —— **198**	
6.2.2	Laser-patterned carbon —— **202**	
6.3	Heteroatom-containing carbon —— **210**	
6.3.1	N-containing carbon —— **211**	
6.3.2	S-containing carbon —— **214**	
6.4	3D printing of carbon —— **215**	
6.5	Waste-derived carbon —— **217**	
6.6	Future of carbon technology —— **220**	
6.6.1	Interdisciplinary approach —— **221**	
6.6.2	Biodegradable electronics —— **221**	
6.6.3	Low-cost and sustainable manufacturing —— **221**	

A   Additional questions (all chapters) —— 225

Bibliography —— 227

Index —— 241

1 Micro- and nanodevice technology

1.1 Background

A device is defined as any technological invention that has a specific use or function. When the primary functional element of a device is ≤100 µm or 100 nm in dimensions, the device is known as micro or nano device, respectively. Most of the manufacturing processes employed at micro- and nanometer length scales are different from the large-scale or the so-called traditional manufacturing processes. At such small length scales, the inherent properties of the material become dominant, defects and impurities play important roles, and unconventional manufacturing tools need to be used. As we move from µm to nm, bottom-up manufacturing processes, such as self-assembly, become more viable compared to top-down techniques.

Though micro- and nanoscale fabrication is significantly different from the traditional metal and ceramic processing, certain techniques, such as drilling, have been scaled down to operate in the micro-meter regime as well. Scaling-down does not work in the nanoscale, but certain fundamental concepts and basic definitions can very well be translated from large-scale to the micro- and nanoscale as well. It is clear that for understanding the design and fabrication of miniature devices, one needs to not only understand the fundamental scientific principles applicable at these length scales, but also the properties of specific materials that are used for device fabrication. The goal of this book is to describe the use of carbon materials in micro and nano device manufacturing. Often, the material itself is synthesized during device fabrication when it comes to carbon.

Carbon is an element, but it cannot be treated as a single or unique material. There are numerous allotropes (different forms) of carbon, each with its own set of properties. Some forms of carbon are synthesized in the form of nanomaterials, whereas some are bulk crystalline or amorphous solids that are reduced to very small scales. Unlike other materials, different forms of carbon may be compatible with entirely different manufacturing process. Most of the carbon forms used in device applications are synthetic. It means that they are produced from a precursor material, typically at high temperatures. In many instances, this carbon production is carried out during the preparation of the pattern, which is ultimately used in the device. The fabrication conditions as well as precursors determine the type of carbon one would obtain. There are certain similarities among all carbon materials, such as their light weight, low chemical reactivity (unless intentionally activated), and thermal stability in inert environment. However, many other properties, including electrical conductivity, surface area, electrochemical stability and thermal conductivity, etc., differ from allotrope to allotrope. Note that diamond and graphite are both carbon materials, but their properties are significantly different. Many other carbon materials, which fall in between diamond in graphite in terms of their net hybridization or crystal structure, exhibit properties accordingly.

https://doi.org/10.1515/9783110620634-001

Nanoscale carbon materials have their own unique features. Bulk and individual properties of nanodiamonds and nanographites are not the same as bulk forms of these crystalline solids. Needless to say, carbon-based micro- and nanofabrication is unique combination of material development and processing. The available literature on this topic is often dedicated to either material development or device manufacturing. But unlike other elements, in the case carbon it is essential to study material and manufacturing aspects simultaneously. The purpose of this book is to provide the reader with a single source of all micro- and nanofabrication concepts specific to carbon, along with the fundamental science of carbon materials.

1.1.1 Need for micro/nano devices

A natural question is, what is the need for such small devices when a millimeter or centimeter scale device can perform a similar function. Indeed, a large number of commercially available devices are of the dimensions that can be held in a human hand. Support electronics in a device, for example electrical contact pads, cables or fluidic chips, generally span over a few centimeters. But the idea is to reduce the size of the functional element to micro or nano for the following reasons: (i) a high surface area exposes many active sites in the material, which can more efficiently perform a desired task; (ii) a high surface-to-volume ratio renders the material lightweight and low in density; (iii) material properties can be tuned without changing the elemental composition by simply converting it into a nanomaterial/device; (iv) typically, defects (which are high energy configurations) cannot stay inside extremely small material units, given a higher mechanical strength per unit volume; (v) overall material requirement in micro- or nanoscale devices is much lower compared to their larger counterparts.

Different fields of science and technology benefits from miniaturized devices in different ways. For example, biomedical engineering may benefits from nanomaterial-based contrast agents during medical imaging; new drug carriers based on pre-defined functional materials; micro- and nanofiber-based tissue growth templates, and many supporting materials and platforms for *in-vitro* cell culture and testing of various drugs. Chip-based micro-/nanodevices have already established themselves in the field of computing and artificial memory storage. Several semiconductor devices utilize micro- and nanofeatures, and miniaturized transistors have been in use even before the advent of what we call nanotechnology today. In the energy section, high surface area of micro-/nano-devices drastically improve the performance of batteries and supercapacitors. Nanoscale catalysts or enzyme-mimic can be used for highly specialized tasks, for which they have been engineered and synthesized. The material design is also nowadays supported by artificial intelligence, machine learning, and other computational techniques, which can predict the properties of a certain combination of size, shape, and material, thus allowing the scientists to come up with highly specialized functional materials.

Nanomaterials can be used in a solution form, in a lithographically patterns fashion (chip-based), or supported by microscale patterns. For example, a catalyst in the form of a nanomaterial can be supported on a porous carbon material having micrometer-scale porosity so as to enhance the exposure of individual nanounit with the target chemical for an enhanced interaction between them. In some cases, nanomaterials are printed or grown on to a substrate and used as bulk materials. Many other examples, which will be discussed at relevant places in this book, will elucidate how miniaturized materials and devices have taken the current technology to another level of performance in the last couple of decades. In this context, various carbon materials, including nanomaterials, and both micro- and nanoscale fabrication processes will be discussed in detail.

The current chapter is dedicated to introducing the fundamental concepts behind micro- and nanofabrication, along with a brief description of devices in which carbon materials are commonly used. A general idea of device physics and performance parameters is also provided. Chapter 2 covers an introduction to carbon materials, their hybridization-based classification, and some general properties. Chapters 3 to 5 contain details of chip-based, nanomaterial-based and fiber-based carbon devices along with their manufacturing processes, respectively. Some emerging carbon materials and technologies for next-generation electronics are covered in the last chapter.

1.2 Micro- and nanofabrication: basic concepts

Micro- and nanofabrication is the name given to all manufacturing processes that are capable of producing features smaller than 100 µm and 100 nm, respectively. These features can be passive or functional depending upon the application requirements. The branch of engineering that deals with the fabrication, process optimization, and testing of micro-scale features and devices is known as microsystems engineering. At the nanoscale, the fabrication of features utilizes the principles of fundamental science rather than traditional engineering. Preparation of nanoscale features (also known as nanostructures), nanomaterials as well as their applications are collectively studied under this branch nanotechnology. Other terminologies, such as nanoscience, nano-engineering, and nanomanufacturing, are also used, but they are less common.

1.2.1 The idea of scale

Micrometer (10^{-6} m, denoted as µm) and nanometer (10^{-9} m, denoted as nm) are units of distance, which are often used to indicate the size of miniaturized features. As shown in Figure 1.1, micro- and nanometer are three orders of magnitude apart. This is same

10^{-9}	10^{-8}	10^{-7}	10^{-6}	10^{-5}	10^{-4}	10^{-3}	10^{-2}	10^{-1}	10^{0}	10^{1}	10^{2}	10^{3}
1	10	100	1	10	100	1	10	100	1	10	100	1
nm	nm	nm	µm	µm	µm	mm	mm	mm	m	m	m	km

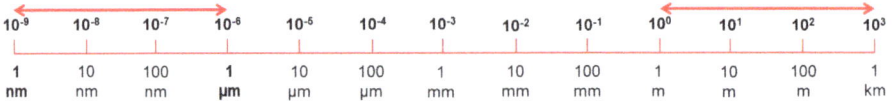

Figure 1.1: Units of distance represented in terms of orders of magnitude (powers of ten). On a logarithmic scale (base 10), the distance between 1 nm and 1 µm would be the same as that between 1 m and 1 km.

as the difference as in the case of a meter and a kilometer. A pattern or feature is considered micro- or nanoscale if at least one of its dimensions is ≤100 µ or nm, respectively. This may not apply to two-dimensional patterns, such as thin-films or coatings, since their thickness is considered negligible compared to the length and width. Here, if the film thickness is ≤100 nm, one could call it a nanofilm/coating. However, since such coatings have been extensively used for hundreds of years (e. g., in electroplating of utensils), it is rather uncommon to consider them a part of nanotechnology unless their nanoscale thickness is responsible for certain specialized property. Although micro and nano sizes are significantly distant, the techniques and materials employed to make miniaturized patterns are somewhat similar. Many such techniques were initially developed to make micrometer scale patterns, and were further refined or modified for submicron features. Subsequently, they were extended to the nanometer regime. Self-assembly or bottom-up processes are more commonly used for nanoscale materials. One must remember that while microscale patterns exhibit properties similar to the bulk material that constitutes them, this is not always valid for nanomaterials.

Similar to large-scale fabrication, different material classes, e. g., metals, polymers, composites, ceramics and semimetals are compatible with different micro- and nanofabrication techniques. For all material types, the chemistry plays an important role in micro- / nanofabrication. For example, polymers are often patterned using their ability to cross-link (form bonds across molecules so as to harden the material). Metals and metal-oxides are deposited from their vapor phase in an atom-by-atom fashion. A range of nanoparticles are chemically synthesized using suitable reactants. Semiconductors such as silicon are processed employing a variety of chemical etching, film growth, and vapor deposition techniques. When in its crystalline form, diamond, is an electrical insulator, whereas glass-like or porous carbons may be good electrical conductors, despite being physically amorphous. They demand unconventional and often nonintuitive fabrication techniques. Carbon nanomaterials are generally synthesized using organic precursors. In many instances, a combination of additive and subtractive techniques are used to obtain carbon having the desired properties, along with the planned shape or pattern. Manufacturing techniques suitable for specific carbon materials are described in chapters 3–6. Here some general definitions pertaining to micro- and nanofabrication are detailed.

1.2.2 Important definitions

1.2.2.1 Material to device

Any manufacturing process begins with a randomly shaped material and finally results in a well-defined pattern. The dimensions, accuracy, and properties of this pattern or shape should be within a certain range or tolerance limit, set prior to material processing. Though the fundamental principles remain the same, there are minor variations when in the context of fabrication of miniature patterns. Some relevant definitions are provided below.

Material: Any substance (element, compound, mixture, alloy, etc.) that is tangible, displays a specific set of properties, and is generally (but not necessarily) used for making objects is called a material. In micro- and nanofabrication, gaseous substances, for example methane, are also used for material synthesis or deposition. These gases are not tangible, but the deposited (or grown) films are. In some cases, a solid substance is first sublimated, and then its vapor form is again used as a precursor to yield another element or compound. Such processes are less common at large scale. Various terms, such as construction, manufacture, fabrication, and synthesis, are used to denote "making." Fabrication is generally associated with soft materials, such as polymers, whereas synthesis indicates the use of chemical processes in manufacturing. In concurrent literature pertaining to micro- / nanofabrication, these terms are used exchangeably.

Pattern, shape, or feature: At micro- and nanoscale, the individual manufactured units are called a pattern, shape, or feature rather than a part. Pattern may also refer to the entire design or array of micro- / nanoscale features. The term "structure" is also frequently used to denote a manufactured shape. However, one should not confuse it with the crystal structure of the underlying material. In this book, the term pattern is most commonly used, but other terminologies may be occasionally present when they are contextually more appropriate.

When an individual micro- / nanofeature or an array of such features is used as a functional unit to performs a certain function, such an assembly is known as micro- / nanodevice. Device is typically a combination of multiple components, which may be larger (e. g., mm or cm-scale). Nonetheless, device can be called micro or nano, as long as the functional element is micro- / nano meter scale. A schematic to illustrate the aforementioned concepts is presented in Figure 1.2.

> A device entails multiple materials, manufacturing processes and packaging techniques. They need to be benchmarked in terms of their output, for example, power generation capabilities. Integration and compatibility of various materials, performance-oriented design and manufacture, information processing, and output management is collectively studied under *device physics*. A detailed study of device physics is more pertinent to semiconductor devices rather than those described in this book.

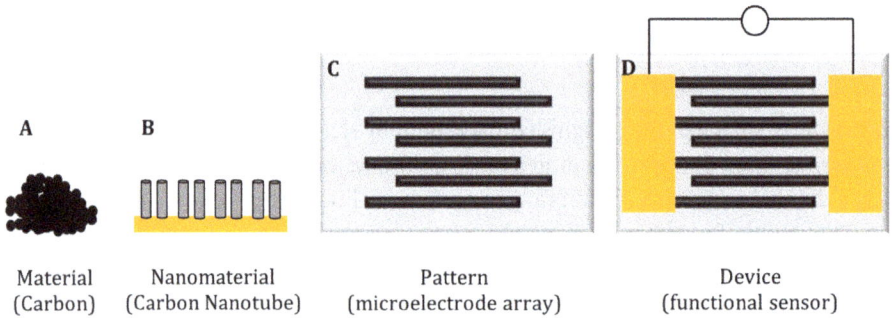

Figure 1.2: Illustration showing (A) material (bulk substance), (B) nanomaterial (collection of nano-scale structures), (C) structure (patterned material or nanomaterial), and (D) device (functional structure). Color code: Black: Carbon; Yellow: Metal.

1.2.2.2 Nanomaterial

A nanomaterial is a collection of well-defined nanoscale units of any material called nanostructures. These individual units, that are responsible for the overall material properties, are not bonded to each other by any strong chemical bonds. They may, however, be agglomerated or joined together via weak Van der Waals forces. Properties of nanomaterials are usually different from the bulk solid made of the same constituents owing to the differences in surface energy, defect distribution, and in some cases, quantum effects. Both static and dynamic behavior of materials exhibits size and shape dependence when it is in the form of a nanomaterial.

Although most materials are composed of small (atomic to nanoscale) structural units, they are only designated nanomaterials when these nanoscale structural units are discrete rather than a structural component in a continuous solid. For example, a single flake of graphene is a nanomaterial, but graphene-like layers in a graphite crystal are not considered a nanomaterial. Graphene-like layers indeed attain their special properties only when they are in the form of single or few layer arrangement; not when they a part of a large graphite crystal. Other than flakes, nanomaterials can be in the form of particles, fibers (only the fiber diameter needs to be ≤100 nm), pyramids, polygons, tubes, and other possible shapes. They may be synthesized in liquid media or grown on a solid support known as substrate. Figure 1.2B is a schematic diagram showing nanotubes grown onto a solid substrate similar to trees growing in a forest. Layered materials, such as graphene, can also be grown or deposited in a similar fashion. Particles having their diameters ranging between 10 and 100 nm are known as nanoparticle, whereas ≤10 nm are generally called quantum dots.

1.2.2.3 Microstructure

Microstructure refers to the arrangement of structural units in a material (element, compound or alloy), when observed with a very high magnification. These structural

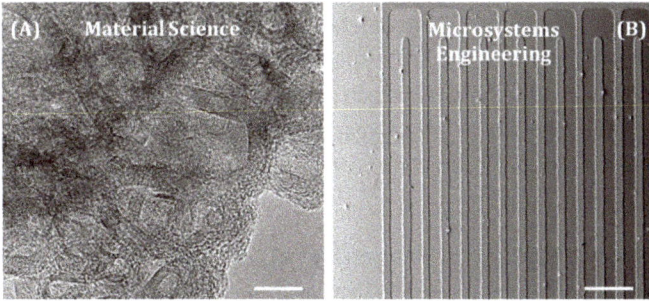

Figure 1.3: Implication of the term microstructure in (A) material science (arrangement at molecular level) and (B) microsystems engineering (a pattern that has at least one dimension in the micrometer scale). Image (A) is a transmission electron micrograph of glassy carbon; scale bar: 20 nm, and (B) is a scanning electron micrograph of a microscale structure made of glassy carbon; scale bar: 100 μm.

units encompass different phases (crystalline, amorphous), defects, and other plausible arrangement of atoms or molecules owing to residual strain. Experimentally, microstructure can be revealed by recording high-resolution images of a material through an electron microscope (see Figure 1.3(A)). Traditionally, microstructure was observed using optical microscopes at 25× or higher magnifications. With the enhanced capacity of electron microscopes today, it has become possible to magnify the images up to 2 million times. With this, one can indeed observe a few molecules down to 0.05 nm resolution. Nonetheless, any conclusion regarding the microstructure of a material is only drawn when a sufficiently large area, which can be considered a true representative of the overall material structure, is taken into consideration. This characteristic is more useful for understanding the properties of bulk continuous solids, rather than nanomaterials. In microfabrication, sometimes microscale features (see Figure 1.3(B)) are also called microstructures (or in one word: microstructure). The term structure here is used to indicate a manufactured or patterned shape, which is a common practice in the discipline of manufacturing. One must pay attention to the intended use of the terminology for specific cases.

Microstructural investigations are an essential part of carbon material development. Unlike other materials, carbon exhibits a wide range of microstructures—from entirely crystalline to highly disordered. Some carbon forms may also contain heteroatoms such as H, which influences the arrangement of crystals and noncrystalline fractions in the material. The fact that carbon already has the microstructure is strongly intertwined with the properties of carbon makes it highly interesting to investigate the microstructure of each one of its various forms. One can often deduce the properties of carbon based on this information. It is of paramount importance to understand what type of defects are present in a given carbon, as the so-called defect of one carbon form can be the constituting units of another. It has been shown that defects in graphene and similar carbon materials improve performance involving electron transport, since the defects tend to expose the edge atoms of graphene. In the case of nongraphitizing carbons, the extent of crystallinity and the layer spacing between carbon sheets are responsible for both electrical conductivity and electrochemical behavior. Several electron micrographs showing the microstructure of a carbon material will be used in this book to elucidate the structure-property correlation of various carbon materials.

1.2.2.4 Manufacturing synonyms at micro- / nanoscale

The term manufacturing refers to making something in large quantities in a reproducible manner within and the given tolerances. Industrial manufacturing is carried out using tested and well-established materials and methods. When a researcher develops a new process or a novel material that can potentially lead to a device, it is not the industrial manufacturing stage. At this stage, several trials and errors and process optimization take place. Since micro- and nanomanufacturing are only a few decades old technologies, a large number of them is still at research or laboratory stage. Nonetheless, some robust micro- / nanofabrication have already made it to the commercial scale, and this list is rapidly getting longer.

Although no hard and fast rules are followed regarding the nomenclature of micro- / nanofabrication techniques in the literature, typically there are three synonyms of manufacturing at small scales: fabrication, micromachining, and synthesis. In addition, the terms micro- and nanopatterning are frequently found in the literature. *Fabrication* generally indicates processing of soft materials, such as polymers. Machining of relatively hard materials (e. g. silicon) at micro-scale is called *micromachining*. Chemically assisted preparation of nanomaterials is known as *synthesis*. Vapor deposition of material, liquid-assisted coatings, or generally any manufacturing technique that involves chemical reactions or breakdown for creating new material can be designated a synthesis. Often, a combination of all or some of these pathways is essential for making one device. Altogether, one can simply designate the entire process as micro- / nanofabrication.

1.2.2.5 Nanoscience and nanotechnology

Design, synthesis, characterization, and understanding of the properties of nanomaterials are defined as nanoscience. The goal is to reveal the information about their structure and properties caused by the small size and/or specific shapes. Other than material analysis, investigations related to certain scientific phenomenon occurring at nanoscale is also covered under the domain of nanosciences. Often these nanoscale effects are specific to their respective size, i. e., the behavior of the same material would differ significantly as bulk.

Nanotechnology, on the other hand, is the idea of utilizing the properties of nanomaterials for a certain function, for example, sensing. This entails acquiring a good control over the properties of nanomaterials, and their integration with the larger-scale device features, such as electrode pads, which can be connected to the external circuitry. Therefore, nanotechnology primarily includes manufacture and implementation of nanoscale structures, devices, and systems by controlling their shape and size. Nanoscience is inevitable for nanotechnology applications and vice versa.

1.2.3 Top-down and bottom-up manufacturing

Similar to large-scale manufacturing, material removal processes are known as top-down, whereas material addition (including growth from a precursor) techniques are called bottom-up approaches (see Figure 1.4). Solid carbon materials, such as graphite and glass-like carbon, are more compatible with top-down manufacturing and nano-materials (e. g., graphene, CNTs) are generally more suitable for bottom-up or additive processes.

| Top-down manufacturing | Bottom-up manufacturing | Pyrolytic carbon cone (top-down) | Graphene-resin cone (bottom-up) |

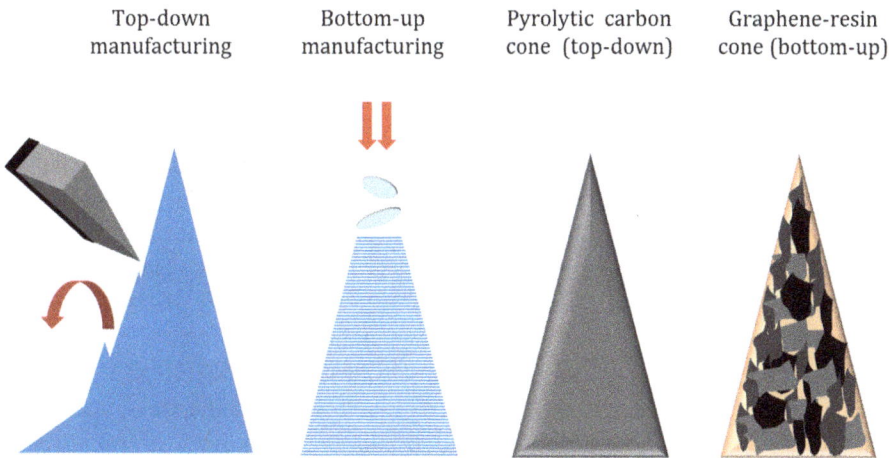

Figure 1.4: A top-down manufacturing approach (left) featuring material removal for making a conical shape. Adding layers of material (right) for obtaining the same shape is an example of bottom-up manufacturing.

Examples of top-down manufacturing at micro- / nanoscale include bulk and surface micromachining, drilling, etching, lithography etc. On the other hand, electroplating, chemical synthesis (e. g., of fullerenes and carbon quantum dots), self-assembly of molecular monolayers, and printing fall within the bottom-up manufacturing approaches. Using laser-assisted cross-linking, it has also become possible to 3D print polymer patterns at micro- / nanoscale, which can be subsequently converted into carbon patterns. As the size of the intended feature approaches the nanoscale, bottom-up processes become more cost effective and convenient. Many device-compatible features are prepared by use multiple processes, which are a combination of both manufacturing pathways. All relevant techniques will be described in each chapter, along with the type of carbon material and intended application.

1.3 Common micro- and nanoscale devices

In this section, a brief introduction to those micro- / nanodevices in which carbon materials are most extensively used. Energy generation and storage devices are on top of this list, both due to the compelling need for inexpensive energy technology and the suitability of carbon for their fabrication.

Production and storage of various conventional and unconventional forms of energy is of paramount interest worldwide. Energy research has two main aspects: energy generation and energy storage. Energy generation is in principle the conversion of one energy form into another (e. g., mechanical to electrical). However, in common terminology generation may be used to indicate that the device can be used as a source for a certain type of energy based on its output. The most common form of energy required for everyday utilities as well as industrial processes is electrical energy. It is traditionally produced using turbines connected to a generator. Turbine blades can be of a few centimeters to tens of meters in size, depending upon the type and magnitude of mechanical forces applied for their rotation. The source of the mechanical forces may come from a stream/jet of water, wind, hot gas, steam, or their combinations. When the turbine is rotated using the force or impact of a water jet, these systems are known as hydroelectric power generation systems. Wind turbines, which have the largest blades among all types of turbine, are rotated by the mechanical forces generated by the natural winds. Gas turbine is similar to a turbo (jet) engine used in the airplanes, where a fossil fuel is burnt to produce a mixture of hot gases that rotate the blades of the turbine at very high speeds. Steam turbines can be further divided into thermal, geothermal, nuclear, and waste (biomass)-derived power. The turbine ultimately rotates due to a flow of steam onto its blades, but the heat used for boiling the water to produce the steam is different. In thermal power plants, this heat source is coal; in geothermal power, it is the natural heat available under the Earth's crust that causes the water to boil; in nuclear power, the energy generated by nuclear fission of uranium or thorium is used as the heat source in the boilers, and finally, in biomass reactors the combustion of waste biomass is used as the heat source. Turbine-based electricity generation using coal, fossil fuels, and water is typically considered a conventional way. The use of wind, nuclear, geothermal, and biomass for power generation is relatively recent, so they can be considered unconventional. Note that water, wind, nuclear reactions, and geothermal energy are renewable energy sources. Biomass is not renewable, but since it enables the utilization of waste, it is also gaining popularity as an unconventional energy source.

There are some energy generation methods that are free of turbine and generator, for example, solar power. In such systems, a driving force is artificially created to induce the flow of electrons in a material, which is utilized as electrical energy. In solar energy generation systems, a n-type semiconductor material is initially activated by photons (energy from the Sun). This change in material property induces a driving force when the n-type semiconductor is in contact with a p-type material. This causes a flow of electrons in a predesigned circuit. It is clear that solar energy generation is strongly dependent on material properties rather than mechanical forces or rotary action. In addition to semiconductors such as silicon, various organic materials with a HOMO-LUMO gap are being developed for solar cell fabrication.

Another turbine-free energy generation method is based on the use of fuel cells. Fuel cells convert chemical energy into electrical via electrochemical routes. The most extensively utilized chemical energy source is hydrogen at present. Fuel cells contain an anode, a cathode, and an electrolyte, and they operate by means of redox reactions. Importantly, both anode and cathode entail catalysts (known as electrocatalysts due to the electrochemical nature of the reaction) that promote the evolution of hydrogen and oxygen, if they are produced *in-situ*. Without these catalysis, the generation of the fuel (hydrogen) itself will become very energy consuming, and the overall process will not be very efficient. One can understand that unlike batteries, fuel cells are primarily energy generation (i. e., conversion) devices. However, they can also be used for energy storage when combined with fuel producing devices, such as electrolyzers [214].

The aforementioned methods are capable of continuous production of electrical energy, but it may not be possible or practical to immediately consume all of it. Whether it is a laptop computer, mobile phone, electric vehicle, or a miniaturized sensor, often consumable forms of energy are only utilized at a time and place different from their generation. Electric vehicles cannot be constantly charged while driving. Solar panels only produce photocurrents during the day time, but the power requirements also exist during the night time. Needless to say, the storage of electricity is essential for the effective utilization of generated electricity. Battery is the most popular example of an electrical energy storage device that can store electrical energy in the form of chemical energy. Other examples include capacitors and supercapacitors, which will be explained below. Other than electrical, heat is another form of energy that can be stored chemically (e. g. in molten salts). Mechanical energy can be stored by designing specialized shapes, the simplest example being a spring. In the context of this book, the primary type of energy storage to be discussed is electrical, which is carried out in devices based on electrochemical processes.

Properties of functional material in devices as well as the characteristics of the device itself are studied under the field of device physics. Device physics is most commonly studied for semiconductor materials and associated devices. Topics studied range from quantum mechanical properties of semiconductor material to various electronic transitions, band theory and mobility of charge carriers, and behavior of the device under different bias conditions. The idea of the physics of carbon devices is different from semiconductors, particularly because each carbon material features a unique set of electronic properties. Additionally, behavior of a single nanoscale carbon unit (e. g., a flake of graphene) is significantly different from its bulk (e. g., graphene printed with a binder on to a substrate). Electronic transitions, density of states, work function, etc., have been theoretically and experimentally studied for graphite and some disordered carbons, but even minor variations in their preparation conditions (e. g., in heat-treatment temperature) can lead to differences in their electronic properties.

Nanoscale features, spread over a few square micrometer or millimeter area exhibit a very high value of usable surface. With their functionalities, they can be used for making high-performance devices for a large number of applications. Even without nanomaterials, simply microscale features in a certain geometric area increase the overall available surface manifolds as compared to a flat (feature-less) material. Given that these properties are useful for several applications areas, the most common ones are described below. Another reason for discussing these device types is the fact that different forms of carbon are playing a major role in the development of these technologies. Note that the description below is only introductory in nature. One could refer to more detailed review articles and books on each topic for further understanding.

Depending upon the type of device, different parameters are used for benchmarking, standardization, and performance testing. In most devices, multiple materials and components are used. First and foremost, a compatibility between these elements needs to be established. This is important at the time of fabrication as well as utilization. There is often either electrical or mechanical or both types of contacts. Even in the case of a perfect physical contact, the electron transport of the two materials can be significantly different. This leads to what is known as a contact resistance. In other words, contact resistance is defined as the electrical resistance caused by the interface between two materials. Work function is the energy required to pull one electron from any material, which, of course, is different for every material. Nonetheless, two materials may have a their work function values very close to each other. In such cases, the contact resistance may be negligible and the contact are called Ohmic. An Ohmic contact is the one that follows the Ohm's law. Interfaces between two different carbon materials can be Ohmic, but it is often hard to form an Ohmic contact between a metal and carbon.

1.3.1 Miniaturized battery

Electrochemical cells and batteries (multiple connected electrochemical cells) observed in the daily life are at least large enough to be held in a human hand. One can also see heavier and larger batteries in cars, electric grids, etc., but as the size of portable electronic devices is decreasing altogether, much smaller batteries with a high energy and power density has become essential. A micro-scale battery consists of active materials on both anode and cathode, along with a liquid or solid electrolyte. For portable electronics, solid phase electrolytes, such as various conductive polymers, are preferred. Often a conductive powder or paste of carbon (e. g., carbon black) is added to the electrode material for increasing its electrical conductivity.

In the case of modern rechargeable batteries, such as Lithium-ion battery (LIB), the anode material is often a conductive form of carbon (graphite, carbon fiber, porous carbons, etc.). In principle it can be replaced by another conductive material, in fact, Li metal itself. Cathode is generally made up of an inorganic layered material, which enables a reversible insertion of Li in and out of it without much change in its own crystal structure. Conductive carbon powder or paste may be added to the cathode material. The electrolyte contains a Li salt. The two electrodes are prevented from short circuit by a separator. Other components of the battery include current collectors, protective casing, etc. During battery charging, Li ions move from cathode to anode, and vice versa. The insertion of Li (i. e., intercalation) also takes place on anode, which is facilitated by the layered structure of graphite. If the ions are slightly larger than Li (e. g., Na), non-graphitizing carbons that feature a larger layer spacing are more suitable. A simplified diagram of a rechargeable LIB having a graphite anode and a layered sulphide cathode is shown in Figure 1.5.

> Carbon materials can be used as both anode and cathode material, depending upon the design and function of the battery. Anode fabrication is more common, particularly in the case of standard Li-ion batteries. High surface area carbons can also be added in the cathode fabrication material to increase its electrical conductivity, or to prevent its agglomeration and dendrite formation.

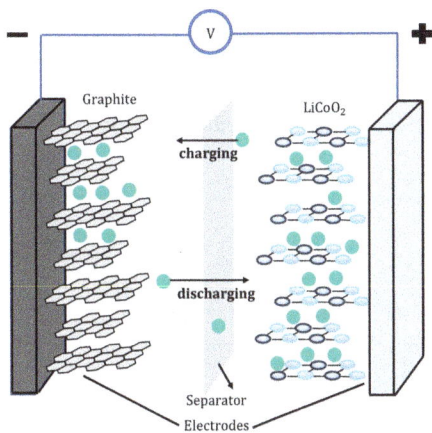

Figure 1.5: A schematic diagram showing Li-ion battery. Copyright ©2013 Royal Society of Chemistry. Reproduced with permission from [85].

1.3.1.1 Important device parameters: battery

Below are some parameters used for evaluating a battery. These must be taken into consideration while designing a battery or selecting an electrode material.

Voltage rating: Voltage rating (volts) indicates the nominal voltage at which the battery is supposed to operate.

Capacity rating: Capacity rating (ampere-hour) alludes to the amount of charge that a battery can deliver at the rated voltage. Generally, the capacity of a battery is proportional to the quantity of the electrode material.

Energy capacity: Energy capacity (watt-hour) is the measure of the total amount of energy that the battery can store. It is obtained by multiplying voltage and capacity ratings of the battery.

C-rate number: C-rate is the rate at which a battery discharges with respect to its capacity. C-rate number indicates the discharge current (current in Ampere at which the battery is being discharged) divided by its nominal capacity (in ampere-hour). The inverse of C-rate number provides the time a battery will require for a completely discharging (in hours).

Efficiency: Battery efficiency (in %) is calculated by taking the ratio of total storage input and the total storage output at system level. Input and output are taken in the units of energy (kWh).

Cycle lifetime: Cycle lifetime of battery, usually stated as an absolute number, is the number of charge/discharge cycles after which the battery capacity drops below 80 % its nominal value.

In addition, a battery's robustness is determined based on its mechanical and thermal stability, its resistance to corrosion due to the growth of any internal chemical species, or caused by external factors, such as humidity. Some of the parameters described above are also influenced by environmental conditions. These effects play an important role in determining the most suitable electrode material for a given environment.

1.3.2 Supercapacitors

Supercapacitors are devices that store charge with a much higher capacitance value compared to standard capacitors. Based on their charge storage mechanism, they are classified into two types: electrical double-layer capacitors (EDLCs) and pseudocapacitors. Some supercapacitors also exhibit a combination of these two mechanisms, and are therefore known as hybrid supercapacitors.

EDLCs store charge electrostatically in the form of an electrical double-layer (formed by positive and negative charges) at the electrode/electrolyte interface. A higher surface area reachable to the electrolyte is the most important characteristic of the electrode material. Many high surface area carbon materials, such as porous and activated carbons, are extensively used for EDLC fabrication. Common models used to describe the EDL phenomenon are Helmholtz, and subsequently developed Gouy–Chapman and

Gouy–Chapman–Stern models. Electrochemical double layers are therefore also known as Helmholtz double layers.

In the case of pseudocapacitors, there is a faradic component in charge storage. This implies that there is a partial charge transfer between the electrode and the electrolyte. This typically includes redox reactions (indicating a highly reversible surface redox system), and intercalation of ions in the material (known as a fast electrolyte ion intercalation). In redox systems, the redox-active electrode material is reduced to lower oxidation state while charging, which is coupled with adsorption of cations from the electrolyte. During discharge, this reaction is reversed. In the case of (diffusion controlled) ion intercalation, the electrolyte ions migrate in and out of the atomic layers of electrode materials during charging and discharging. This is similar to that of Li-ion batteries, but the process is characterized by a much higher reversibility.

Carbon materials are more extensively used in EDLCs, but some may feature partial presudocapacitance. High surface area activated carbons are the most common EDLC electrode materials. Supercapacitors mainly consist of two electrodes: an electrolyte and current collectors. A schematic showing a comparison between an EDLC and a pseudocapacitor is provided in Figure 1.6 [85].

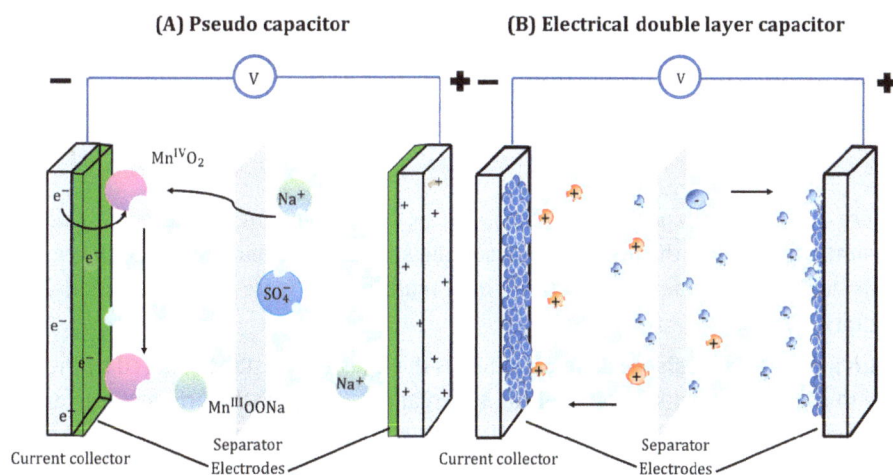

Figure 1.6: A schematic showing a comparison between an EDLC and a pseudocapacitor [85]. Copyright ©2013 Royal Society of Chemistry. Reproduced with permission from [85].

1.3.2.1 Important device parameters: supercapacitors

Total (or cell) capacitance: The total capacitance (C_T, in F) of a supercapacitor expresses the total electrical charge stored in the device at a given voltage change, i. e.,

$$C_T = \frac{\Delta Q}{\Delta V},$$

(1.1)

where ΔQ (in coulomb) and ΔV (in volt) are the stored charge and voltage change, respectively.

Specific capacitance: Specific capacitance (C_s, Fg^{-1}) is the measure of the ability of a supercapacitor device to store electric charge per unit mass. Capacitance may be normalized with respect to mass (gravimetric capacitance), volume (volumetric capacitance), area (areal capacitance), or even the total (cell) weight of the device. Specific capacitance is generally the gravimetric capacitance value, but for research purposes one may vary the normalization parameter. In the case of activated carbon-based commercial supercapacitors, C_s is typically considered to be an between $100\,F/g$ and $70\,F/cm^3$ in organic electrolytes. It is calculated using the cyclic voltammograms as follows:

$$C_s = \frac{A}{2mk\Delta V},$$ (1.2)

where A is the area covered by the CV discharge curve; m is the mass of material in g; k is the scan rate in mV/s, and ΔV is the potential window in V. Different scan rates are used to ensure the wetting of the entire electrode surface. Usually a linear increase in the electrochemically available (wetted) area can be observed with a decrease the in scan rate. Since specific capacitance is strongly dependent on the surface area of the material (typically bulk activated carbon), it is a common practice to carry out surface area analysis using gas adsorption isotherms and compare the results with the cyclic voltametry (CV) curves. The following relationship is used for this purpose:

$$\frac{C}{A} = \frac{\epsilon}{d},$$ (1.3)

where C in F is the capacitance, A (m^2/g) is the surface area of the electrode, ϵ is the permittivity, and d is the distance between the surface of the electrode and the center of the ionic layer (same as the distance between capacitor plates in a conventional capacitor).

Equivalent series resistance or ESR: ESR is the internal resistance offered by the device to charging and discharging of the supercapacitor. This is the primary reason for a reduced efficiency of the device and depends strongly upon both electrode and electrolyte materials.

Other parameters, such as voltage rating, are similar to those of a battery, at least in the commercial available devices. There are several ways of calculating these parameters while evaluating a new material for its supercapacitor performance. Details of all experiments and calculations are not covered here. Readers may refer to literature specific to electrochemical devices [267, 174].

1.3.3 Fuel cells

A fuel cell is an energy conversion device that runs on a continuous supply of hydrogen, ammonia, carbon monoxide, or light hydrocarbons; along with an oxidant such as oxygen or hydrogen peroxide etc. [253]. Hydrogen and oxygen (typically from air) are the most popular fuels in the present-day technology. In a hydrogen-based fuel cell, hydrogen is oxidized at the anode, which results in protons and electrons. On cathode, two phenomena take place: (i) oxygen is reduced to oxides, and (ii) water is formed by combining hydrogen and oxygen. The reaction that takes place at the anode is known as hydrogen oxidation reaction (HOR), and that on cathode is called oxygen reduction reaction (ORR). Catalysts may be used for increasing the rate of both HOR and ORR. There is an electrolyte between the two electrodes. The entire assembly is known as the membrane-electrode assembly (MEA). The MEA allows only ions to traverse through it and reach the cathode, which pushes the electrons to take the path of the external circuitry. This is indeed the power output in the form of Direct Current from the device. MEAs may have an acidic or alkaline pH, which is principally determined by the type of the electrolyte. For example, Nafion is a commonly used acidic electrolyte in the MEAs. Water formation reaction at cathode is exothermic, hence some heat is also generated in this process. Any extra hydrogen is sent back to the fuel tank, and water and heat outlets are connected to the cathode. A schematic representation of a basic fuel cell is shown in Figure 1.7. The chemical reactions [5] occurring at anode and cathode are as follows:

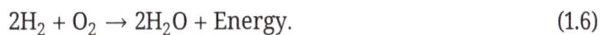

$$2H_2 \rightarrow 4H^+ + 4e^-, \tag{1.4}$$
$$O_2 + 4H^+ + 4e^- \rightarrow 2H_2O, \tag{1.5}$$
$$2H_2 + O_2 \rightarrow 2H_2O + \text{Energy}. \tag{1.6}$$

In terms of design, a fuel cell is similar to battery or other electrochemical devices. It consists of an anode and a cathode along with an electrolyte sandwiched between them.

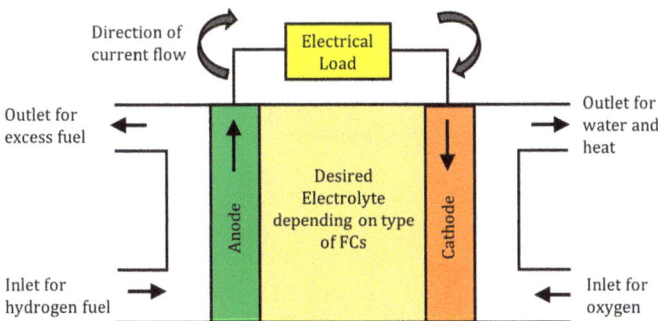

Figure 1.7: A basic schematic showing working principle of a hydrogen fuel cell. Redrawn with minor modifications from [5].

As a result, often a comparison is made with batteries. However, it must be noted that the standard fuel cells neither store electrical energy, nor are they rechargeable. In a fuel cell, a continuous supply of hydrogen (or another fuel) on anode and that of oxygen on cathode is essential for electricity production.

Large-scale fuel cells are used to power electric vehicles. One fuel cell pack is an array of multiple individual cells, which are connected so as to increase the overall voltage output. Hydrogen for these cell-assemblies are stored in high pressure cylinders, placed somewhere in the vehicle. Hydrogen for commercial purposes is largely produced from fossil fuels. Greener alternatives for hydrogen production include electrolysis (splitting of water using electricity), a process known as steam-methane reforming, and gasification of biomass [221]. Cylinders are simply refilled prior to their use in the vehicles. Hydrogen is a highly flammable gas, and special cylinder bodies are required for its storage and transportation. Extensive research is ongoing in this field and indeed lightweight carbon-based composite materials are a preferred choice for making hydrogen cylinders. This topic is beyond the scope of this book.

In high-pressure cylinders, hydrogen is physically stored. To avoid the use of cylinders, various chemical storage methods are being developed. For this, specialized materials, such as metal-organic frameworks are developed, which can hold hydrogen molecules within their structural configuration and release it on demand. Another option is to use electrolyzers, which are basically water-splitting devices. They can theoretically be considered the opposite of a fuel cell. The two main reactions on different electrodes of an electrolyzer are oxygen and hydrogen evolution reactions (known as OER or HER, respectively). Similar to HOR and ORR, electrocatalysts are used for accelerating HER and OER. There are several metal-based electrocatalysis available, but some nitrogen-containing carbon materials have also recently shown electrocatalysis activity for OER [47].

Like other electrochemical devices, high surface area carbon materials are a preferred choice for the fabrication of fuel cell electrodes or as catalyst support. In addition to their high electrochemical stability, in the case of fuel cells their corrosion resistance, under both acidic and alkaline media, is of tremendous value.

> **!** Carbon fiber reinforced polymers (CFRPs) and carbon fiber reinforced carbon are advanced manufacturing materials that can withstand very high loads and pressure. Among these, CFRPs are used for manufacturing pressure vessels, such as those for hydrogen storage. CFRCs, being one of the most expensive manufacturing materials, are more commonly used in high-end aircrafts and parts of spacecrafts that need to operate under extreme P and T conditions.

1.3.3.1 Important device parameters: fuel cell

Unlike other devices, micro- and nanoscale fuel cells are yet to be mass produced, and many of them are in their research phase. Large fuel cells, e. g., those used in electrical vehicles are generally characterized based on their operating temperature, pressure, and relative humidity range. If one is developing electrocatalysts for HER, OER, HRR, or

ORR, their testing requires standard electrochemical characterization via cyclic voltam-metry, impedance spectroscopy, and chronoamperometry [47].

Some additional parameters are briefly described below.

Air stoichiometry: Air stoichiometry is the ratio of supplied oxygen to required oxygen for a stoichiometrically balanced chemical reaction. This parameter is important for proton exchange membrane fuel cells.

Efficiency: Efficiency η of a fuel cell can be calculated thermodynamically as follows:

$$\eta = \frac{nFE_{\text{cell}}}{\Delta H} \times 100, \tag{1.7}$$

where n is number of electrons transferred during the chemical reaction, E_{cell} is the cell potential, and F is Faraday constant.

Sustainability and emission: Sustainability of a fuel cell is measured by the emission of greenhouse gases. In a hydrogen fuel cell, water is the only byproduct. However, if hydrogen is produced using fossil fuels, the generation of greenhouse gases must be factored in.

1.3.3.2 A comparison of different energy storage devices

To compare the mechanism and attributes of battery, capacitor, supercapacitor, and fuel cells, two primary quantities, energy density (Wh/L), and power density (W/L) are con-sidered. Power is the amount of energy (work that can be done) per unit of time. The implication of "density" here is the production or consumption of energy or power per unit mass (or volume). Therefore, energy density is defined as the available energy con-tent per unit mass (or volume) in a device. Power density, on the other hand, is the mea-sure of the rate at which energy can be released from the device. When the calculations are based on per unit mass, they are known as gravimetric, whereas volume-based cal-culations are known as volumetric. Both schemes are used depending upon the type of materials used. Other related terms are specific energy (Wh/kg) and specific power (in W/kg). These gravimetric quantities indicate the energy and power that can be delivered by the device per unit of its mass. A comparison of different energy conversion and stor-age devices in the form of the aforementioned quantities is called a Ragone plot, which is shown in Figure 1.8 [253].

Evidently, fuel cells are high-energy systems; batteries display an intermediate power and energy, and the supercapacitors are high-power systems. Batteries exhibit some overlap in terms of these characteristics with both fuel cells and supercapacitors, which is influenced by their design and the material used for their fabrication. For further details on electrochemical devices, one can refer to several books and research articles, e. g., a review by Winter et al. [253].

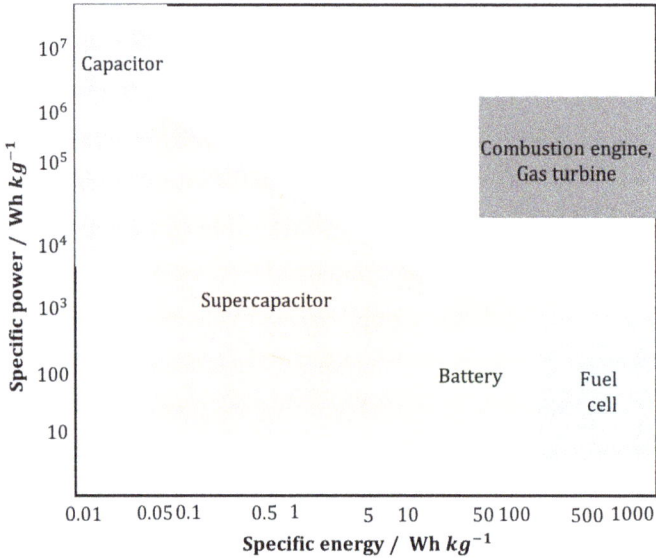

Figure 1.8: A simplified Ragone plot for various electrochemical energy conversion/storage devices. A comparison with conventional internal combustion engines, turbines, and capacitors can also be seen. Reprinted Adapted with permission from [253]. Copyright ©2004 American Chemical Society.

1.3.4 Sensor

Sensors are devices that detect any changes in their surrounding environment, or in other words, any external stimuli. Their detection method is strongly influenced by the material used in the sensing element, because it is the sensing element that changes certain characteristics that are used to measure the response of the device. For example, if there is a wire made of a conducting material that has a certain standard value of electrical resistance. Now this wire is placed in a chamber filled with some gas, which reacts with the wire material. After this reaction, the wire's electrical resistance changes. By observing whether or not there is any change in wire's resistance, the presence or absence of that gas (analyte) can be confirmed. This would be a very simple *yes/no* type sensor. If the quantity of the gas needs to be measured, one would need to fill various known quantities of the analyte gas in a controlled fashion in the chamber and measure the change in resistance. As the concentration of the gas continues to increase, the resistance will continue to change (say, decrease). But after a certain concentration, the chemical reaction between wire and gas may slow down (for example, due to a compound layer formation on wire surface). Now the change in resistance may not be linear, and at some point, there will be no change at all. All of these factors need to be extensively studied, and the detection limits of a senors on both higher and lower side need to be fixed before analyzing an unknown sample of the given analyte. Not only this; if the change in wire's electrical response can be caused by any other chemical present in that environment (real-time sensors will not have the controlled environment like the chamber), that needs to be established as well.

The aforementioned example is simply based on change in electrical response of the sensing element, and it is a chemosensor, because the analyte is a gas, which is a chemical. There are several different types of sensors, including strain sensors, biosensors, touch/tactile sensors, environmental sensors, optical, and electromagnetic radiation sensors etc. Each one of these categories can be further expanded based on the large number analytes that can be detected. There may even be overlaps in different types of sensors. For example, an environment pollutant, such as a dye mixed in drinking water, is actually just a chemical. So the sensor can be called both environmental sensor and chemosensor. There are highly specialized sensors with ppm-level detection limits that have applications in practically every field of modern science and technology.

The functional element of a sensor either detects a change in the system, or changes at least one of its own properties based on a change in the system. This change is then measured in the form an output signal, for example, change in electrical resistance or impedance. Here the term *system* implies the environment within the detection range of the device. *Change* could be an increase or decrease in the concentration of an analyte (chemical, biomolecule, etc.). Change can also be in the form of intensity variation for a certain signal or electromagnetic radiation. The transfer function, which defines the relationship between the output and input signals is plotted for various values of the input. The resulting plot is known as the calibration curve. For example, in a chemosensor, a calibration curve is drawn by plotting the known concentrations of an analyte against the sensor response in an incremental manner. The slope of this calibration curve in the linear response region is known as the sensitivity of the device. In physical sense, it indices how accurate and robust is the method of detection.

Sensors can be classified in a variety of ways. They can be divided into physical, chemical, and biosensors based on their applications. In fact, each of these sensor classes can be further categorized according to their sensing mechanism or the type of response. Let us take the example of biosensors. They encompass a bioreceptor, a transducer for converting the signal to another detectable form, and finally, the signal processing unit for achieving the device output. The sample may be in the form of a biological fluid, such as blood or urine. This sample first interacts with the receptor, which can be an enzyme, antibody, DNA, or even quantum dots or functional nanomaterials. After binding or attaching with the receptor, its quantity can be measured with the help of the transducer. The transducer can be based on electrochemical, colorimetric, optical, thermal, or simply a mass or volume sensitive entity. It converts the required information (e. g. mass) from the bioreceptor in a format detectable by the signal processing unit. At this end, software-assisted processing and readout methods may be implemented. One can classify such sensors either based on the type of receptor, transducer (or sensing mechanism), or output signal.

Sensor can be single or multimodal. A multimodal sensor is capable of detecting multiple analytes simultaneously while maintaining decoupled sensing mechanisms.

Most of the next-generation sensors, such as those used in self-driven vehicles, digital manufacturing units, various health monitoring devices, and even smart phones, are expected to collect, transmit, and process different types of data at the same time. Most of them are flexible in terms of the substrate used, as wearability is one important requirement. At present, the most common practice to fabricate multimodal sensors is to combine multiple sensing units in-plane (tiling), or in the form of layers (laminating) [258]. Carbon nanomaterials and their composites, along with piezoelectric, thermometric, and triboelectric materials are widely used in multimodal sensors.

Carbon materials are useful in sensor device fabrication owing to the following properties: (i) good electrical conductivity, (ii) high electrochemical stability, (iii) chemical and structural stability even under harsh environments, (iv) ability to be functionalized with various chemicals and biochemicals, (v) biocompatibility, and (vi) manufacturability. In this book, the focus will be on the last aspect, where micro- and nanomanufacturing with carbon will be described.

1.3.4.1 Important device parameters: sensor

Signal to noise ratio: Signal to noise ratio for any device is calculated as follows: Refer to Figure 1.9 to facilitate understanding sensitivity and limit of detection. The curve in Figure is plotted between the output signal (y-axis) and the input variable (x-axis). In other words, this is a plot of the transfer function of the device. For example, in the context of a colorimetric chemosensor, the output may be the intensity of the measured wavelength (or the desired color), and the input variable may be the concentration of the analyte. The curve is then known as a calibration curve.

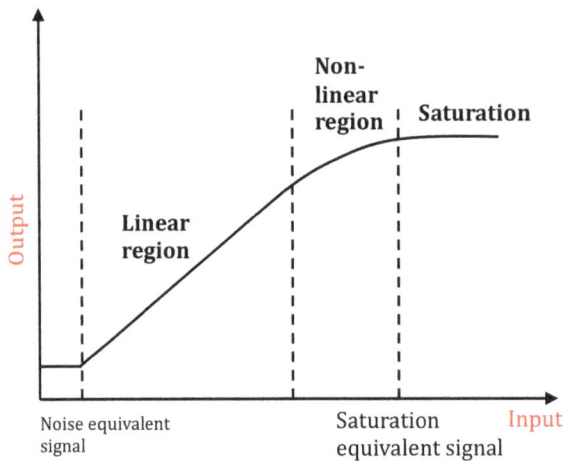

Figure 1.9: A typical calibration or response curve of a sensor device.

As it can be observed, the output increases with an increase in the input up to a certain point. This is called the linear range of the sensor device. If the input parameter/value is further increased, the response may become non-linear owing to various factors, such as saturation of the signal. Nonetheless, if the nonlinearity relationship is known, or the sensor only needs to operate at a fixed input with minor deviations, it can be used in this region as well. Some other device parameters are defined below:

Sensitivity: Sensitivity of a sensor is the measured change in output with respect to per unit change in the input. It is the value of the slope (dy/dx) of the transfer function (or calibration) curve (see Figure 1.9).

Limit of detection: The lowest point of the linear region of the transfer function curve is called the lower limit of detection (or lower limit of quantification). Lower limit of detection is the smallest input signal that can be differentiated from the background or noise. This must be true for at least 99 % readings. In other words, lower limit of detection for a sensor is the smallest input signal that can be distinguished from the noise with a 99 % confidence level. The upper limit of detection, in general, is the upper end of the linear response region of the sensor. However, if the sensor operates in the nonlinear region, the upper limit may be accordingly defined. The entire input range where the sensor can reproducibly function is known as its dynamic range.

Specificity: Specificity represents the ability of a sensor device to only respond to a single target analyte. An idea sensor would strictly detect only the target analyte, nothing else, irrespective of the number and quantity of any other entities present in the system. Specificity is often an important performance parameter in chemo- and biosensing. Enzyme-, antibody-, and aptamer-based sensors are examples of highly specific devices, as they feature a lock-and-key type binding mechanism with the analyte.

Selectivity: Selectivity generally alludes to the ability of a sensor to respond to a group of analytes or specifically to a single analyte without detecting anything else. In the case of a single analyte, selectivity is similar to specificity. But nowadays the term selectivity is more commonly used in the context of array-based or multimodal sensors [189]. In such sensors, each sensing element interacts differently with an analyte to generate its fingerprint. The same can be done for another analyte using the same sensing element, which will have a different fingerprint. Based on these data, a library of analytes is generated. One can selectively detect the target analyte once its fingerprint is matched with the database [189]. Multimodal sensors, such as electronic noses and tongues, are expected to be highly selective.

Cross-sensitivity: Cross-sensitivity is an undesirable characteristic in a sensor; it indicates the extent of interaction with the potentially interfering entities present in the system, other than the target analyte. For example, dopamine sensors tend to react with ascorbic acid, commonly present in the same biological environment, which indicates the cross-sensitivity of the sensor.

Linearity: Linearity of a sensor expresses the relative deviation of an experimentally determined transfer function plot (see Figure 1.9 from an ideal straight linear region).

Response and recovery time: Response time is the time required for a sensor to respond to one step change in the input value. Recovery time is the time a senor takes for the returning to its initial value after a step change in input from a given value to zero.

Resolution: Resolution is the smallest difference in two input values that can be differentiated by a sensor.

All aforementioned parameters are to be measured at a given temperature, pressure, and relative humidity. For further reading, readers may refer to literature associated with the specific sensor type [23, 189].

> **!** **Problem:** The change in output voltage of a sensors is 1 mV when the concentration of the analyte is increased by 100 μL. The device cannot measure concentrations below 1 μL. If the difference between two concentration values is <0.5 μL, the output values are not reproducible. Calculate its sensitivity, lower limit of detection and resolution.
> **Solution:** sensitivity = 1 mV/100 μL = 0.01 mV/μL or 10 V/L
> Lower limit of detection: 1 μL
> Resolution: 0.5 μL

1.3.5 MEMS and NEMS

MEMS and NEMS are acronyms for micro- and nano-electro-mechanical-systems, respectively. They belong to the class of micro-/nanodevices that utilize electrical as well as mechanical properties of the functional element. By definition, the device should produce an electrical response as a function of mechanical movements, or a mechanical response as a function electrical property changes. However, over the years the definitions of MEMS/NEMS have been extended to devices that may produce only electrical or only mechanical signals, as long as their functional element is in the micro- or nanoscale. Generally, passive shapes (e. g. cell-culture platforms) should not be called MEMS/NEMS, irrespective of their size or fabrication process.

Some examples of MEMS are microactuators, micro-pressure sensors, and gyroscopes. NEMS are the nanoscale variants of these devices. Most MEMS/NEMS are integrated onto chips (small flat pieces such as a mobile phone *sim* card), which is performed using processes borrowed from the CMOS (complementary metal oxide semiconductor) technology. MEMS can be used as sensors, actuators, or other electronic devices. Accordingly, the device performance evaluation parametrization is carried out. MEMS is simply the name given to the designs that (i) have their functional element <100 μm, or (ii) provide an electrical response to mechanical stimulation, or vice versa. The same goes for NEMS (functional element dimension <100 nm). Simple examples of MEMS include gyroscopes and atomic force microscopy tips. As such, no specific parameters can be described for testing MEMS and NEMS, because the testing depends upon the application. Measures such as manufacturing tolerances, surface roughness, accuracy, and precision may be applied for evaluating their fabrication process.

> *MEMS/NEMS* represent a class of miniaturized devices, whereas *MEMS/NEMS technology* may sometimes refer to a set of micro- and nanofabrication techniques aimed at manufacturing small-scale devices. This technology can also be utilized to fabricate passive shapes and patterns in the micro / nano size-range. ⚠

1.3.6 Biomedical devices

Any device that interacts with a living or nonliving biological entity, or a signal produced by it, can be considered a biomedical device. Biosensors, bioelectronic medicine, and numerous drug delivery systems come under this category.

Biosensors are often recognized by their function, for example, a neural sensors detects biochemical signals arising due to the activity of neural cells and so on. Biosensors differ from other physical and chemical sensors in terms of their biocompatibility and robustness under harsh biological environments. These are features that need to be considered in addition to the standard sensor parameters, such as sensitivity, selectivity, and linearity.

Other than biosensors, solid-state cell culture platforms and scaffolds are also often regarded as biomedical devices in the contemporary literature. Strictly speaking, only those patterns or shapes can be called devices that produce some kind of an output signal. But some scaffolds may be functional, e. g., it may be possible to provide an electrical stimulus to the cells during their growth. Hence, they are sometimes referred in the literature as devices. Nonetheless, they are important microscale patterns that are extensively used for artificial cell and tissue culture.

Cell culture refers to the process of growing and nurturing biological cells outside of their natural environment. This is referred to as *in-vitro* analysis. Tests performed inside a living animal body (such as drug monitoring or implants) are called *in-vivo*. Prior to conducting *in-vivo* test, new research experiments related to various aspects of biomedical engineering as well as medicinal and cosmetic chemistry, are performed on the artificially cultured cells. Animal models are not only expensive, they are strongly discouraged due to various ethical concerns. There is also a high probability of data variation due to the physiology of individual animals. Cultured cells are used (i) to study the developmental patterns of various diseases, (ii) as a testing platform (model) for new drugs and chemicals, (iii) to test the effect of foreign elements, such as microorganism on biological systems, (iv) in evolutionary biology, (v) to understand basic cell biology and biochemistry, (vi) to produce genetically engineered proteins, and (vii) to study gene therapy and genetic engineering, etc.

Cell culture is performed under controlled conditions, typically in petri dishes, where the required nutrients (glucose, minerals, antibiotics) are supplied to the cells in the form of a solution known as the cell culture medium. The cells can be simply distributed or suspended in the medium, where they, depending their nature, can grow in clusters. In some studies it is required to provide the cells with a solid support to evaluate their collective behavior. Two-dimensional supports are known as culture substrates, whereas the three-dimensional cell-housing is generally designated a scaffold. Substrates and scaffolds allow the cells to interact with each other, which is closer to their natural biological environment.

There are several different types of cells in the human body. They originate from certain primary cells, known as the stem cells. Various types of stem cells are present during the early development of the human body from an embryo as well as in the tissues of the adult human. When the conditions are right, stem cells can either divide or differentiate. In cell division, one parent stem cell creates two daughter cells of the same type for the purpose of self-renewal. In differentiation, new cell types are produced. For example, a neural stem cell can differentiate into neurons and various glial cells. Similarly, hematopoietic stem cells in the bone marrow differentiate to form red blood cells, white blood cells, and platelets.

To fabricate a cell culture substrate or scaffold, first and most important property is the cytocompatibility of the material used. Cytocompatibility is the measure of cell's compatibility with the material, which is tested by recording the response of the cells towards the material (e. g. inflammatory reaction). Scaffolds should also feature mechanical compatibility with the cell or tissue. In most cases, the scaffold material should have its stiffness in the same range as the tissue. An essential quality of the scaffold or substrate is to allow the culture medium to reach the cells and ensure that the waste materials produced by the cells can be removed. Often porous materials and membranes are chosen for this purpose. In most cases, proteins from the extracellular matrix, such as laminin, are coated onto the surface of the scaffold to provide the cells with a more *in-vivo-like* environment.

Cells, through the culture substrate or scaffold, may be provided with an electrical stimulus combined with the measurement of their activity. For example, neural cells may produce certain biochemicals (e. g., dopamine) under the influence of an external stimulus. The quantity of this biochemical can be measured by integrating a recording platform. Altogether, this would be a complete biomedical device. Other biomedical devices include bioimaging platforms, point-of-care diagnostic tools, and device for medical assistance. In this book, only the device pertinent to carbon technology are detailed. Similar to MEMS/NEMS, no specific device performance parameters can be defined for biomedical systems owing to their versatility. In all cases, however, biocompatibility and biosafety aspects must be thoroughly evaluated.

> **!** Cell culture is a general term that refers to creating artificial conditions for the survival and growth or any living cell or tissue. These cells may be extracted from a living organism, or from a previously established cell line or cell strain. Cells can be cultured in a liquid medium, but in some cases, a solid support is often provided to better mimic their natural environments. 2D supports are generally known as substrates, whereas the 3D ones are called scaffolds. The terminology, however, is exchangeably used.

1.3.7 Function-specific performance evaluation

In addition to the standard device performance parameters, it is essential to also characterize the device against the state-of-the-art in that particular field and compare the pros and cons. For example, in the case of a EDLC-type supercapacitor, it is necessary to benchmark your new carbon material with respect to activated porous carbon. On the other hand, if you are using your carbon material as an electrocatalyst, the performance should be compared with existing metal catalysts. Often carbon materials show an inferior performance compared to metals when it comes to their electrical conductivity or catalytic activity. Their corrosion resistance and electrochemical stability in a wide range of pH values acts as a compensating factor. Manufacturability of carbon materials is usually scalable and typically does not require extensive wet processing steps or harmful chemicals. So despite a lower electrical conductivity, they not only compete with but often supersede the performance of metals.

The are two primary challenges associated with carbon material development, even if a high degree of control is achieved in the process. These are (i) high energy consumption due to elevated temperatures involved in material development, and (ii) biodegradability. Although carbon materials are biocompatible and generally nontoxic, their high stability makes them nonbiodegradable. This is a problem in the case of excessive inhalation of carbon nanomaterials, and also when there is a carbon-based implant in the human body, which is expected to degrade after performing its job. Nonetheless, both of these challenges are being addressed by material and design modifications; they will be discussed in the coming chapters.

1. What is a nanomaterial? Are nanomaterials crystalline or amorphous?
2. You are given a material that features (i) high surface area due to microporosity, and (ii) electrical resistance in kilo-ohm range. What are the possible device applications of this material?
3. What are the primary components of an electrochemical sensor?
4. Is it possible to fabricate a complete device using just one material? Explain with an example.
5. What does the term "microstructure" mean to (i) material scientists and (ii) microsystem engineers?
6. What is the difference between microfabrication and micromachining?
7. Classify the following as top-down or bottom-up manufacturing techniques: photolithography, drilling, metal casting, 3D-printing, plastic extrusion, 2D-printing, self-assembly of biomolecules.
8. Why are bottom-up manufacturing processes more suitable than top-down in nanoscale fabrication?
9. What is the difference between sensitivity and lower limit of detection of a sensor?
10. Can a sensor be used for input values for which the response is beyond the linear range? If yes, what will be the requirements of such a sensor?
11. Define transfer function of a sensor.
12. What are energy storage devices, and how are they different from energy generation devices?
13. What are the similarities, if any, between a supercapacitor and a battery?
14. Why, in your opinion, are porous carbon materials preferred for EDLC fabrication?
15. What are No-ion and Li-S batteries? How are they different from Li-ion batteries in terms of both materials and mechanism?
16. Why are commercially available hydrogen fuel cells still very expensive? Suggest a few options for their cost reduction.
17. In which ways can carbon materials be used in the fabrication of a fuel cell?
18. A certain type of carbon quantum dots display a change in color when exposed to a heavy metal ion in an aqueous solution. Which type of sensor (based on detection method) can be fabricated using this material? Explain how will you draw the calibration curve, taking any concentration range of the input.

2 Carbon materials for devices

Carbon materials range from nanoscale to crystalline and disordered solids. In addition, there are microscale carbon forms, such as carbon fibers and lithographically patterned micro-scale features, that exhibit bulk properties but with a much higher surface area. Industrial carbon materials, such as porous carbons, come in the form of pellets, granules, and particles. Carbon dots, which are only a few nanometer in size, exhibit interesting properties that may be significantly different from large-scale materials.

2.1 Carbon and silicon

Carbon and silicon are group IV elements in the periodic table, and many of their properties are somewhat similar. In both elements, sp^3 hybridization leads to strong sigma bonds, resulting in interpenetrating face-centred cubic crystals. These two elements also form a very stable compound, silicon carbide, which also features a diamond-like crystal structure. Silicon has undoubtedly been an important technological material in the last few decades. A direct comparison between the two elements and their industry prospects is therefore natural.

What differentiates carbon from silicon is its unique ability to form double bonds, despite the possibility of a very stable crystal form (diamond) entirely based on single bonds. This characteristic of carbon is attributed to the small size of its atom. Heavier elements in the same group have their atomic orbitals more diffused, making it difficult to form π bonds, which result from the partial overlap of two orbitals (from two different atoms) perpendicular to the bond axis. In other words, after the formation of a stable σ bond, the two orbitals should be close enough to form an additional π bond. It is easier for most of the small atoms (e. g., second period elements in the periodic table), not just carbon. However, this ability combined with the electronic configuration gives carbon a unique position among all elements, which enables the formation of a range of allotropes.

The reasons for silicon's popularity in the semiconductor industry primarily include (i) its availability in highly pure crystalline form, which can be synthetically grown, and (ii) formation of a solid-state oxide, which can serve as an insulating layer in the devices. If we compare the oxides (particularly, di-oxides) of silicon and carbon, we find that silicon dioxide is a solid, whereas carbon oxide is stable as a gas form. Due to a high stability, carbon dioxide molecules can exist in gaseous state, whereas silicon dioxide tends to condense and form a solid. Again, the stability is owing to the π bonds accompanying σ. Carbon allotropes having predominantly sp^2 hybridization also feature a remarkable stability. In the last three decades, since the discovery of fullerenes, it has also become widely known that not only pure sp^2, but also carbon allotropes featuring a hybridization state between sp^2 and sp^3, come with an amazing stability. This behavior is not observed in any other element.

https://doi.org/10.1515/9783110620634-002

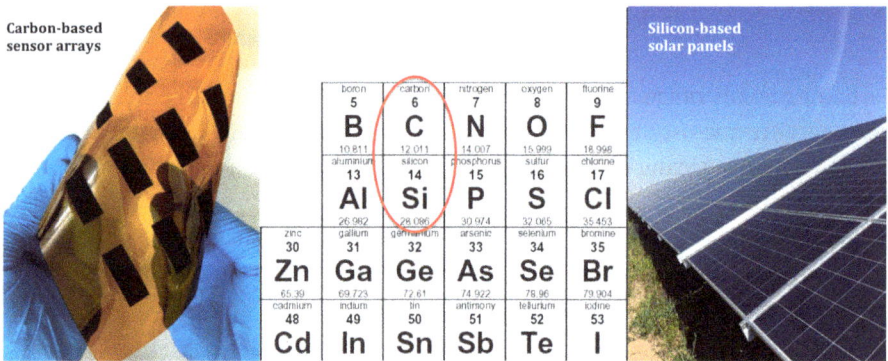

Figure 2.1: An illustration comparing carbon and silicon, their position in the periodic table and device applications. Periodic table image is taken from Wikimedia Commons and Solar Panel from pxfuel.com.

In today's technology, carbon and silicon both have their unique positions and applications. They are comparable in certain aspects, but they are not competitors in the market. In fact, silicon is used in combination with carbon, for example, as the substrate material in many device applications. Details of such devices and fabrication techniques are provided in Chapter 3. This technological coexistence of the two elements can be attributed to the fact that silicon is more useful in semiconductor or photovoltaic applications that demand an extremely high purity, crystalline structure, possibility of doping and insulating thin-films. Most forms of carbon, on the other hand, are electrical conductors electrochemically stable, hence appropriate for other device classes, such as batteries and sensors. The manufacturability of the two elements at micro- and nanoscale is also different. Silicon is often used as chips with well-defined crystal orientations. Carbon materials are almost never used as chips, except in the case of highly oriented pyrolytic graphite (see Chapter 4 for details). Whereas silicon is presynthesized, cut into wafers and then used in devices; carbon can potentially be synthesized from its precursors during the micropatterning. Carbon's utilization and fabrication strongly depend upon the pattern size, shape, material's crystallinity, and often the preparation method. In other words, there is no single fabrication strategy that can cover the entire carbon-device patterning. Explanation of these techniques is the primary objective of this book. But prior to that, it is essential that carbon materials, their hybridization states, and their origins are clearly understood.

> Erich Fitzer, a well-known name in industrial carbon technology, provided an excellent comparison of carbon and silicon during his lecture in the 13th Biennial Conference on Carbon held in Irvine, CA, USA in 1977. The contents were later compiled as a paper and published by the journal *Carbon* in 1978 [64]. The lecture gives an insight on the parallel development of carbon and silicon as technological materials, which has led us to today's device technology.

2.2 Carbon materials in miniaturized devices

Carbon materials used in device applications can be broadly divided into two types: carbon nanomaterials and bulk (crystalline/disordered) carbons. Nanomaterials are often present in the form of powders or suspensions that contain discrete nanoscale units of a material. This material can be an element or a compound with a well-defined set of properties. The size (e. g., diameter) must be <100 nm. When it comes to carbon materials, some of its nanoforms, such as graphene, also serve as the building block for its bulk crystalline form graphite. Similarly, fullerenes and other curved carbon fragments are found in glass-like and possibly other nongraphitizing carbons. Nonetheless, when any nanomaterial becomes a part of a continuous material, it loses the special properties that were caused by its size, geometry, and very high surface-to-volume ratio. Therefore, the properties of graphene are not the same as graphite. Among carbon nanomaterials, graphene and its derivatives, CNTs, CNFs, fullerenes, carbon black are the most common materials for devices. Though most of them are synthesized via bottom-up processes, the patterning of the device may entail top-down approaches as well. Some carbon nanomaterials are shown in Figure 2.2. Note that it is only a specific arrangement of graphene sheets (ABABA..) that results in graphite.

Among bulk solid carbons, two main classes of carbon materials are used. These are known as graphitizing and nongraphitizing carbons, which are explained in detail in Chapter 3. Pure graphite itself is also used in devices. In fact, this is one of the oldest

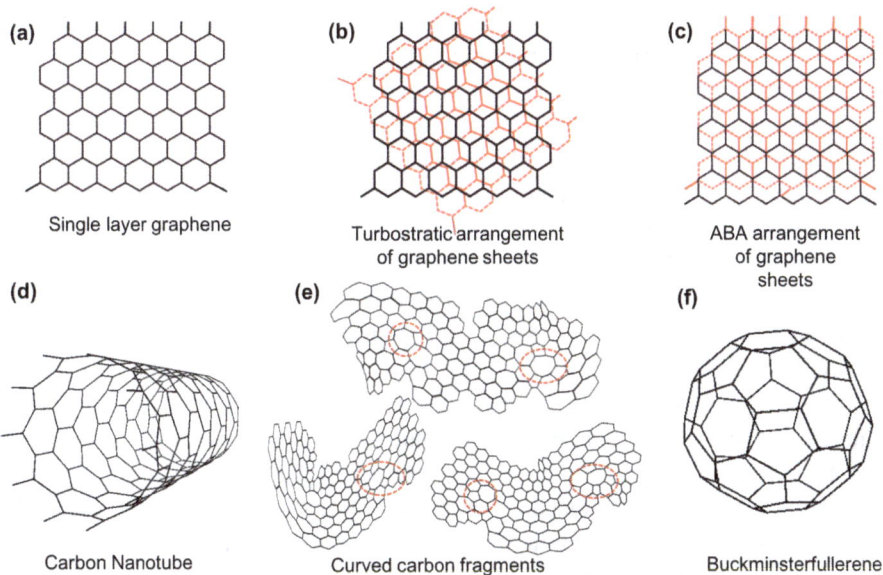

(a) Single layer graphene

(b) Turbostratic arrangement of graphene sheets

(c) ABA arrangement of graphene sheets

(d) Carbon Nanotube

(e) Curved carbon fragments

(f) Buckminsterfullerene

Figure 2.2: An illustration showing carbon nanomaterials and building blocks for bulk solid carbons. Image reprinted from [50].

material that is still in use in batteries. Graphite can be used as a powder with a binder, or in the form of a polycrystalline solid. Its manufacture is commercially carried out using a raw material known as needle coke. Needle coke is a processed form of a petroleum coke that contains stacked carbonaceous fragments, which can yield ordered graphite material on further heat-treatment.

Carbon fiber(s) is yet another type of carbon material widely used in devices. Fibers made of carbon can be from a few micrometers to several centimeters long, depending upon their fabrication process. In terms of diameter as well, there is a wide range from just a few nanometers to about 10 μm. Accordingly, their properties can either be similar to nanomaterials or bulk solids. In all cases, however, the morphology and aspect ratio of the fiber play decisive roles in property determination. Fibers having micrometer diameters and a high degree of uniformity are well-known for their high tensile strength and Young's modulus. They are used in manufacturing applications, typically in the form of a composite. These large-scale manufacturing applications will not be discussed in this book. Rather, the focus will be on devices where patternability and high surface area of carbon fibers are exploited.

Other than the aforementioned forms, some new additions to the carbon family include carbon dots and laser-patterned carbon. Some nitrogen-containing carbon forms are also gaining a lot of popularity, particularly in device fabrication owing to their special functionalities, which are caused by the presence of nitrogen in their structure. These materials, along with their preparation methods will be discussed in Chapter 6.

Whereas carbon nanomaterials generally feature an expected and defined physicochemical behavior, properties of different bulk carbon materials are strongly intertwined with their respective microstructures. Over 95 % carbon materials used in device applications are graphite-related and primarily sp^2-hybridized. Diamond and related materials are used for their mechanical and thermal properties, often in the form of a coating. Nitrogen-vacancy center diamonds are utilized for their optical as well as magnetic characteristics in sensor devices. Within graphite-related carbon materials, one can again find a large variation in terms of microstructures, which depends upon their preparation method and precursor. Pure graphite occurs naturally, but can also be prepared via synthetic routes, e. g., by heat-treatment of needle coke and certain polymer. Some synthetic carbons can be converted into graphite when heated to very high temperatures, whereas others cannot. Many are disordered, i. e., they contain graphitic crystallites embedded into an amorphous carbon matrix. The crystallite size and the separation between the layers that constitute the crystallites also differ. In many carbon materials, there is micro, macro, and meso porosity. Some of them may have a well-formed network of the crystallites, whereas others may feature disconnected, particle-like graphitic regions. Altogether, this very complex microstructure makes it difficult to understand their properties at a first glance. In a complex device, carbon needs to be used with other materials, such as metals. In such events, the interaction of carbon with them becomes an important parameter for an optimum device function. Difference in work function of carbon and metals, high electrical resistance at ohmic contact points, and surface energy difference are some such parameters that have been extensively studied for various carbon-metal combinations.

To understand carbon materials, it is essential to appreciate the element carbon and possible hybridization states during its bonding, especially with one or more carbon atom. It is indeed the stable co-existence of different hybridization states in the same carbon form, which makes it possible to have many carbon materials *in-between* graphite and diamond. This will be elaborated with examples in the following sections.

2.3 Materials and their properties

To learn about any material, one needs to understand it at three levels: atomic, molecular, and crystal. The atomic structure gives information about the number of electrons, energy levels at which these electrons are distributed, nucleons and isotopes, and accordingly the weight and size of a single atom. The second level is the molecular structure. When electrons are available in the valence shell, atoms tend to bond with their neighboring atoms. The neighboring atom may or may not be of the same species (material). Two-atom molecules, such as H_2, are simpler to understand. In such cases, two atomic energy levels, or orbitals, form new molecular orbitals. This facilitates a more energetically favorable distribution of electrons. This implies that the properties of the material that were resulting from the distribution of electrons in the atomic orbitals may not be valid anymore. We now have to look at the material at the molecular level and relate its properties to the available electrical charges. Charge on a molecule result from the excess or lack of electron. This can also be influenced by other neighboring molecules or atoms. Complex multiatom molecules, such as proteins, may have inherent charges, given they do not have simple distribution of electrons into molecular orbital, such as that of H_2. They experience various attraction and repulsion forces and often also steric hindrance. Polymers are large molecules with repeating chemical units known as the monomers. Their properties significantly differ from the constituting atoms. Interestingly, despite the fact that the molecular orbital theory supports the formation of a C_2 molecule (two covalently bonded carbon atoms), it is highly unstable and has only been observed in carbon vapors for extremely short duration. Molecules encompassing 60 or more carbon atoms adopt a football-like shape. They are known as fullerenes, which will be discussed Chapter 4.

The third level is the crystal structure that is often simply called the structure of a material. Unlike atomic or molecular arrangements, crystal structure represents a macroscopic physical attribute, which is responsible for a number of physicochemical properties of a material. When several atoms or molecules adopt an order by organizing themselves in the form of repeating units, the resulting configuration is known as a crystal. Simple crystals are those composed of the same type of atoms, i. e., elements. Crystals of compounds contain heterogeneous neighboring atoms, and the size of atoms of chemically different atoms plays an important role in determining the crystal type.

In an ideal crystal, all atoms have a well-defined distance between them. This means that all bond-lengths (between same or different atoms) are identical. Materials that do not form a crystal and feature variable bond-lengths are called amorphous. In a physical sense, crystals always break along a plane, whereas the amorphous materials shatter on breaking. All powders, in principle, are amorphous at a macroscopic level, because they do not have a packed structure. Based on the arrangement of small crystallites, solid materials can be divided into crystalline (single- and polycrystals) and amorphous. Certain materials contain small crystallites embedded into an amorphous matrix, which may or may not be connected to each other. Such materials are known as disordered solids. Many carbon materials fall in this category. An additional type is nanomaterial, where nanoscale structural units, that may be crystalline or amorphous, are present in their discrete form. A schematic representation of different material classes based on crystallinity is given in Figure 2.3.

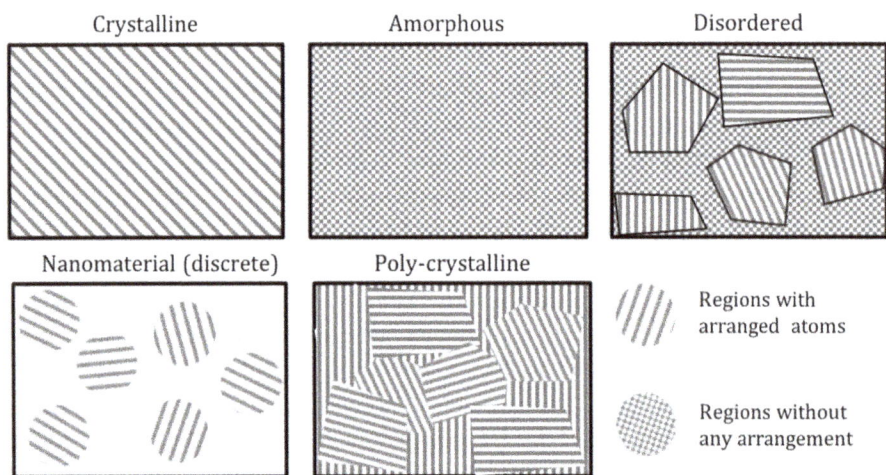

Figure 2.3: An illustration showing crystalline, amorphous, disordered, and nanomaterials.

For most materials such as metals and alloys, crystallinity is an indicator of electrical conductivity. Amorphous materials, such as glasses, tend to be insulators. But in the case of carbon, its crystalline form diamond is an insulator, whereas many disordered carbon materials (which are physically amorphous) feature a good electrical conductivity. Graphite, which is yet another crystalline form of carbon is a conductor, which is called a semimetal owing to its valence and conduction band touch at the Fermi level. Many powders consisting of carbon nanomaterial may also feature electrical conductivity, if the individual units (say particles) in that powder are conductive and form a percolated network. Various crystalline and noncrystalline carbon materials have specialized nomenclature proposed by the IUPAC if they have a recognized set of structure and properties. If the microstructure of a carbon material is unknown or undefined in the *IUPAC Gold Book*, it may simply be referred to as *carbon*.

> IUPAC stands for the International Union of Pure and Applied Chemistry. The Compendium of Chemical Terminology published by the IUPAC is informally known as the IUPAC Gold Book. In case of any confusion regarding the correct nomenclature of carbon materials, IUPAC Gold Book should be referred. Updated versions of this compendium are made available periodically, which contain new forms of carbon, such as advanced nanomaterials.

2.3.1 Isotropic and anisotropic materials

Isotropic materials are those that feature same properties in all directions. Anisotropic materials, as the name suggests, have preferred orientation and accordingly, their properties. Materials with a layered structure, such as graphite, are excellent examples of anisotropy.

Properties of a material are defined as the response of the material to a certain field under standard conditions. These standard conditions are typically established in the form of temperature, pressure, humidity, or other relevant environmental conditions. The field could be in the form of any external force, such as electrical, mechanical, thermal, and so on. For example, when a voltage is applied across a crystal, certain current passes through it. The current is the material's response, and it depends upon the electron flow patterns characteristic of that material. This is initially proportional to the voltage, but if we keep on increasing the voltage value, it may become nonlinear. This is indeed the case most often. A proportionality constant, such as electrical resistance, can be deduced for the linear range. For understanding the nonlinear behavior of the material, several factors such as its electronic configuration, crystal structure, vibrational motion of the molecules, etc., need to be investigated. Any property P of the material in the linear response region can be defined as follows:

$$R = PF, \qquad (2.1)$$

where F is the force/field and R is material's response to it. The value of P can be represented as

$$R = R_0 + \frac{dR}{dF}F, \qquad (2.2)$$

where the term R_0 is only for those properties that can exist without any external field (e. g., magnetism in iron). R_0 is neglected for properties such as electrical or thermal conductivity, as they are only observable when the external field is present. For an isotropic material, R is a scalar quantity independent of the direction of F. However, for anisotropic materials, such as graphite, R and hence P are dependent on the direction of F with respect to their crystal orientation.

Among carbon materials, pure graphite displays highly anisotropic properties due to the arrangement of its so-called basal planes ($00l$ planes) on top of each other in a layered fashion. However, most other forms of synthetic bulk carbons show a high degree of isotropy. Even if these materials contain small crystallites of graphite-like carbon, these crystallites are randomly oriented in all directions, thus resulting in isotropic properties as a whole. A schematic showing examples of anisotropic and isotropic carbon materials is presented in Figure 2.4.

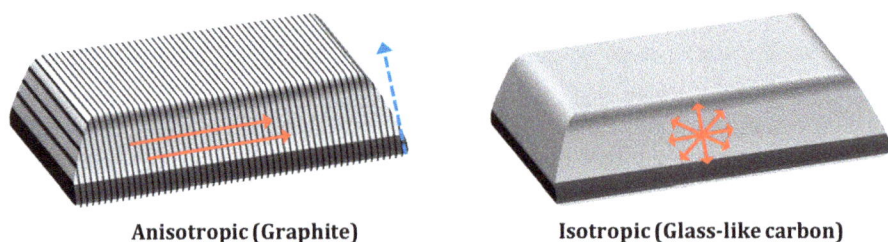

Anisotropic (Graphite)	Isotropic (Glass-like carbon)

Figure 2.4: An illustration showing anisotropic and isotropic carbon materials. The red (solid) arrows show the direction of material's response. In the case of anisotropic graphite, properties along the blue (dashed) arrows are significantly different from the direction perpendicular to it. Graphite and glass-like carbon here are only two representative examples.

2.4 Atomic structure of carbon and hybridization

Carbon (chemical symbol: C) is the sixth element in the periodic table and has an atomic weight of 12.0107 amu. There are various known isotopes of carbon having atomic weights between 8 and 22. Three stable isotopes, which are found in abundance in nature are ^{12}C, ^{13}C, and ^{14}C. ^{13}C is nuclear magnetic resonance or NMR active, whereas ^{14}C is the radioactive carbon used in carbon dating. The average atomic weight of carbon is therefore calculated based on the abundance of these three isotopes as follows:

$$[(12 \times 98.9) + (13 \times 1.1) + (14 \times 0.0001)]/100 = 12.0107.$$

Note that the resulting number is not exactly 12.0107 owing to the (negligible) natural presence of other carbon isotopes as well. The three primary isotopes are shown in Figure 2.5.

> ! ^{13}C-NMR spectroscopy is a well-known analysis method in chemical sciences. It provides useful information about the $C - C$ network in organic compounds that can reveal essential structural information.

The electronic configuration of carbon is $1s^2 2s^2 2p^2$. As shown in Figure 2.6, after filling two pairs of electrons in s orbitals, the remaining two electrons are distributed in the p_x and p_y orbitals as unpaired. This is the natural or ground state of carbon, which is rather

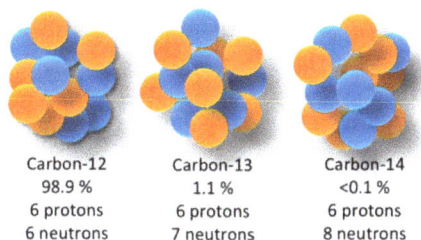

Carbon-12	Carbon-13	Carbon-14
98.9 %	1.1 %	<0.1 %
6 protons	6 protons	6 protons
6 neutrons	7 neutrons	8 neutrons

Figure 2.5: Three most commonly found natural isotopes of carbon.

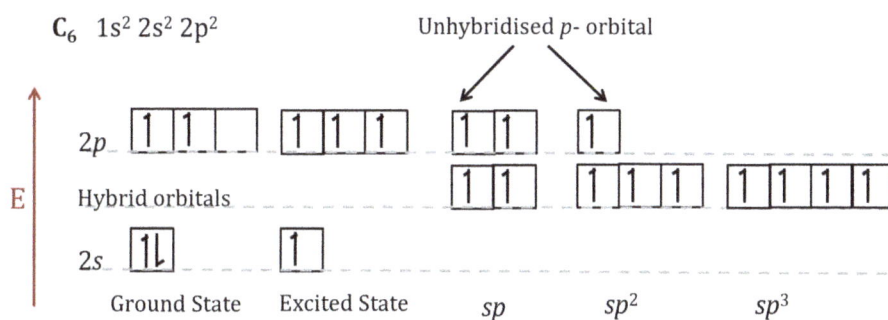

Figure 2.6: Distribution of electrons and hybrid energy levels in carbon during bonding.

rare in carbon materials. This is because during bond formation, the energy levels of $2s$ and $2p$ orbitals readjusted such that new hybrid orbitals are formed. The term hybrid is used because the new orbitals have their energy levels between the s and p. Accordingly, their properties (that are influenced by their distance from the nucleus) are considered a hybrid of $2s$ and $2p$ characters. All hybrid orbitals are at the same energy level, as shown in Figure 2.6. The shape of sp-hybridized orbitals is linear, sp^2 orbitals is trigonal planar, and sp^3 orbitals is tetrahedral. sp hybridizations are commonly observed in carbon compounds, whereas the other two are dominantly present in elemental carbon, graphite and diamond being the principal examples of sp^2 and sp^3 carbon, respectively. However, a new class of carbon materials known as carbyne has been discovered that contains elemental carbon with sp hybridization. Numerous efforts are being made in the direction of bulk production of carbyne experimentally. Yet another interesting hybridization state that can be adopted by carbon atoms in their bulk solid forms is sp^{2+n}. Clearly, it is a state that is energetically between sp^2 and sp^3. But its visualization is sometimes difficult. These sp^{2+n} states with a variable n are common in fullerenes and other curved carbon fragments.

Carbon atoms prefer to form covalent bonds since their valence shell (principal quantum number, $n = 2$) contains four electrons. This implies that carbon atoms don't have the tendency to either donate or accept electrons. Carbon undergoes covalent bonding with metals, nonmetals, and with itself. The most notable and unique feature of carbon is its tendency to form large chains and sheets by self-bonding, which is occasionally also denoted as self-polymerization. This so-called polymerization, however, has no specific monomers. These extensive carbon sheets or layers can be synthesized by both top-down and bottom up processes.

The concept of hybridization of atomic orbital is fundamental to the modern understanding of the structure of carbon materials. Hybridization can be defined as the merging or fusing or two or more atomic orbitals. In quantum mechanical terms, hybrid orbitals can be seen as the linear combinations of the wave functions of the atomic orbitals. The orbitals participating in hybridization have at least two different energy levels (i. e., azimuthal quantum numbers), but all resulting hybrid orbitals have the same energy. This enables all bonds formed by that atom to be at the same energy level, making it more stable.

Let us start understanding hybridization in carbon with sp^3 state, which enables bonding of all four valence electrons. The shape of hybrid sp^3 orbital is pyramidal, and one atom is distributed in each hybrid orbital. After forming four covalent bonds with a similar (sp^3-hybridized) carbon atom, a network of carbon is formed in the pyramid-like shape, as shown in Figure 2.7A. The most common example of this type of bonding in carbon is diamond.

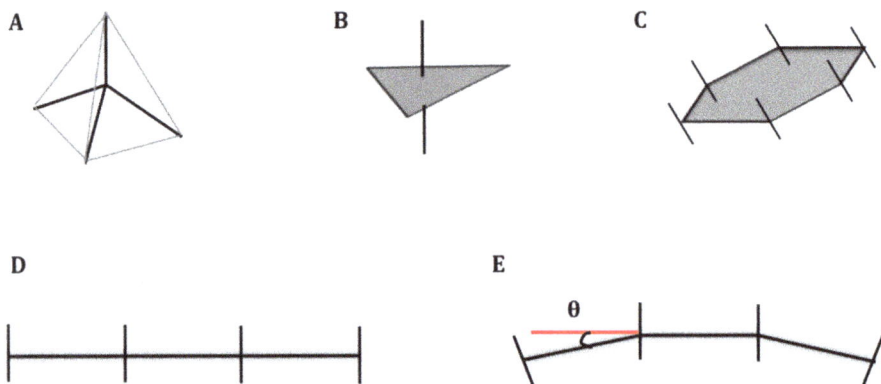

Figure 2.7: Different geometries forms by differently hybridized carbon atoms: (A) Pyramid geometry of sp^3 hybridized carbon, (B) Trigonal planar geometry of sp^2 hybridized carbon, (C) A hexagonal carbon geometry formed by six planar sp^2 hybridized carbon atoms, (D) Size view of the hexagon shown in (C), (E) A curved carbon sheet (side view) formed by a bent hexagon, a result of sp^{2+n} hybridization. Angle θ indicates the deviation from planar geometry.

The second well-known hybridization state is sp^2. Majority of technology relevant carbon materials are at least partially sp^2-hybridized. Here, there are three hybrid orbitals with one electron present in each. Naturally, it is these three that participate in bonding, leaving one unpaired electron. This one electron is indeed responsible for not only the wide range of interesting properties of carbon materials, but also in certain bonding phenomena that are not purely sp^2. The geometry of sp^2-hybridized orbitals, and the resulting carbon crystals, is trigonal planar with the unhybridized p_x and p_y orbitals located on the top and bottom of this plane (see Figure 2.7B). A single planar sheet formed by sp^2 bonding among six coplanar carbon atoms is known as graphene (Figure 2.7C). When multiple graphene layers are arranged on top of each other in an ABABA.. fashion, the material is known as graphite.

The crystal structure of graphite is hexagonal, but not closed packed. As shown in Figure 2.8, alternating layers having a pure sp^2 hybridization and a network of carbon hexagons in a planar fashion, are stacked upon each other. These are considered alternating because of their position in the crystal. One corner atom of the hexagon of layer A is located right in the center of the hexagon of the B layer. The A and B layers are separated by each other by a distance of 0.335 nm. The distance between two A (or two B) layers is 0.670 nm. One unit cell of graphite, as shown in Figure 2.8, covers two A layers and one B layer. Each of these layers is known as the basal plane, as it constitutes the base of the unit cell. The vertical c– parameter of the crystal therefore equals 0.670 nm. The A and B layers are only bonded by weak Van der Waals forces containing the electron cloud from the unhybridized p-orbital. Hence, the crystal geometry is not closed packed; it is just hexagonal. In the case of a HCP crystal, parameter $c = 1.633a$,

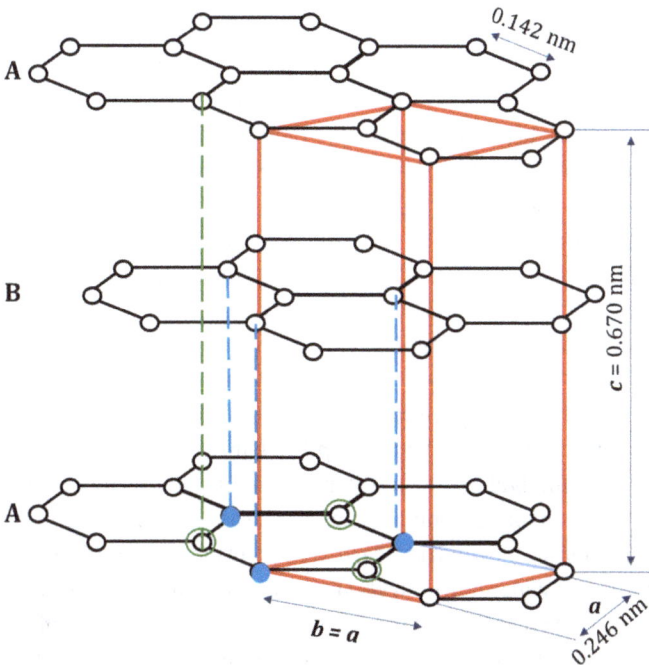

Figure 2.8: Schematic representation of a graphite crystal. The box connecting two A layers (red color) represents the unit cell. In one hexagon in the bottom-most plane, two types of nonequivalent carbon atoms are shown with blue (solid) and green (open) circles.

whereas in graphite $c = 2.723a$. Here a is the length of the lattice vectors lying in the basal plane having a length equal to 0.246 nm (see Figure 2.8). The bond length between adjacent carbon atoms is 0.142 nm. Space group that represents a flat h-graphite crystal is $P63/mmc$.

Although all atoms in a hexagon or even in the entire graphite crystal are chemically identical, according to their distance from the nearest neighbor in the crystal, there are two types of atoms. In the bottom basal plane shown in Figure 2.8, these nonequivalent atoms are represented by red (solid) and green (empty) circles. One can also observe that the unit cell connects four red (or green) atoms. This is essential for achieving an identical repeating unit. Further details of nonequivalent atoms in one basal plane is described in Chapter 4 within the explanation of graphene crystal.

The $sp^{(2+n)}$ hybridization is better understood based on the resultant s- and p-character of the hybrid orbital in a multiatom molecular structure of, for example, of a spherical shape. It is important to keep in mind that orbitals are nothing but a probability density of an electron. A hybrid orbital defines a new region where the electron can now be found. This shape of this region is influenced by the parent orbitals, and is mathematically determined by taking the linear combination of the wavefunctions associated with the parent atomic orbitals (LCAO). The probability of finding an electron in an orbital can be a fraction. This implies that an electron can spend partial time between two different orbitals. The "net" hybridization of the entire molecule is calculated, which depends upon the s- and p-character (which is in turn decided by the average time spent by electrons in these orbitals) may result in a fractional hybridization.

Have a look at Figure 2.7 parts D and E. D is the cross-sectional illustration of a single hexagon of a graphene sheet. The vertical lines indicate p_x and p_y orbitals located above and below the bonded carbon plane. If this sheet is given some curvature, as shown in part E, it will make an angle with the horizontal plane. The angle $(90+\theta)$ will be the angle between this bent plane and the direction of the unhybridized p-orbital. This direction is sometimes denoted as π-orbital axis vector or POAV. Angle $(90 + \theta)$ is known as the angle of pyramidalization [87], which is used to quantify the curvature in carbon molecules. If one now back-calculated the net hybridization of a carbon molecule (spherical or just curved), it will fall between that of sp^2 and sp^3. The angle of pyramidalization for pure sp^3 hybridized structure is 109.5° and for pure sp^2, it is 90° $(\theta = 0)$.

The most notable carbon structure that features $sp^{(2+n)}$ hybridization is Buckminsterfullerene or C_{60}. These are football-like hollow carbon entities made-up of 60 carbon atoms, which result in a net hybridization value of $sp^{2.28}$. Other similar molecules, such as C_{60}, are also called fullerenes. Cylindrical carbon nanoforms, such as CNTs, also exhibit a very small value of n owing to the curvature and minor structural strain at tube ends. In some texts, all curved carbon structures are considered members of the fullerene family, and hence denoted as fullerene-like or fullerene-related structures. C_{60} has a diameter of 0.71 nm. It is composed of twelve pentagonal and twenty hexagonal carbon rings arranged such that no two pentagons share an edge. In general, such structures can be unstable, but C_{60} is surprisingly stable. The average bond length in

C_{60} is 1.4 Angstrom, which is an average of double and single bonds present between the hexagon-hexagon and hexagon-pentagon pairs, respectively. C_{60} is electron acceptor with a strong electron affinity of 2.65 ± 0.05 eV [247]. This is due to the fact that C_{60} contains conjugated ring systems (with resonating electron) but is nonplanar. Curl, Smalley, and Kroto were honored with the Nobel Prize in 1996 for their discovery (synthesis, characterization, and analysis) of fullerenes.

The idea of pyramidalization was based on the concept that since the π-orbital is not exactly orthogonal to the bonded or σ-orbitals, it results in a local pyramid-like shape. The pyramidalization angle, $90+\theta$, is therefore also simply written as $\theta_{\sigma\pi}$ (see Figure 2.9). The atomic orbitals in nonplanar conjugated carbon molecules readjust their hybrid energy levels to maximize the overlap between all bonds [87].

The discovery of fullerenes and analysis of the hybridization states of carbon atoms in such molecules marked the beginning of a new era in carbon science. Curved carbons, which were widely considered defect-containing metastable forms of carbon, were finally regarded as stable allotropes after Buckminsterfullerene's discovery. This discovery also helped in understanding nongraphitizing carbons, which comprise of carbon fragments featuring a range of curvatures.

What does this mean in terms of hybridization? Note that once the π-orbital is not orthogonal to σ orbitals, it is not the same hybrid π-orbital as in graphite (pure sp^2 hybridization). According to the POAV2 calculations suggested by Haddon [88], this π-orbital is now that hybrid orbital which forms the angle $\theta_{\sigma\pi} > 90°$ with all three σ-orbitals. This phenomenon is known as the rehybridization of the carbon atom, since it can be seen as an already hybridized orbital further hybridizing with the s-orbitals. If the number of carbon atoms, and hence the geometry of a spherical carbon structure is known, one can mathematically deduce $\theta_{\sigma\pi}$. In the case of C_{60}, $\theta_{\sigma\pi} = 101.6°$ (see Figure 2.9 C). Based on this, one can calculate the contribution of the π-orbital in the hybridization.

Coming to the sp hybridization in the context of elemental carbon: Until 1960s it was believed that this hybridization state could only exist in carbon compounds. Although there exist scattered reports on synthesis of one-dimensional carbon molecules since the late 19th century, the existence of carbyne is still debated. Nonetheless, more and more publications, including both theoretical and experimental work, are supporting the idea of the existence of carbynes, and the carbon community now has a wide acceptance for this form of carbon.

Carbynes are linear carbon molecules with alternating single-triple or double-double carbon bonds. These two chemical structures, schematically shown in Figure 2.10, are believed to exist simultaneously (in resonance). These carbon chains further assemble themselves in the form of a hexagon. However, this hexagonal structure is made up of chains rather than individual atoms. The chains are assembled via van der Waals forces (sideways), rather than covalently. This arrangement of carbyne chains and the cross-section of this assembly is also represented in 2.10. Often, it is

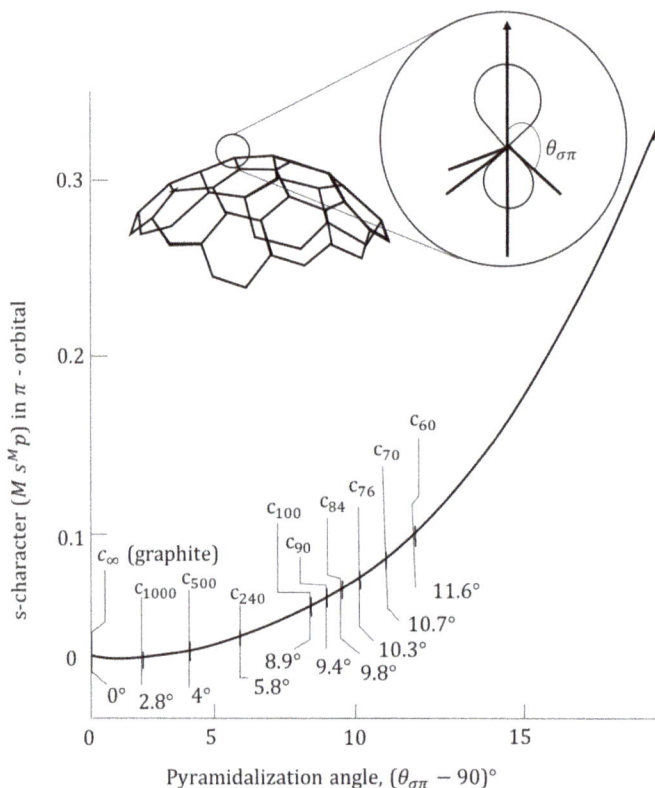

Figure 2.9: An illustration of rehybridization and the angle of pyramidalization in fullerenes. Copyright ©2017 Royal Society (for scanned images). Reprinted with permission from [88].

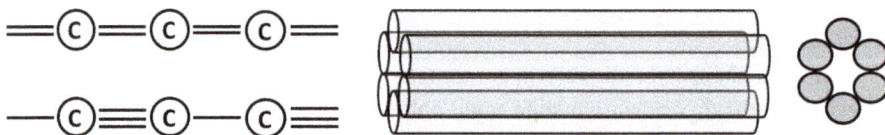

Figure 2.10: Proposed structure of carbyne molecules. The first image illustrate possible chemical bonding with either all double or alternating single-triple bonds between carbon atoms. Second and third sketches illustrate the assembly of carbyne in the form of a hexagonal bundle (front and cross-sectional views).

the diffraction patterns arising from these hexagons that are detected during electron microscopy and that confirm the presence of carbynes. Similar to many other carbon forms, the term carbyne is used to denote a class of carbon materials rather than an individual chemical entity. The chain length, arrangement, presence of functional groups, and varying chemical analogues can lead to the formation of different types of carbynes. For a more detailed study, the readers may refer to the book edited by Heimann, Evsyukov, and Kavan [98].

The existence of carbyne or any elemental *sp*-hybridized carbon in bulk has been debated for years. But many recent publications suggest a possibility of their production in pure forms via, for example, chemical synthesis. In this context, synthesis of *sp*-carbon wires is noteworthy.

2.5 Allotropes of carbon

The word allotrope, adapted from Greek, means *other turn* (or to take other turn). In the context of materials, it is interpreted as *another form*. As evident from the discussion in the previous section, carbon appears in nature in many different forms due to its ability to conform to different hybridization states. These different elemental forms are called allotropes of carbon. The term allotrope is strictly used with respect to elements, not compounds. By the traditional definition, all forms of a material, disordered or crystalline, can be classified as different allotropes as long as they are stable and occur naturally. However, recent discoveries of carbon nanomaterials as well as the possibility of the existence of plenty of other carbon forms has compelled the carbon community to revisit the very definition of allotrope.

There is yet another related terminology, polymorph, which comes from ancient Greek as well, which implies *many forms*. Polymorph is used to designate different physical (typically crystalline) forms of the same substance, irrespective of whether it is an element or a compound [30]. The exact meaning of polymorph changes depending upon the research field. In biology, polymorphism indicates alternative phenotypes of the same animal. In Object Oriented Programming, the idea of accessing different objects with the same interface is referred to as polymorphism. In medicine, this terminology indicates certain data in electrocortiogram. Though in material science different carbon forms are exchangeably referred to as polymorphs, allotrope is a more appropriate description since (i) allotrope may also include disordered carbon forms, whereas polymorph is conventionally employed for crystalline forms only, and (ii) all atoms in one system must feature the same coordination number for it to be considered a polymorph. Although this works for pure diamond and graphite crystals, many other carbon forms have differently hybridized atoms in the same material, which may locally change the coordination number and bonding behavior. Such materials are better described as allotropes.

For a very long time, diamond, graphite, and occasionally, amorphous carbon were regarded as the three major allotropes of carbon. The reason was their natural availability and distinct chemical properties. Among the so called amorphous carbons, coal, charcoal, and any carbon material with disordered crystallites were included. In 1967, a cubic, diamond-like carbon, Lonsdaleite (also called hexagonal diamond) was identified on the rim of the crater formed by the Canyon Diablo meteorite strike on Earth. The material was assumed to be formed by the colossal impact of a graphite-carrying meteorite on the ground that transformed the sp^2 hybridization of the atoms into sp^3, while still retaining the hexagonal crystal geometry. Hexagonal diamond could also be prepared in the laboratory by pressure and shock treatment of graphite. This metastable material was considered a *polymorph of diamond* in the contemporary literature. Similarly, a variation of natural graphite known as rhombohedral graphite, which features sp^2-hybridized atoms but ABCABC-type crystal structure, was designated a *polymorph of graphite*. In different publications, these carbon forms were later denoted as the new allotropes of carbon. By definition, these materials are polymorphs (due to their crystalline structure) as well as allotropes (because they are different forms of carbon). The naturally available quantities of these carbon forms were large enough for their consideration as allotropes.

> **!** Crystal description of rhombohedral or *r*-graphite is best expressed by the space group $D_3d^5 - R^3m$ with lattice parameters $a = 256.6$ pm and $c = 1006.2$ pm. However, one can adopt both rhombohedral or hexagonal system for representing its unit cell. The lattice constants differ in each case. For more information one can refer to a chapter by M. Inagaki in the Handbook of Advanced Ceramics [110].

Graphite has been synthesized artificially from different precursors, including SiC, petroleum cokes for over a century. Between 1960 and late 1980s, the process of thermal cracking (or pyrolysis) of hydrocarbons followed by deposition of carbon from this chemical vapor was developed and optimized to yield large quantities of pyrolytic carbon. With this, it became possible to synthesize many different versions of graphitic carbon. This process was further refined to yield bulk graphene layers (earlier known as pyrolytic graphite), carbon fibers and tubes. In the late 1980s, fullerenes were discovered and analyzed for their properties as well as hybridization mechanism. This is also the approximate timeline when nanotechnology became a recognized branch of science and technology, and an extensive research started on the nanoscale carbon materials. Some other examples of carbon materials that gained popularity around and after the 1990s include carbon foam, carbon whiskers, carbon walls, carbon peapods, nanographite, nanodiamond, etc. Some previously known carbon materials, such as carbon black, glass-like carbon, and porous carbon were also revisited and structurally tuned to be used in micro-/nanodevices, such as microbatteries and sensors. The nomenclature of these materials also experienced inconsistencies, and it became a challenge to segregate all forms of carbon without considering them allotropes. Today, majority of carbon community accepts graphite, diamond, disordered carbons, and fullerenes (including nonspherical curved carbons) as carbon allotropes. It is still hard to find consensus on the nomenclature of those carbon forms that can be derived by a slight structural modification of a previously existing form. Nonetheless, it is clear that each new carbon material having a unique structure and set of properties must be identified and assigned a clear nomenclature. IUPAC is generally ensuring this and adding new carbon materials into its list on a regular basis.

2.5.1 Classification of carbon allotropes

In this book carbon allotropes are classified based on their hybridization, which in turn influences their structure. This scheme is illustrated in Figure 2.11, where the primary allotrope associated with each hybridization is shown in the first row. Primary allotrope is simply the signature material in each category that represents that specific hybridization. Secondary allotropes are the ones that were added to the list at a later stage. Though some of them may indeed have a high purity in terms of hybridization, there are many reasons for calling them secondary. For example, they may exhibit a very poor (or no) natural abundance. They may only be a structural unit of another allotrope (e. g., graphene or nanographite). Or simply, the research work pertaining to them is currently limited.

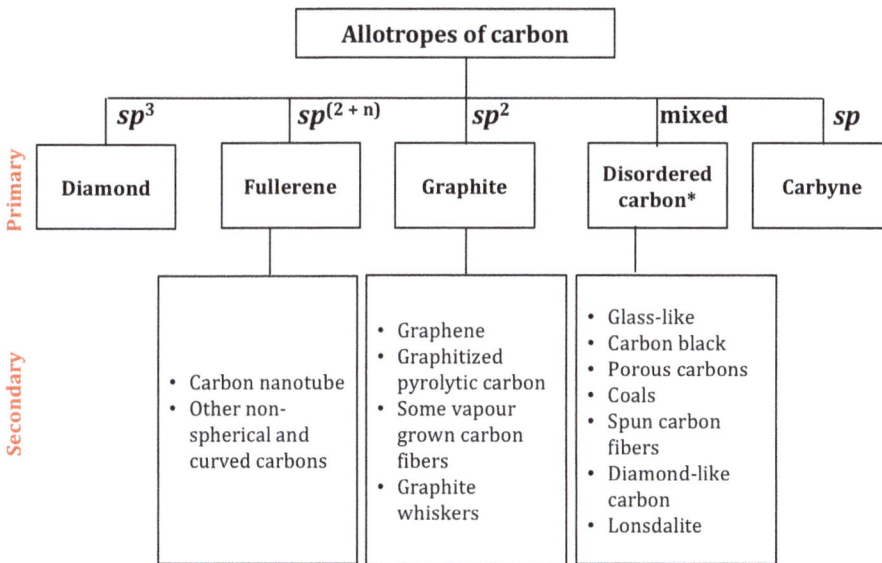

Figure 2.11: A hybridization-based classification of the allotropes of carbon. More carbon materials can be added to secondary allotropes.
*Disordered carbon is not a single material but an indicator of a class of carbon materials.

There are a few more important things about this classification. Other than the three main hybridization states, sp, sp^2, and sp^3, the sp^{2+n} state has been included to accommodate the entire curved carbon family. Fullerene (particularly C_{60}) may be designated as the primary allotrope in this class, because it was indeed the stability and extraordinary properties of C_{60} that generated interest of the scientific community in curved carbons. As stated earlier, CNTs are also curved, and they feature a net hybridization that is not pure sp_2. If they are capped (bowl-like shape at one or both ends), they may even contain pentagons. Thus they are also placed in this category. Often there are curved carbon sheets as well as small fragments that are found as defects in many pyrolytic carbons. Depending upon their manufacture, they may have strong or weak curvatures, with a significant structural nonuniformity. Such curved fragments as well as spherical fullerenes are also detected in polymeric nongraphitizing carbons. But since these materials contain a measurable fraction of sp^2-hybridized carbon, they have been assigned a mixed hybridization state.

Coming to the terminology: mixed hybridization. The term mixed is not used for individual carbon molecules or their hybridization states, but rather to express that the overall (bulk solid) material contains small regions of differently hybridized carbon. For example, a porous carbon material contains both crystalline and amorphous regions. The crystallites are typically <5 nm in their diameters, and they are not perfect graphite crystals. They contain defects in the form of non-six-membered rings or voids, and hence their arrangement is turbostratic. Nonetheless, the individual layers that constitute such crystallites may feature sp^2 hybridization. In the amorphous parts, the atoms can be sp^2-, sp^{2+n}-, or even sp^3-hybridized. In fact, there may exist clusters of six-membered carbon rings, but not in any particular order. Such regions exhibit a significant residual strain in their structure. As a bulk solid, porous carbons are physically amorphous, but many of them have a certain electrical conductivity owing to the presence of crystalline regions. This is the case with many carbon allotropes. These are placed in the mixed allotropes in Figure 2.11. Details of the structure and microstructure of such carbons will be detailed along with their manufacture and applications in upcoming chapters.

> **!** Lonsdalite, also known as hexagonal diamond, was given this name in the honor of crystallographer Kathleen Lonsdale. This form of carbon naturally occurs in meteorites and is also synthetically produced by hot-compression of graphite. Some nanoscale materials having this type of structure have also been developed. They are commonly referred to as carbon-wurtzoids.

2.5.2 Composition diagram

An interesting representation of carbon materials based on hybridization was suggested in the editorial of a book on carbynes and carbynoids [98]. A simplified version of this diagram, here referred to as the composition diagram, is shown in Figure 2.12. In this representation, the three primary hybridization states are placed in three corners of a triangle. Carbon forms having sp^{2+n} hybridization, of course, appear between sp^2 and sp^3. Carbynes sit in the corner, representing pure sp hybridization, whereas graphite and diamond hold the sp^2 and sp^3 carbon positions, respectively. What is interesting about this diagram is the fact that the properties of the materials match with their positions. For example, fullerenes, which can be arranged between diamond and graphite according to their increasing p-character, ultimately collapsing into diamond, are semiconductors. Structural aspects can also be easily explained with this method of under-

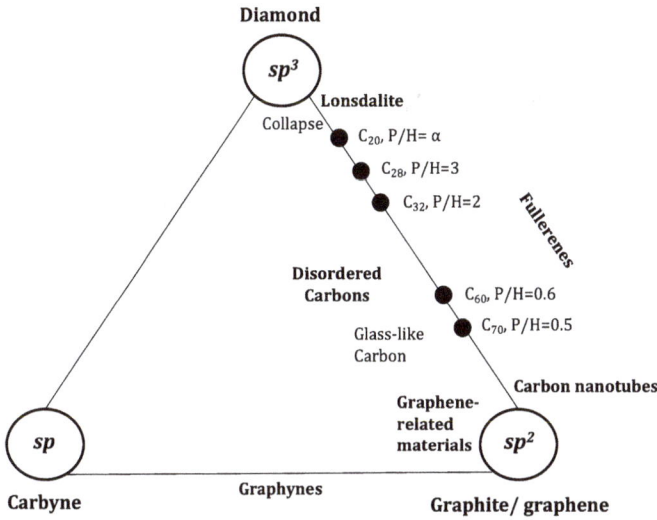

Figure 2.12: A simplified diagram showing the distribution of carbon allotropes based on their hybridization states. P/H indicates pentagon to hexagon ratio in fullerenes.

standing carbon. Evidently, there are still many structural possibilities among carbon materials. Some such materials have been proposed, and their existence in carbon vapors has been hypothesized. But due to the lack of any experimental data, those are still considered hypothetical materials. Nonetheless, this diagram tells us how little we know about carbon, despite knowing so much!

2.5.3 Allotrope conversion

Stable allotropes of carbon generally don't adopt another allotropic form. Graphite is the thermodynamically most stable allotrope of carbon, which technically means that given a choice, every carbon material will try to convert into graphite. But it is not the case in practice, because often there are energy barriers that prevent other allotropes from dissociating into carbon atoms or molecules. Crystalline forms of carbon, such as diamond, are metastable, but they can remain stable for hundreds or years in nature. Fullerenes and nanotubes are also very stable, and they don't naturally change their forms or convert from one allotrope to another. However, allotropic conversion in carbon is observed under extreme conditions, such as very high temperatures, irradiation with high-intensity beams, and very harsh chemical environments.

Let us take the example of diamond and graphite. Natural graphite and diamond are two most studied crystalline allotropes of carbon. Graphite is typically found within the first 150 km below the Earth's surface. Diamond, on the other hand, is typically present deeper than 150 km. Graphite has the lowest energy among all carbon forms. At standard pressure and temperature ($T = 298.15K$, $P = 1$ atm), the enthalpy (or heat) of formation (ΔG_f^o) of graphite is 0 kJ/mol (standard state of the element), whereas for diamond it is 2.900 kJ/mol. This implies that the formation of graphite is thermodynamically more favorable compared to diamond, however, it does not take place automatically due to a very high activation energy of the reaction. In other words, the reaction is thermodynamically favorable, but not kinetically. We will look at both of these aspects here. First, considering the thermodynamic aspects, the state function (ΔG_{rxn}^o) and the reaction constant K for graphite to diamond conversion is calculated as follows:

$$\Delta G_{rxn}^o = \Delta G_{gra}^o - \Delta G_{dia}^o$$
$$\Delta G_{rxn}^o = 0 - 2.900 = -2.900 \text{ kJ/mol}$$

A negative value of ΔG_{rxn}^o indicates that the equilibrium favors the process. The reaction constant, K can be calculated as follows:

$$K = e^{-\Delta G_{rxn}^o / RT}. \tag{2.3}$$

From here, we get $K = 3.22$. The positive value indicates that the reaction is feasible. Now we look at the same reaction from the kinetics point of view. Kinetic calculations take the activation energy and reaction time into consideration, whereas thermodynamics only deals with the energy difference between the start and end products in their respective states of existence. Let us try to find the enthalpy (heat) of reaction for the graphite to diamond conversion with the help of combustion reactions for both of these allotropes. This data is known, and pertains to kinetically favorable reactions. At constant P, we have

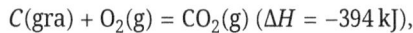

$$C(\text{gra}) + O_2(g) = CO_2(g) \ (\Delta H = -394 \text{ kJ}),$$

and

$$C(\text{dia}) + O_2(g) = CO_2(g) \ (\Delta H = -396 \text{ kJ}).$$

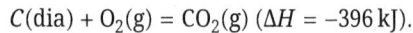

By combining these reactions we get

$$C(\text{gra}) \longrightarrow C(\text{dia}) = +2 \text{ kJ}.$$

Evidently, the enthalpy of reaction has a positive value, which implies this is an endothermic process (gain of heat by the system), which renders it kinetically unfavorable. This is due to the fact that the activation energy of this process is very high. This is due to the high cohesive energy (energy to convert a crystal into infinitely separated atoms) of diamond (717 kJ/mol). However, if this energy can be provided by heat or any

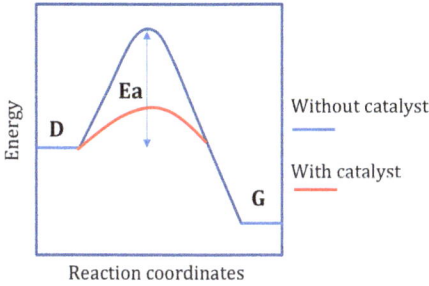

Figure 2.13: Activation energy plot for diamond to graphite conversion. The blue (top) line indicates reaction without a catalyst and the red (bottom), with the catalyst.

other means, diamond can quickly convert into graphite. The activation energy (see Figure 2.13) can be experimentally calculated using the Arhenious equation as follows:

$$K = Ae^{\frac{-E_a}{RT}}, \tag{2.4}$$

where K is the reaction constant, A is pre-exponential factor, E_a is the activation energy, R is the universal gas constant, and T is temperature in Kelvin. The activation energy can be lowered by using a catalyst in the reaction, as shown in Figure 2.13.

Another example of allotropic conversion is the transformation of graphene flakes into fullerenes. This can be achieved via arc discharge, laser ablation, e-beam irradiation, and other high-energy processes. In this process, single sheets of graphite (i. e., graphene) are subject to very high intensity of a certain form of energy, which causes them to form smaller flakes containing defects in the form of non-six-membered rings. Optimum formation of pentagons is extremely important, as they are needed for inducing the curvature in a flat sheet. Once a strong curvature (bowl-like shape) is achieved, it is favorable for this fragment to adopt a spherical fullerene geometry. A detailed theoretical and experimental study on this type of allotropic conversion was reported by Chuvilin et al. [39].

2.6 Phase diagram

A *phase* is defined as a region in a material that is chemically homogeneous, mechanically separable, and is in the same physical state, at a given pressure and temperature. Phase is a thermodynamic quantity that can be defined for any substance, i. e., element, compound, or ionic mixtures. In the case of solids, difference in atomic or molecular arrangements result in different phases. For example, amorphous and crystalline, or different crystal arrangements (e. g. FCC and BCC) make separate phases. Phases occur naturally at the given combination of P, T, and material composition.

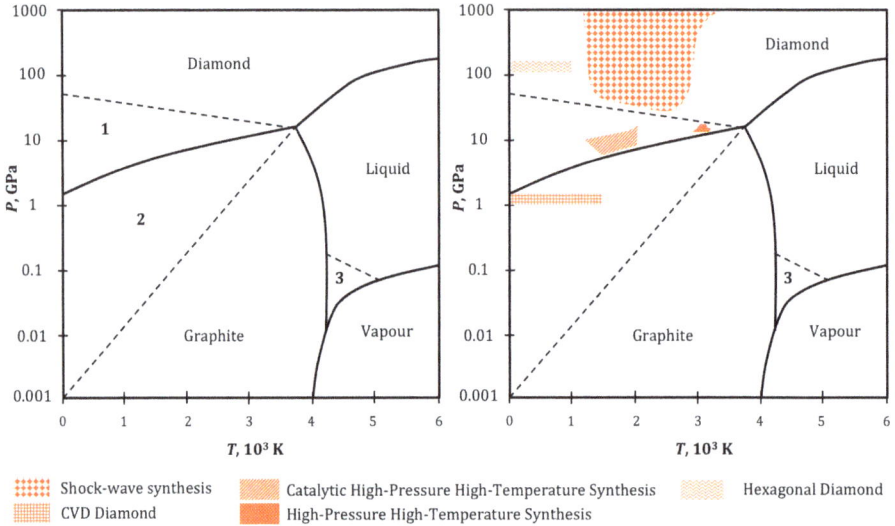

Shock-wave synthesis   Catalytic High-Pressure High-Temperature Synthesis   Hexagonal Diamond
CVD Diamond   High-Pressure High-Temperature Synthesis

Figure 2.14: Phase diagram of carbon. First illustration shows a basic phase diagram, whereas the second one has overlaid information. Region 1: coexistence of diamond with metastable graphite; 2: coexistence of graphite with metastable diamond; 3: possible presence of metastable liquid carbon along with carbon vapours.

A phase diagram elucidates regions of existence of various phases/relationship between the phases that are in a microstructure-level equilibrium in a system consisting of one or more materials, typically as a function of P, T, and composition. In the case of an element, for example carbon, the composition need not be mentioned, since the material is always 100 % carbon. The most widely accepted phase diagram of carbon is shown in Figure 2.14. This is the simplest phase diagram of carbon with no overlaid information.

As one can observe, graphite and diamond are the two prominent crystalline forms or phases or carbon. At temperatures >4000 K, as well as a high pressure, liquid carbon and carbon vapors exist. The lines that separate the two phases indicate that there is an equal probability of the existence of two phases at that point. Often, such phases are in microstructural equilibrium. At the triple point (where three lines meet), three phases can coexist, giving us zero degrees of freedom (as per the Gibb's phase rule). In this diagram, region 1 indicates the coexistence of diamond with metastable graphite, and 2 shows graphite along with metastable diamond. Yet another region, 3, in a triangular shape is the combination of thermodynamic variables where metastable liquid carbon should be present. Evidently, there is no sudden transition from diamond to graphite, or the other way around. A very large number of conditions can induce the phase transformation between the two materials. Carbon materials, such as hexagonal diamond (Lonsdalite) or rhombohedral graphite are metastable states, which fall in these regions. You will note that the P (in GPa) and T (in multiples of 1000 K) are much higher than what is experienced in daily life. Therefore, phase diagrams are based on both theoretical calculations and experimental validation. Computational tools are often used to drive the phase of extreme pressure and temperatures. Now we will discuss some more related questions about this phase diagram.

Where do we show disordered carbon on the phase diagram?

Disordered (and amorphous) phases are not shown on the phase diagram since it is believed that for every P/T pair, there exists a thermodynamically stable crystalline form, which defines the region. Since carbon materials may feature crystallinity at the nanoscale (short-range order) while still being in a physically amorphous state as bulk solids, they are not marked in the phase diagram. Disordered carbons are considered metastable, despite the fact that many of them may survive for very long periods of time without microstructural modifications. Polymer-derived carbon materials, such as glass-like and porous carbon, are shown under regions 1 and 2.

Do equilibrium lines change in a catalytic conversion processes?

Any point in a phase diagram indicates how many material phases are possible at that P and T, irrespective of their origin. For example, graphite can be converted into diamond using catalytic processes. The resulting material is metastable diamond with residual graphite. Such materials can be, and have been, shown on the phase diagram. One must, however, understand that this is only an additional information to give an idea of some of the various metastable materials that may exist in the region. The phase diagram is not incomplete without this information.

Where are the nanomaterials on the phase diagram?

Nanomaterials such as CNTs and graphene may exhibit crystallinity at the nanoscale; their bulk physical state (powders) is amorphous. As phase diagrams only represent crystalline phases, hence nanomaterials are excluded. If one applies pressure to the nanomaterial powder, it will deform. If the pressure is applied in a closed confinement, the spaces between the nanounits will decrease and the material will experience "packing," but not in crystalline sense. Hence, nanomaterials in bulk cannot be considered distinct phases of a bulk crystalline solid. Nonetheless, some phase diagrams for carbon nanomaterials have been proposed based on their cluster sizes and phase transitions [212, 239].

What is the role of phase diagram in micro- / nanofabrication?

Phase diagrams provide information about the microstructure of a material at a given P and T, and this information can be used for selecting the right fabrication process. Many carbon synthesis methods entail deposition of the material from chemical vapors.

An optimum combination of P and T is essential for not only increasing the deposition window, but also to control the crystallinity in the carbon deposits. Though vapor phase carbon is a highly complex system, certain cues can be deduced from the possible phase distribution. For example, one can envision a sp^2- or sp^3-dominant carbon. Phase diagrams are also very useful during carbon thin-film or coating fabrication. Certain carbon materials such as diamond-like carbon tend to be a combination of sp^2- and sp^3-hybridized carbon atoms, which have indeed also been placed onto the phase diagram for understanding metastable carbons. Nonetheless, one needs to careful about the exact nucleation point of one carbon phase in another while developing a new material. This information may have minor variations compared to the phase diagram since nanoscale dynamics of materials in their ultrapure states may not match with their bulk natural phases. Altogether, phase diagram is an important chart of carbon's different forms, which is fundamental to their origins.

2.7 Comparison of different forms of carbon

It is clear from the sections above that carbon is available in various forms for device applications. Certain forms of carbon are traditionally used for large-scale applications, for example, manufacture of water purification columns or as lightweight composites. But their variants have been adopted in micro/nano-fabrication. The increasing need for micro- and nanodevices in today's technology has compelled researchers to investigate the compatibility of carbon with other materials and tune their properties accordingly. Needless to say, several novel carbon forms have been developed, and some previously overlooked forms of carbon have regained attention. A general comparison of properties of some well-known carbon allotropes used in micro- / nanodevices is tabulated in Table 2.1. These properties may vary depending upon the specific fabrication and carbonization parameters.

In addition, another comparative analysis of different carbon materials for one selected device application—neural sensor—is provided in Table 2.2. Neural sensors are chosen for this comparison because they are highly sophisticated microscale devices that require an electrochemically and mechanically stable biocompatible material, which can conform to the tissue or organ where it is implanted. The state-of-the-art commercial technology still utilizes noble metals, such as Pt, but there are certain advantages, such as corrosion resistance, durability, compatibility with imaging techniques such as MRI and an improved longevity, which have encouraged the use of carbon for the fabrication of neural sensors. Since a number of carbon materials are used in the same application with more or less the same dimensional and surface characteristics, the table can provide the reader with a good overview of how different carbon allotropes behave in a similar environment. Properties pertaining to specific carbon materials and their dependence on fabrication process and heat-treatment will be discussed in respective chapters.

Table 2.1: Comparison of the properties of different carbon materials used in device applications. Reprinted with minor modifications from [51].

Carbon material	Fabrication method	σ @ RT (S.cm^{-1})	YM (MPa)	SSA (m^2.g^{-1})	Ref.
Graphite	Commercial: Graphite Grade JP932	10^4	11.5×10^3	–	[1]
Graphene	Mechanical/chemical exfoliation of graphite	10^2–10^3 (theoretical) ~2700 (experimental)	10^6	~2629	[166, 179, 274]
MWCNT	Arc discharge/CVD	2×10^4 (Single tube)	3.70×10^6 (Single tube)	64.8	[57, 172, 232]
Electrospun carbon fibers	Electrospinning + carbonization (1000 °C)	2×10^3	228×10^3	<0.7 (can be increased on activation)	[159, 112, 249]
Vapor grown carbon fibers	CVD (1100 °C)	$\sim 8 \times 10^3$	237×10^3	2.73	[229, 230, 100]
Glass-like carbon (bulk disordered solid)	Commercial (Polycarbon USA/Hochtemperatur-workstoffe GmbH, Germany)	2×10^2	20–40×10^3	–	[19, 76]
Polymeric carbon (micro/nano patterned)	Photo patterning + carbonization (900 °C)	$\sim 4 \times 10^2$ (nanowires)	25.27×10^3 (micro pillars)	–	[160, 208]
Activated carbon	Carbonization + activation	25.13×10^{-3}	35–55×10^3 (PAC/starch composite)	~1200	[3, 137]
Laser-patterned carbon	IR laser irradiation of polyimide sheet	10–30	1.02	200–300	[264, 47, 154]

Table 2.2: Comparison of different carbon materials used for fabricating neural sensor devices. Reprinted with minor modifications from [51].
Acronyms: PI-Polyimide, CVD–Chemical vapor deposition, CNT-Carbon nanotube, LP-C-Laser-patterned carbon, CF: Carbon fiber, RIE: Reactive Ion Etching.

Carbon materials	Substrate	Fabrication process	Electrochemical stability window	Advantages	Remarks, if any	Ref.
Glass-like carbon	PI	Photolithography followed by pattern transfer	−0.8 V to 1.2 V pH:7	Biocompatibility, High signal to noise ratio compared to Pt during *in-vivo* experiments	Expensive fabrication process including pattern transfer	[240]
Graphene	Parylene C	CVD, wet transfer of graphene. Electrode patterned by RIE.	−0.6 V to 0.8 V pH:5.5	*In-vivo* imaging of the cortical vasculature via fluorescence microscopy	Expensive fabrication process including wet transfer of graphene	[181]
CNT	Self-supporting CNT film without substrate	CNT growth by CVD	−0.8 V to 1.0 V pH:7	Wide electrochemical stability window	*In-vivo* cytocompatibility, a concern	[134]
Carbon fiber	PI	Electrospinning and CF patterning by RIE	−0.9 V to 1.1 V pH:7.4	wide electrochemical stability window	Expensive fabrication like electrospinning and RIE	[241]
LP-C	Parylene C	Direct laser patterning of parylene C	−1 V to 1.7 V pH:7.4	Large electrochemical stability window compared to Pt and other flexible carbon electrodes	Inexpensive and one step fabrication process	[242]

1. State two advantages of carbon over silicon for the manufacture of devices that require a high-surface area material.
2. What is the difference between an amorphous and a disordered material?
3. Which factors can cause anisotropy in a material?
4. Which isotope of carbon is used for carbon dating? Why is carbon a suitable material for dating archaeological structures?
5. Is it possible to have a stable C_2 molecule similar to O_2 or H_2? Why or why not? Draw a molecular orbital diagram to support your analysis.
6. Why is graphite called a hexagonal crystal and not hexagonal closed packed?
7. Why are alternating atoms in a hexagon present in graphite or graphene considered nonequivalent? Do they have different chemical reactivities?
8. What is the difference between carbon and graphite, if any?
9. What is the difference between multilayer graphene and graphite?
10. What is turbostratic carbon?
11. What are nongraphitizing carbons? Name at least two carbon materials that belong to this class of materials.
12. What is the difference between porous and activated carbon?
13. What is diamond-like carbon? Do all carbon atoms on this material feature a sp^3 hybridization state?
14. What is the hybridization state of carbon atoms in a C_{60} molecule? Suggest a way to calculate such intermediate hybridization states of the geometry of the molecule is known.
15. Why do some fullerenes exhibit a semiconductor behavior? Does it have something to do with their hybridization states?
16. What is a carbyne? Is it a stable material and can it be synthesized in laboratories? Support your answers with recent research articles on this topic.
17. What is the difference between an allotrope and a polymorph?
18. Why don't diamonds naturally convert into graphite, despite the fact that it is thermodynamically less stable than graphite?
19. What is the difference between a material's phase and microstructure?
20. Can liquid carbon exist under standard pressure and temperature? Why or why not?
21. Name at least three carbon materials along with possible manufacturing processes that can be used in flexible electronics?
22. Which carbon material has the highest electrical conductivity? Can this material be used in devices in the same physical form in which it exhibits this high value of conductivity?
23. Draw a graphite crystal with lattice parameters and clearly show the following planes: (002), (101), (100), (112).

3 Chip-based carbon devices

3.1 Introduction

For the last few decades, electronic devices containing integrated circuits (ICs) fabricated on to silicon chips (small pieces of a thin silicon wafer) have dominated the world markets and led to an accelerated technological growth. Their numerous applications include components of computer systems, smart phones, electronic note-pads, data storage hardware, accessories for vehicles, and a range of sensors. These devices have become such an integral part of life that it is unthinkable to define technology without them. Silicon, its oxides, and occasionally nitrides, are the primary constituents of the ICs. In addition, some metals and ceramics are used for making various components of these ICs. The IC industry has become highly sophisticated over time, and the size of the chips has drastically decreased from several micrometers to a few nanometers. Smaller structures not only require much less material, they have a high available surface area, faster cooling, light weight, and an overall low maintenance and replacement cost.

Back in the 19th century silicon processing techniques were mainly developed for metallurgical applications, because silicon could be used as an inexpensive alloying agent. These initial applications made it a familiar material to the industries and led to the development of crystal growth and purification processes that are in use even today. In the 1950s when transistors changed the entire course of electronic device technology, the potential of silicon was recognized in a new light. This semiconductor material could be conveniently doped, etched, and most importantly, sealed with its own thermally grown oxides. Some of the first silicon-based transistors were manufactured by Bell Labs in the US, which could be considered the inception of silicon chip-based device technology. The oxides and nitrides of silicon provided it with an edge over other industrial manufacturing materials. It was this IC technology that laid the foundation of the microfabrication. Most of the early microfabrication techniques, such as bulk micromachining, surface micromachining, etching, lift-off, etc., were almost entirely silicon-based. These processes supported 2D-fabrication, which implies that the devices featured only stacked films electrically connected through etched holes, whenever required. Complementary metal–oxide–semiconductor (CMOS) transistors are good examples of such multilayer film type devices.

Since the 1980s, there has been a major emphasis on reducing the footprint of the structures, while still increasing their functionalities. This is only possible if the height of the structures can be increased, or in other words, one moves towards a 3D fabrication from 2D. This is where microfabrication used in ICs has its limitations. Though the number and thickness of films can be increased to a certain extent, the rising demands of fitting a large number of micro- and nano-scale components inside popular electronic gazettes cannot be fulfilled without 3D-microfabrication. This can be achieved by fabricating carbon structures on to silicon chips. For this, one needs to first make polymer structures using lithographic techniques on the chips and subsequently convert them

https://doi.org/10.1515/9783110620634-003

into carbon. Several polymers have already been used as a precursors to carbon for almost a century. The achievement of the last 30 years is the translation of this bulk carbon manufacturing technology into micro- and nanofabrication. Notably, silicon is only used as a substrate in most cases. Carbon devices are rather used for electrical and electrochemical applications than as semiconductors.

Carbon-conversion of patterned polymers can be achieved via top-down fabrication as well as bottom-up processes. In this chapter, mainly the top-down lithography processes will be covered. The basic principle here is to first pattern a carbonizable polymer in the micro-/nanoscale, and then convert it into carbon via heat-treatment. Naturally, this approach can be broken down into two primary steps: (i) finding the optimum fabrication technique for making the desired patterns, and (ii) optimization of heat-treatment parameters for obtaining carbon from this prepatterned polymer. Both steps require different expertise, but a perfect device can only be fabricated when material and design complement each other. So let us start with step (i), i. e., micro- and nanofabrication techniques used for patterning carbonizable polymers on to silicon chips. Other chip materials (e. g., sapphire, high purity quartz and carbon itself) may be used as a replacement for silicon, as long as they can withstand a minimum temperature of 900 °C without any deformation.

3.2 Fabrication techniques for chip-based device

As you may recall from Chapter 1, micro- and nanofabrication techniques are the processes that are capable of manufacturing structures having their dimensions ≤100\μm and ≤100 nm, respectively. Different materials, such as polymers, metals, and ceramics, are compatible with different microfabrication processes. Microfabrication processes have a strong dependence on the type of materials that can be used with them. In fact, many micro- / nanofabrication instruments are specifically designed for a class of materials. A complete device may consist of several structures and typically multiple materials. In such cases, both material addition and removal (bottom-up and top-down) have to be employed in the right sequence. Conversion of polymer to carbon requires high temperatures. It is therefore important to carry out the fabrication steps in such a way that thermally sensitive materials are not damaged during the process.

Unlike large-scale manufacturing, it is often difficult to achieve 3D shapes at small scales by top-down manufacturing. Many lithographic techniques facilitate the increase of height (i. e., in z-direction) of a 2D shape. But patterning anything along the z-axis is not possible. This is mainly because the micro / nanomachining tools are significantly different from those used in traditional manufacturing. For example, <100 μm light or other electromagnetic radiations are generally used as tools. Materials processed this way are in the form of 2D films that are exposed to the light from the top (details below), which prevents the user from patterning anything on the side walls of the structure. The structures are generally fabricated in large numbers over a given area. As a result, they are called patterns, not workpiece(s). Patterns consisting of multiples of identical structures are known as arrays. These structures can be hundreds or thousands in number. As a result, various forms of lithography enable batch fabrication.

Fabrication techniques that allow for the increase in height of the structures to some extent, but cannot be used to pattern the structure along its height, are known as two-and-a-half-dimensional (2.5D) manufacturing techniques. A comparison of 2D, 2.5D and 3D shapes is shown in Figure 3.1. Structures that have a much smaller height compared to the length and width are considered 2D, whereas those with a comparable height but no patterns on the sidewalls are called 2.5D. A truly 3D structure would also have sidewall that are not straight. They can be inclined, curved, and/or may contain certain features. The ratio of height to the largest feature size in the xy-plane is known as the aspect ratio (AR). Though there are no very precise AR values that would qualify a structure or technique to be 2.5D, the general idea is that there is no control over the z-patterning.

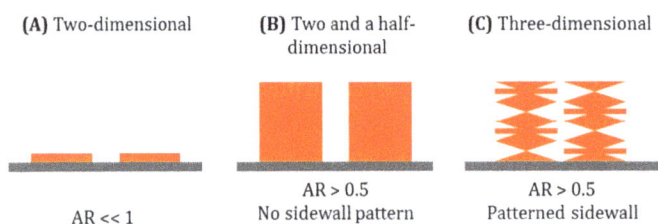

Figure 3.1: Two-, two-and-a-half-, and three-dimensional microscale shapes. AR (Aspect Ratio) is the ratio of height over base diameter for a cylindrical structure.

3.2.1 Photolithography

Photolithography is the most common polymer processing technique that can be used for making carbon structures. This process is used for patterning photosensitive polymers. Some of these polymers are good carbon precursors, i. e., they yield a high fraction of electrically conductive carbon materials at temperatures that are below the melting point of silicon (or the substrate used). The step-wise details of photolithography are described below.

Photolithography is used for making structures with a critical dimension (smallest feature size) larger than 5 µm. This technique is inspired by film-based photography. Before the advent of digital photography, the cameras came with a film roll (see Figure 3.2), which was exposed to the light for a very short time for capturing the image. This exposure caused a wavelength-dependent modification to the film properties. Unmodified parts were then washed away or "developed" inside dark rooms. The films were indeed coatings of photosensitive polymers onto thin plastic sheets. After developing and printing the photograph, the master copy known as the *negative*, could be preserved and used for reprinting. The term "negative" is believed to originate from the fact that

Figure 3.2: A camera along with the film-roll that contains a thin film of a photosensitive polymer. Picture source: www.freeimageslive.co.uk.

the photosensitive polymer was a negative photoresist. The difference between positive and negative resists is described later. First, let us understand the basic principles and process steps.

Basic principle: Photolithography is based on the principle that the UV light can induce chemical modifications into certain materials, such that the UV-exposed parts attain a different solubility behavior compared to the unexposed ones. The can used to selectively degrade of cross-link films of UV-sensitive polymers via a photo-mask containing micro-scale features.

Materials used: Photosensitive polymers are generally known as photoresists or simply resists. The photoresist can be of two types: positive or negative. If the resit degrades under the UV radiation, it is a positive resist. If it hardens due to the cross-linking of the polymer chains, it is known as a negative resist. The reasons for this terminology becomes clear after understanding all the process steps used in photolithography.

Process steps: The steps used in the photolithography process are shown in Figure 3.3. First, a thin layer of resist is spin-coated onto a hard and flat substrate, such as a silicon wafer or a glass slide. Spin coating machines have a holder (known as the vacuum chuck) that can rotate in the xy-plane, generally up to 10000 rotations per minute (rpm).

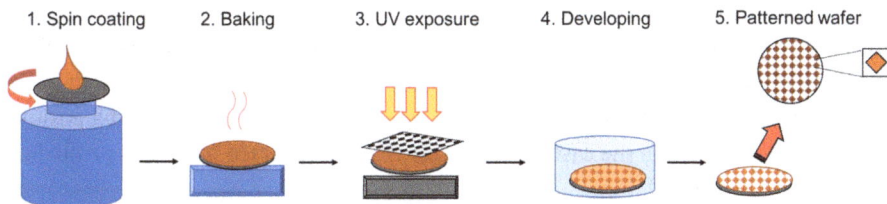

Figure 3.3: Basic process steps in photolithography. 1. Spin coating a resist onto silicon wafer, 2. pre-exposure bake, 3. exposure to the UV light through a photomask, 4. developing the structure by dipping the wafer into a chemical that selectively dissolves the degraded parts (positive resist) or the uncross-linked parts (negative resist), 5. patterned wafer that can be diced into small pieces, or further processed if needed. Negative resists also require a post-exposure bake to enhance cross-linking.

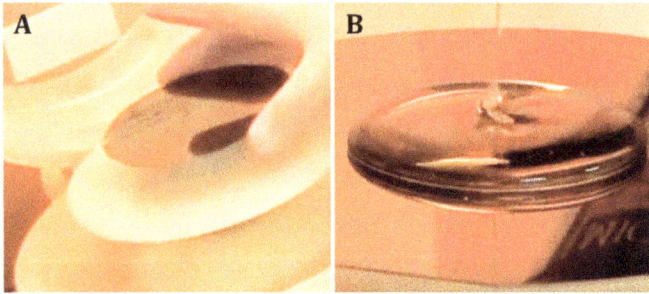

Figure 3.4: (A) A typically spin-coating set-up with a silicon wafer mounted on it; (B) a negative photoresist onto silicon wafer before spin-coating.

The substrate is placed on this holder and firmly fixed by creating a low vacuum using suction. 2–5 mL resist is pipetted onto the substrate, which is rotated at a rpm value optimized for the specific resist viscosity, such that it yields the desired thickness. Images of a spin-coating equipment and a viscous photoresist being poured onto a silicon wafer are shown in Figure 3.4.

For most commercial resists, the rotation parameters and the corresponding resist film thickness are provided by the suppliers in a document called the data sheets. Spin coating is a highly reproducible process. For a given resist-rpm combination, the resulting film thickness will always be the same (very small errors), irrespective of the initial volume of the resist placed on the wafer. Any extra resist is thrown out by the centrifugal forces. The spin-coated resist film is baked to get rid of the solvents and to decrease the film stresses (e. g., ripples, material deposition near the edges) by reducing its surface energy. A film with uneven thickness will not contact the mask at all points, thus leading to irregular UV doses and undesired diffraction of the UV light, which will ruin the final structures. The flat, baked film should have glass-like properties, i. e., it should be well below the glass-transition temperature of the resist to ensure the right UV dose, and also to protect the mask from any damage.

The next step is to selectively expose the resist film to the UV radiation. This is carried out using a photo-mask, which is printed with a UV-opaque ink on a UV-transparent silicate glass (such as quartz). When the critical dimension is <10 µm, the mask needs to be prepared using atomic deposition of metals, such as chromium via electron-beam deposition. The mask carries the exact same pattern as the final structures if a positive resist is used, whereas in the case of a negative resist, a reversed pattern is printed on the mask (see Figure 3.5).

The most critical parameter in photolithography is the determination of the correct UV dose for a given resist. In the case of a positive resist, the UV light basically causes material degradation. Here the films are generally thin. Positive resists are rarely used for carbon device fabrication, mostly because carbon microdevices are generally designed with a certain height. Negative resists are often composed of thermosetting resins with

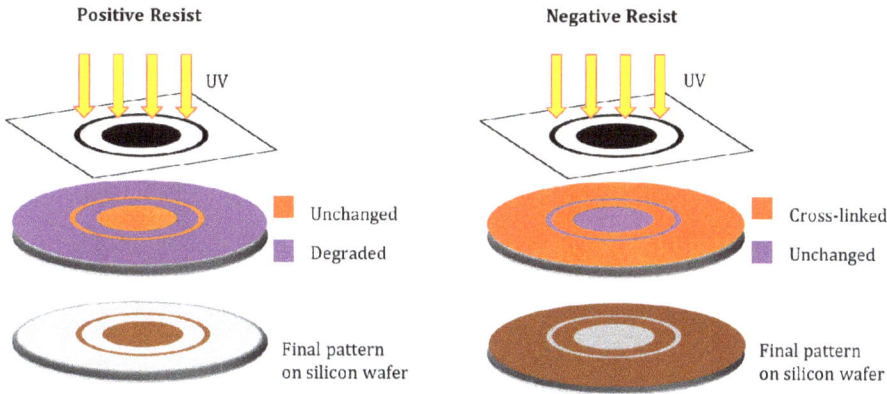

Figure 3.5: Pattern transfer in positive and negative resists after the UV exposure.

epoxy side groups, and are available in different viscosities that can yield films as thick as 150 μm. These resists contain a photoinitiator material, which initiates a chain reaction by producing a photoacid (e. g., via proton or a radical) when exposed to the UV light. The dangling bonds created during this process facilitate cross-linking of polymer chains, thus rendering the overall polymer hard and set. To maximize the cross-liking, a high mobility of the constantly generating radicals is essential. Therefore, although the photoinitiation process starts at the time of UV exposure, its amplification must be assisted by a post-exposure bake. The *post-exposure baking* is an additional step used only in the case of negative resists to propagate the chain reaction, and hence cross-linking. The temperature for this bake is higher than that of the preexposure bake, as the material has already solidified to some extent. As the cross-linking progresses, the material becomes harder. The most common negative tone photoresists used for carbon-based microfabrication are phenol-formaldehyde resins.

The appropriate UV is calculated according to the chemical properties of the resist, e. g., chain length of the polymer, viscosity, concentration of photoinitiator, etc. Many commercially available resists to date have patents protecting the details of their exact chemical structures. As a result, it is a common practice to follow the UV dose recommendations provided by the resist supplier rather than optimizing it. There are some modification, for example, extended post-exposure bake [123] at lower temperatures than recommended, which have shown better results in term of achieving a good resolution. The supplier datasheets generally provide the details of the UV energy per unit area. The exposure time is then calculated by dividing the UV dose by the intensity of the lamp. It is recommended to measure the lamp intensity right before the experiment as the intensity of the UV lamps decreases over time. If the sample is underexposed (lower dose than required), it may not get cross-linked and get washed away during development (next step). If it is over-exposed, the top part may become wider than the bottom, creating an undesirable taper. This type of lithography error is known as T-topping [141].

The next and final step of photolithography is to develop the structures using a suitable solvent. This solvent should dissolve either the degraded resist (positive resist) or uncrosslinked parts (negative resist). Depending upon resist thickness, the time to develop the structures varies. Underdeveloped structures may contain resist in the designed gaps, whereas over-developed structures may peel-off or have their corners damaged.

Finally, the structures are observed under an optical or electron microscope. Subsequently the wafer is diced into small chips. If the goal is to convert these structures into carbon, they can be heat-treated (described in Section 3.3) before or after dicing.

> **!**
>
> **Problem:** Calculate the UV dose (duration of UV exposure in second) for a resist film of thickness of thickness 10 μm. Given the energy required for photoinitiation; cross-linking is 140 mJ/cm² and the intensity of the UV lamps is 7.3 mW/cm².
> **Solution:** Exposure time t (in seconds) = UV dose/lamp intensity
> As the units do not require any conversion, we can directly proceed with this relationship. The value of $t = 19.18$ s. This can be rounded up to 20 s, because during the experiment one can generally not control fractions of seconds.

3.2.2 Nano-imprint lithography

Nanoimprint lithography (NIL) is used for fabricating nanoscale structures in relatively large areas (several centimeter scale). Whereas some of the steps in NIL are similar to photolithography, the main difference is the use of a physical mold, which creates the impression of the structures into a resist film. The resist can then be hardened by simply cooling down. Of course, the polymers that can be used with NIL need to be thermoplastics, as they are expected to soften and harden when the mold temperature is above or below their glass-transition temperature. Further details and process steps are described below.

Basic principle: Patterning relatively large surfaces with nanoscale structures by mechanically pressing a mold having the inverse structures, followed by curing the resist and mold removal. Afterwards, the leftover resist between the structures is removed (in most cases) by dry etching.

Materials used: Thermoplastic resists, heat, or UV curable.

Process steps: NIL can be performed in a polymer or on metal surfaces. If the target material is a polymer, then the first step is to spin-coat it onto a substrate (e. g., silicon wafer) using similar set-up as photolithography. These polymers are generally kept solvent free to avoid any release of volatile materials when the mold is heated. If needed, a baking step can be performed to remove the solvent.

Coated wafer is placed inside the NIL chamber and held tightly on a vacuum chuck. The next step is to press a mold into the polymer film. A good contact between the mold and the polymer is essential to imprint the patterns. Vacuum inside the chamber helps

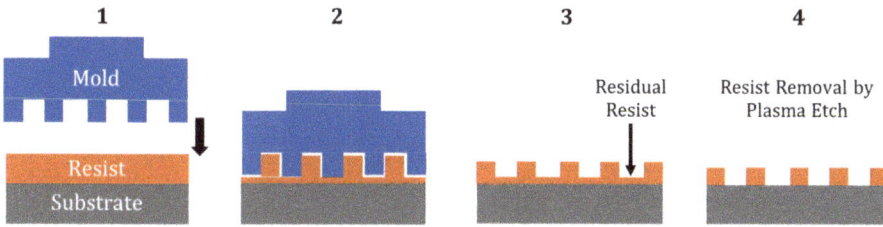

Figure 3.6: Process steps in nanoimprint lithography.

the contact between the mold and polymer. Now the mold is heated to a temperature just above the glass-transition temperature of the polymer. This makes the polymer soft and enables a good insertion of the mold. Following this, the temperature is reduced again, and mold is removed (see steps 1 and 2 shown in Figure 3.6). Naturally, there would be some residual resist in-between the structures, as the mold does not penetrate through the entire film (step 3 of Figure 3.6). This extra material is removed by dry etching techniques, such as deep reactive ion etch (DRIE).

A variation of NIL is photo-NIL, where UV exposure is used in addition to mold insertion for hardening the resist. The calculations of the UV dose vary with film thickness and can be calculated in a similar fashion as in photolithography. UV-assisted cross-linking enables the use of thermosetting polymers in NIL. Note that there is no mask in the process and the mold is made of a UV transparent material. The UV light is only turned on after the mold is properly inserted into the polymer. The energy from the UV light simply cross-links the polymer in whatever shape it has at that moment. Of course, only negative tone photoresists can be used in NIL. These are suitable for carbonization, hence, this process has been used for manufacturing carbon-based nanodevices [186].

If the resists is thermally cured, the process is also known as hot-embossing [254]. In fact, hot-embossing and NIL are two techniques developed in parallel; they have evolved for sophisticated nanoscale fabrication. Recent advancements in NIL include a roll-to-roll continuous fabrication (similar to newspaper printing), where the rolls themselves are used as molds. This process is an excellent tool for creating very large surfaces with nanoscale features.

3.2.3 X-ray lithography

X-ray lithography is one of the first lithographic techniques that is still in use for fabricating very high aspect ratio (100 – 1000) features. This technique was developed in the Forschungszentrum Karlsruhe, Germany, in the late 1980s. The process utilizes X-rays for patterning the material, which can be a polymer, metal, or generally anything that can be degraded with high energy X-rays. The generation of X-rays is carried out in X-ray synchrotrons. Synchrotrons are large particle accelerators with closed circular paths,

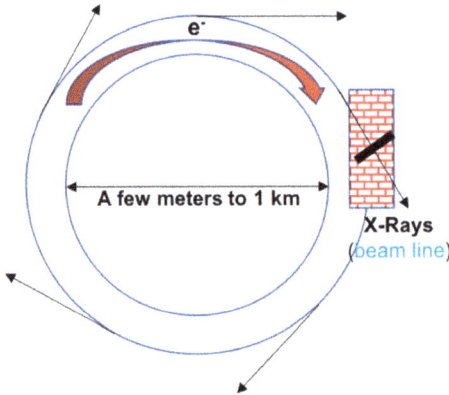

Figure 3.7: A schematic diagram showing the functioning of a X-ray synchrotron.

which can have diameters of up to 1 km. In a closed-loop tube-like cavity, the electrons move while constantly turning to follow the circular path. X-rays are emitted due to this motion, which are then collected in chambers located tangentially. A schematic diagram of a synchrotron is shown in Figure 3.7. These powerful X-rays are exposed on to a resist film through a thick metal mask. The polymer patterns thus fabricated can be electroplated and, after removing the remaining resist, used as a master mold replicating the pattern. Due to the use of synchrotrons as the X-ray source, the process remains expensive and limited to a few facilities. However, the precision of the shape, despite the very high aspect ratio, is unmatched when compared to other lithographic processes. Some resists that are used for X-ray lithography are phenol-formaldehyde-based [120] that can be carbonized.

> Other than X-ray LIGA, synchrotron radiations are used for fundamental science and engineering applications, including material characterization by diffraction analysis, residual stress measurements, high-resolution texture determination, microstructure establishment, elastic constant determination, study of electronic density of states, etc. In addition, many other fields benefit from these high-end facilities, such as laser development, condensed matter physics, and particle physics.

Importantly, X-ray lithography typically encompasses the additional steps of electroplating the resist pattern (Galvanoformung, in German), and subsequently, a mold preparation (Abformung, in German). When these steps are included in the process, it is known as X-ray LIGA. Here *Li* is taken from the German word *Lithographie*, *G* from *Galvanoformung*, and *A* from *Abformung*. In principle, one can stop at the lithography step if only one polymer pattern (e. g., to be used as a master mold) is required. Of course, this will render the overall fabrication highly expensive.

Basic principle: X-rays can be used as a tool to degrade a material by breaking its inherent chemical bonds. This changes the material's solubility, which allows this exposed material to be selectively washed away with common solvents. X-ray penetration is deep with a straight path due to very little diffraction of X-rays.

Materials used: Polymers, metals, and potentially a variety of other substances.

Process steps: Prior to lithography steps, it is essential to understand the multistep mask fabrication compatible with this technique. In the initially developed process, first a silicon wafer (often carbon coated) is used as a substrate for coating a titanium layer of 2 μm thickness. A slightly thicker layer of resist that can be patterned with an electron beam (e. g., poly-methyl methacrylate or PMMA) is then coated on top of it. Using the electron beam lithography, a reverse pattern (opposite of the desired final pattern) is carved onto it. This entire pattern is plated with gold followed by resist removal. This leaves a couple of microns high gold pattern with empty spaces in the shape of the mirror image of the pattern with a titanium backing. This is then cut out of the silicon wafer, generally with a steel frame attached around it (for better handling).

The working substrate is typically a steel plate with a thinned backside coated with 2 μm titanium and a much thicker (>50 μm) resist layer on top of it. This is exposed to the X-rays through the mask followed by resist development. Here, depending upon the resist, its properties and film thickness, an appropriate chemical can be chosen as the developer. Some commercially available resists have a developer prescribed by the supplier. As such, the microscale features are ready at this stage. But one can further perform electroplating of these patterns, generally with nickel and remove the resist altogether. This nickel pattern will then serve as the master mold for further fabrication. If the ultimate goal is to perform molding with this master, the process should be designed so as to make a negative pattern with cavities, where the resist can penetrate and yield a desired pattern. X-ray LIGA can also be used for making molds for NIL and similar processes.

If we compare the feature heights for those fabricated by photolithography and X-ray lithography, there is a significant gain of aspect ratio in the case of X-rays. The designs, however, still remain 2.5D, because there is no sideways patterning. One can get an idea of the relative heights of features made by photolithography of a positive and a negative resist, compared to what can be achieved via X-ray lithography from Figure 3.8.

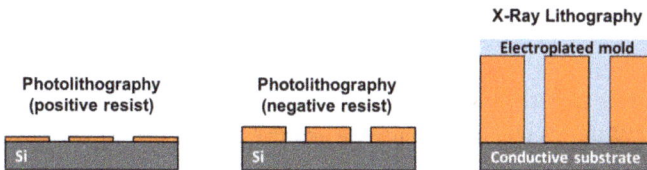

Figure 3.8: A comparison of patterns fabricated via photolithography of a positive tone and a negative tome resist, compared to X-ray lithography.

3.2.4 Two-photon lithography

Two-photon lithography is a micro- and nanoscale 3D-printing technique that has been used with carbonizable resists. With this process, one can get truly 3D carbon shapes, but they are fabricated in a serial fashion (i. e., it is not a batch fabrication technique). This technique is therefore not cost-effective for wafer-scale device manufacturing. But making extremely small individual features for specialized applications, such as conical tips patterned onto silicon cantilevers for atomic force microscopy (AFM), have been possible with two-photon lithography. This process is also based on chemical modification of photoresists. But unlike in photolithography, here the cross-linking of the resist is achieved via a two-photon process. These photons are supplied back-to-back using lasers. Since two-photon absorption by a material is a nonlinear process, very high intensities of laser are required. Resists are generally developed with a high photon absorption for a certain wavelength, as the laser source needs to be selected accordingly. This process is mostly automated as the designs are fed into the computers, which control the laser source. Importantly, this process obviates the need for expensive masks, which is an advantage over photolithography.

Basic principle: Two-photon absorption of a resist that is transparent to certain laser wavelength can drastically modify the chemical properties of the resist, thus making it soluble in certain solvents. The chemical change takes place at the focal spot of the laser, which can be software controlled. Laser can be used for scanning (layer-by-layer) the material according to the features defined in the pattern.

Materials used: Photosensitive polymers.

Process steps: As two-photon lithography is s serial process, there is generally no need to spin-coat a flat layer of resist onto any specific substrate. If the laser is exposed from the backside (i. e., the droplet is facing downwards), transparent substrates, such as glass, are used. One can simply place a drop of a resist (drop casting) onto the substrate, which can be in liquid or solid form. The design is fed into the system, which is connected to the *xyz*-stage as well as the laser source. An appropriate dose of laser is then provided such that there are two simultaneous photons hitting the desired spot. This causes the material to degrade or crosslink, depending upon whether it is a positive or negative photoresist. In this high energy process, often parts of resist can evaporate. The developer is selected accordingly, and in some cases, it may not be required at all. Partial evaporation of resist, however, may lead to porosity or other damages in the patterned shape compared to the initial design. As a result, it requires further optimization of laser parameters. With this technique, one can achieve pretty much any shape, including metamaterials, suspended, and freestanding shapes and other complex patterns. The fundamental process principle is illustrated in Figure 3.9.

In addition to the aforementioned techniques, 3D printing has also been used for making carbon patterns (detailed in Chapter 6). Other microfabrication techniques used for chip-based carbon devices include the use of porous alumina molds, soft

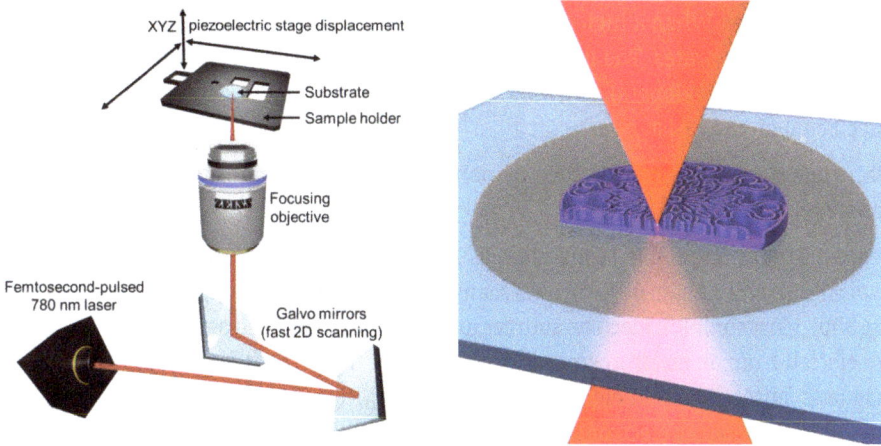

Figure 3.9: Fundamental principle of two-photo lithography. Image from: Bunea et al. [33].

lithography, electrospinning, and other support techniques, such as etching, plasma treatment, and sputtering. Review articles that enlist various microfabrication and heat-treatment methods for fabricating carbon shapes for chip-based devices can be referred [49, 203].

The quality of carbon materials is measured by their electrical, electrochemical, and mechanical properties. In addition the density, surface area, and their ability to maintain a given shape during carbonization are also important parameters. The exact requirements may vary from application to application. To understand polymeric carbons, one needs to learn how the carbon-conversion process works. This high-temperature process is sensitive to even minor differences in parameters and is chemically very complex. In the coming sections, the carbonization process specific to chip-based devices is explained in detail.

3.2.5 Support techniques

3.2.5.1 Electron beam lithography

Although electron-beam lithography (EBL) is an independent micro- / nanofabrication technique used for various tasks, in the context of carbon-based microfabrication it is often used as a support technique, for example, for fabricating masks for photolithography. It is a relatively expensive technique as it is based on direct writing using an electron beam, which can only operate under high vacuum conditions and is a serial and time-consuming process. It is used for making chrome masks, which typically require <10 µm feature sizes. Masks having larger critical dimensions may be printed onto PET sheets employing sophisticated printing techniques. Basic concepts of EBL are described below:

Basic principle: When a high-energy beam of electrons is incident upon a polymer, it degrades or evaporates. The degraded material can be further removed using specialized chemicals (developers). Software-assisted direct-writing is employed for scanning the e-beam on the resist.

Materials used: EBL resists, such as poly-methylmethacrylate (PMMA) and polycarbonate (PC).

Process steps: A suitable resist, such as PMMA, is spin coated onto a rigid substrate (typically Si) and a beam of electron is scanned over it as per the supplied design (see Figure 3.10). Most common electron sources are LaB6 or metals such as W, Zr, etc.. Electrons are ejected from an electron emitter (cathode), and the beam is focused and guided via magnetic lenses. To avoid electron scattering, very high- or very low-energy electrons are used. High-energy electrons can penetrate deep without scattering. Low-energy electrons have too low energy to scatter much. After patterning with EBL, metal film is deposited on top of it using sputtering/evaporation. The resist is then removed (process known as lift-off) and metal pattern is obtained on the substrate. Master molds can also be made using EBL, which can be used for other processes, such as X-ray lithography.

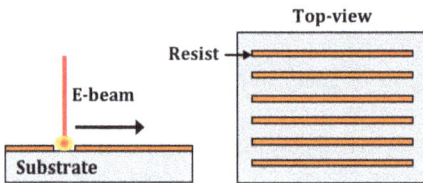

Figure 3.10: A schematic diagram showing the working principle of electron-beam lithography.

3.2.5.2 Metal deposition and etching

At many instances during micro- and nanofabrication, deposition of a thin layer of metal is required for making electrical contacts. To pattern this layer for making well-defined electrical contact lines or pads in the microscale sizes, a photoresist layer is first patterned using, for example, photolithography or EBL. On top of this, a layer of metal (Cu, Au, Cr, etc.) is deposited using techniques known as metal sputtering or evaporation. The resist is removed afterwards, which carries away the metal deposited on top of it. The remaining, patterned metal stays on to the substrate.

The opposite of this is etching. Etching simply implies removal of a material that can be carried out using chemicals (wet processes) or plasma (dry processes). Plasma is the ionized state of matter. It can be generated using high voltages in a vacuum chamber. The target gas that needs to be ionized is filled inside a purged chamber, and then very high voltages are applied. Oxygen and argon plasma are some common types that are used in microfabrication. More reactive plasma, such as XeF_2, may contain other inert gases or their compounds for an improved radical concentration and reactivity. In Figure 3.11,

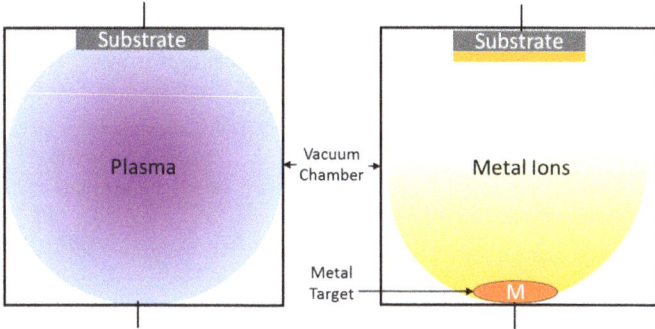

Figure 3.11: A simplified schematic showing the fundamental principles of material etching inside a plasma chamber, and an atom-by-atom deposition of metal from a target known as sputtering.

representative sketches of both plasma etching and metal deposition via sputtering are shown. More details are described below:

(a) Plasma etching:

Basic principle: When a material is placed in contact with a high intensity plasma, it is etched away via the dry etching process. Different materials have a different etch rate. This property is often utilized in microfabrication.

Materials used: Polymers, metals, carbon, or, in principle, thin-layers of any material.

Process steps: A gas is ionized under high voltage inside a vacuum chamber for creating a plasma environment. A substrate with certain parts masked with another material (typically a lithographically patterned sacrificial polymer) is placed inside this chamber. Plasma induces dry etching of the material, which is unselective. The material is removed from both polymer and the underneath (metal) layer. However, the patterns masked/protected underneath the polymer patterns are prevented from etching. Plasma etching is also used for cleaning the surfaces prior to microfabrication, as it etches away any residual chemicals or polymer traces from the substrate.

(b) Sputtering:

Basic principle: Under low pressure or vacuum conditions, metal atoms are removed from a target (e. g., a plate of metal or alloy). If such atoms are filled inside a chamber and there is another electrode (deposition substrate) serving as an attractive force for the metal atoms or ions, the metal selectively deposits onto it.

Materials used: Target metals or alloys, including Cu, Cr, Au, Ti, and various alloys. Substrate can be any material that can withstand the pressure and temperature conditions inside the plasma chamber.

Process steps: A metal is evaporated at low pressure and/or high temperature from a target plate. The metal in its vapor phase is attracted towards the substrate, attached onto an electrode. Substrate typically contains lithographic patterns, but a thin-film of metal can also be deposited as a uniform coating without any specific features.

> **!** Various other support techniques, such as wire bonding, soldering, surface milling, dicing, profilometry, and electron microscopy, are commonly used in micro- and nanofabrication. Though each one of them is not described here in detail, a microsystems engineer is expected to have a working knowledge and fundamental understanding of these techniques.

3.3 Conversion of polymer patterns into carbon

After fabricating micro- or nanoscale polymer patterns, the next step is to convert them into carbon via a controlled heat-treatment process. Polymers have been used as a source of various synthetic carbon materials for over 100 years. Synthetic graphite, glass-like carbon, porous and activated carbon, carbon fiber, and even carbon/carbon composites are all prepared by essentially the same type of heat-treatment process, with certain parameter variations. The difference in the case of micro- and nanoscale shapes is the fact that they are attached to a substrate (generally silicon), and they have a much higher surface-to-volume ratio compared to bulk industrial carbons. The polymers employed in microfabrication generally have a higher purity and a greater extent of cross-linking compared to industrial carbon precursors. As a result, carbonization of microfabricated polymers feature certain aspects that are different from large-scale polymer carbonization. The chemical complexity of the process, however, remains the same, which is fundamentally based on removing all noncarbon atoms from an organic (hydrocarbon) material.

The aforementioned heat-treatment process can be broadly divided into three regions: (i) pyrolysis (generally up to 700 °C), (ii) carbonization (700–2000 °C), and (iii) graphitization (>2000 °C), as shown in Figure 3.12. These three temperature segments can be further segregated based on the type of bond cleavage and/or formation. These phenomena are based on the chemical structure of the precursor. In cases where the pre-

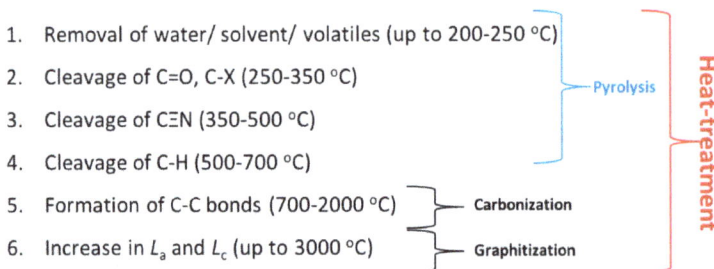

1. Removal of water/ solvent/ volatiles (up to 200-250 °C)
2. Cleavage of C=O, C-X (250-350 °C)
3. Cleavage of C≡N (350-500 °C)
4. Cleavage of C-H (500-700 °C)
5. Formation of C-C bonds (700-2000 °C)
6. Increase in L_a and L_c (up to 3000 °C)

— Pyrolysis — Carbonization — Graphitization — Heat-treatment

Figure 3.12: Chemical changes occurring at various temperature ranges during heat-treatment of solid or semisolid organic materials in the absence of oxygen. These are approximate values and specific materials may show slight variation.

cursor itself is a mixture of various hydrocarbons (e. g., petroleum pitches and cokes), the process can become very complex. Heat-treatment for microfabricated shapes is typically conducted at 900 °C. This is because most micro and nanoshapes have a substrate that may not be able to withstand elevated temperatures. Nonetheless, it is important to understand the entire heat-treatment process to know which type of carbon is expected from a polymer at a certain temperature.

3.3.1 Pyrolysis

Pyrolysis is defined as the thermochemical decomposition of organic molecules in the absence of oxygen. A process can only be called pyrolysis when the breaking of bonds takes place strictly due to thermal energy, rather than chemical reactions with external oxygen (i. e., burning). Pyrolysis is different from both decomposition and combustion. The term "decomposition" generally refers to the natural degradation of a material (organic or inorganic), which may or may not have been caused by heat, whereas pyrolysis is necessarily a thermal process performed in a controlled environment. Since pyrolysis is conducted at elevated temperatures, it is also likely to be confused with combustion (burning). In fact, in some cases, such as large-scale waste treatment, pyrolysis is may intentionally be performed in the presence of some oxygen (from air). Depending on the amount of the supplied air, the actual process can be a combination of pyrolysis and combustion, or purely combustion. Nonetheless, by definition pyrolysis itself is an oxygen-free process, which allows for an exclusively thermochemical decomposition of the organic material [49]. The word "pyro" is derived from the Greek word "pyr," which means fire. As a result, the term pyrolysis has been used to describe various processes involving high-temperatures or fire. Some examples include pyrometallurgic treatments applied for metal refining, and the process of creating *Bhasm* (Saknskrit word meaning ash-like residues) used in *Ayurvedic* medicines. Occasionally, firing an inorganic mixture for the purpose of ceramic manufacturing is also designated as pyrolysis.

Pyrolysis is not limited to solid materials. Any organic molecule, including light hydrocarbons, such as methane, acetylene, alcohols, and other gaseous or liquid material can undergo pyrolysis at high temperatures [62]. In such cases, a carbon smoke is formed inside the reaction chamber, where small fragments of carbon are present. They can be collected onto a substrate, typically with catalytic properties. This entire process (pyrolysis followed by carbon deposition) is known as the chemical vapor deposition (CVD) of carbon, and is extensively used for carbon nanomaterial fabrication. Additional details of this process are provided in Chapter 4.

For chip-based structures composed of solid organic polymers, the heat-treatment is carried out in a tube furnace, as shown in Figure 3.13. A continuous flow of an inert gas (e. g., nitrogen, argon, helium) or a vacuum is maintained in the furnace to avoid any oxygen. The gas flow also pushes the byproducts out of the furnace and prevent them from depositing onto the structures. Most organic materials become unstable at high

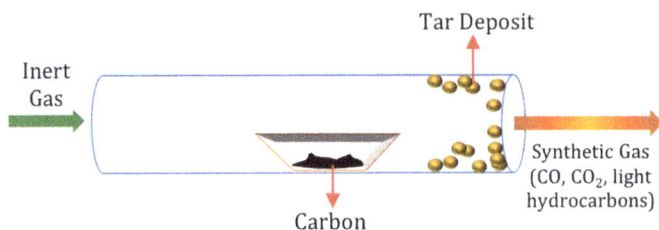

Figure 3.13: Pyrolysis products of solid organic materials heated in a tube furnace under inert gas flow. Organic matter (natural or synthetic) converts into carbon after the release of hydrocarbons that are either condensable and heavy (tar), or light and volatile (synthetic gas).

temperatures and start to form smaller fragments, which are more stable at that temperature. With further temperature increase, these fragments get further fragmented, and so on. During these chemical reactions some very light volatile molecule, e. g., CH_4, may also get released. Some relatively high molecular weight organic materials with tarry or oily texture may also leave the reaction mixture in vapor form. Altogether, the initial polymer converts into three components: solid (residual carbon), liquid (tarry droplets often deposited on furnace tube), and (iii) gas (volatile materials that leave the furnace).

> The term *organic* doesn't necessarily mean biodegradable, biosafe, or free of harmful chemicals (as in organic foods). Any hydrocarbon in solid, liquid, or gas form is organic. Carbon in its pure elemental form is inorganic, as there are no C–H bonds present in it.

In addition to manufacturing carbon patterns (large or micro- or nanoscale), there are three major application areas of pyrolysis: (i) carbon nanomaterial synthesis through the CVD process, (ii) large-scale waste-treatment, and (iii) analysis of organic materials via gas chromatography or mass spectrometry, as shown in Figure 3.14. In some industrial manufacturing processes, more than one type of precursor (solid and gas) may be used, which leads to carbon with lower porosity. In the context of microfabrication with carbon, only polymer-to-carbon conversion is discussed in detail here. For other applications, readers may refer to the review article by Devi et al.

3.3.1.1 Mechanism of polymer pyrolysis

When an organic polymer is heated above its degradation temperature, carbon-heteroatom bonds start to cleave, which is followed by the formation of the new C–C bonds. As shown in Figure 3.12, the weaker links in the polymers, generally the bonds of carbon with halides and oxygen cleave up to 350 °C. This is followed by C–N bond breakage. At the end, it is the C–H bond that is cleaved (around 600–700 °C). It is important to note that all the bonds of a certain type do not simultaneously break in the entire material. The polymer degrades via sequential fragmentation. In fact, this fragmentation pattern is studied in another application of pyrolysis: GC-MS analysis. When these fragments

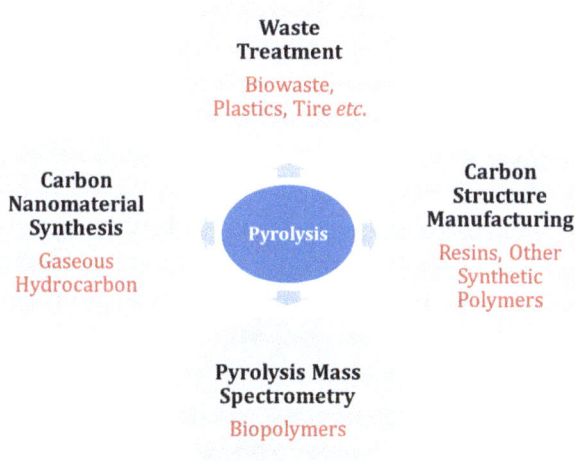

Figure 3.14: Major application areas of the pyrolysis process (black text), with primary organic material used as the carbon source (red text in smaller font).

are produced at a certain temperature, they spark many other reactions in the pyrolyzing mixture. For example, the fragments can undergo further fragmentation, there may be several gases or volatile liquids being produced as byproducts; there may be products of one reaction serving as reactants for another, and so on. As a result, at any given temperature point, it is extremely difficult (almost impossible) to detect all chemicals that are present. Many unstable pyrolysis products are indeed very short-lived, and they may simply be intermediate compounds. The reaction kinetics for these multiple parallel reactions is also different from each other. Altogether, pyrolysis is, chemically, a very complex process. It is not one chemical reaction, but rather a combination of numerous chemical reactions, which may follow a pattern under the same precursor and conditions.

For many precursor materials, for example phenol-formaldehyde resins, the major intermediate materials have been investigated. One can analyze the volatile products generated during pyrolysis using spectrometric techniques. If some pyrolysis products are known, many others can be predicted based on the chemistry of the materials. A schematic diagram showing the main reactions during pyrolysis of a phenol-formaldehyde resin is presented in Figure 3.15. These resins are very important for us, as one of the most commonly used photoresists (commercial name SU8) is of this type. Phenolic resins (details in Section 3.4.3.4) have been used as precursors for some bulk carbon materials for almost five decades.

Figure 3.16 shows the pyrolysis products of yet another resin used in carbon-microfabrication: poly-furfuryl alcohol. Similarly, the fragmentation pattern of naphthalene is shown in Figure 3.17. Only some representative examples of organic materials

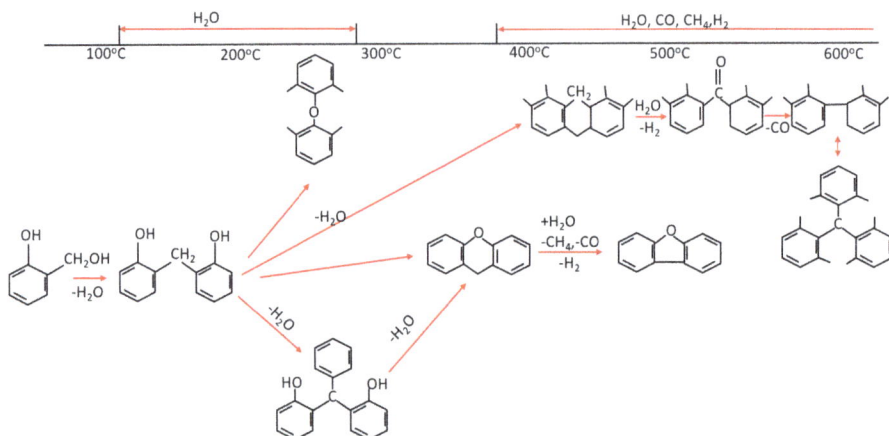

Figure 3.15: Fragmentation pattern of a phenol-formaldehyde resin during pyrolysis. Mechanism originally proposed by Ouchi and Honda [178, 177].

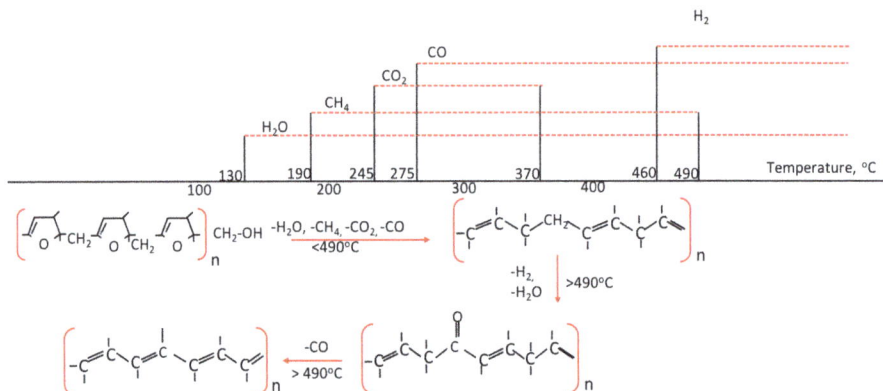

Figure 3.16: Fragmentation pattern of a fufuryl alcohol during its pyrolysis. Copyright ©1970 Elsevier Ltd. Adapted with permission from [67].

have been described here in terms of their pyrolysis chemistry. In practice, each polymer adopts a different pyrolysis chemistry. Even within the same class of polymer, e. g., phenol-formaldehyde resins, the exact pyrolysis chemistry may vary, depending upon the specific analogue. Nonetheless, some basic principles may be followed in the case of similar chemical structures.

The absence of external oxygen in the furnace minimizes the CO_2 and CO formation (i. e., burning). However, if there is any (bonded) oxygen within the polymer itself, it can lead to oxide generation. Pyrolysis products, such as CH_4 and small hydrocarbons (aldehydes, ketones etc.), are volatile. They are released in the form of small bubbles, or in some cases, participate in secondary reactions during pyrolysis. Heavier hydrocarbons are typically condensable at room temperature and form a tar-like product. Such tars de-

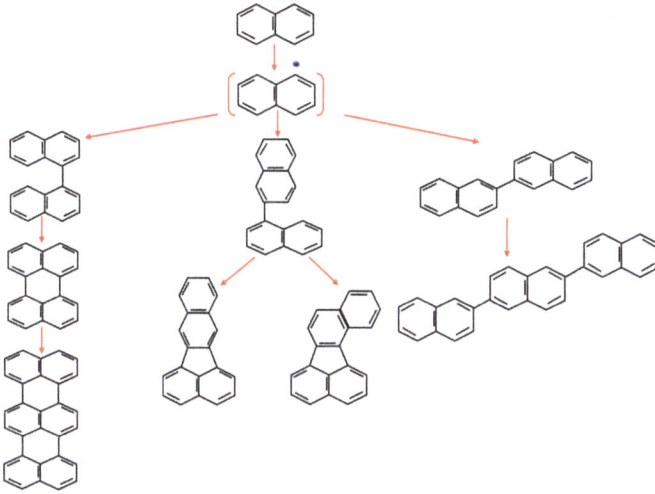

Figure 3.17: Fragmentation pattern of naphthalene during pyrolysis. Image drawn based on naphthalene pyrolysis mechanism proposed by Lang et al. [138].

posit on the walls of the furnace (see Figure 3.13), and can be separately collected if generated in large enough quantities. The solid carbon remains in the sample holder. This material is occasionally also designated as "char," but here we use the term "carbon," since char refers to a specific type of carbon. In the case of pyrolysis of biodegradable waste (biomass), syngas and tars are called biogas and biofuel, respectively.

As mentioned earlier, in addition to solid carbon, pyrolysis of polymers yields tar-like materials (heavy hydrocarbon molecules), and a gaseous mixture of CO, CO_2, and light volatile hydrocarbons, commonly known as the synthetic gas (or just *syn*gas). A schematic representation of pyrolysis performed in a tube furnace is shown in Figure 3.13. If the quantity of polymer is small (such as in the case of microfabrication), the amount of tar and syngas may be negligible. On the other hand, in the case of bulk (several kilogram) of raw polymeric material, tar and syngas are relatively high. They are then separately collected and used as fuels or raw materials in other industrial processes. In the case of carbon device fabrication, the goal is to improve the yield and purity of the carbon residues, whereas during large-scale pyrolysis (often performed for waste treatment), one tries to increase the fraction of tars and syngas. The quality of all products can be controlled by adjusting the pyrolysis conditions.

Among the applications of the pyrolysis process shown in Figure 3.13, the precursor materials for carbon structure manufacturing, waste treatment, and mass spectroscopy are organic polymers. The nature and the quantity of the polymer is different in each case, depending upon the application. Below we will first discuss the history, classification, properties, and applications of carbon materials obtained via the pyrolysis of polymers of a relatively high purity. This type of polymers is used for making devices. Afterwards, we will touch upon the polymeric mixtures present in the municipal waste, which are pyrolyzed but with an intention of obtaining tars and syngas, rather than solid carbon residual.

> **!** A wide variety of polymers have been converted into carbon for various industry-scale applications. Only a few examples relevant to microfabrication are described here. For further reading on this topic, books by Jenkins and Kawamura [115] and the 7th volume of the book series *Chemistry and Physics of Carbon: A Series of Advances* (particularly Chapters 1, 2, and 5) [244] are strongly recommended.

3.3.1.2 Difference between pyrolysis for device fabrication and waste treatment

Pyrolysis aimed at obtaining carbon is a single operation process that is carried out in high-temperature furnaces. Waste pyrolysis, on the other hand, is performed at a much larger scale, typically in continuous mode in flow reactors. For specific complex mixtures of waste materials; batch reactors are also used. The heating and cooling rates are not precisely controlled during waste treatment. Occasionally, even the pyrolysis temperature is determined by flames, rather than being adjusted externally. In the continuous mode flow reactors, there is often no cooling involved. In waste treatment plants, the mixture of waste (sometimes pretreated) is fed into the reactor at the rate of several kilograms per hour. Major differences between the pyrolysis parameters used for carbon structure manufacturing and waste treatment are listed in Table 3.1. Note that here the common biomass pyrolysis carried out in a flow reactor is taken as an example.

Table 3.1: Comparison of pyrolysis carried out for carbon structure manufacturing and continuous waste treatment.

Feature	Carbon-based manufacturing	Waste-treatment
Operation	Single	
Equipment	Furnace	Flow reactor
Temperature	900–3000 °C	Below 600 °C
Cooling	Yes	No
Environment	Inert	May contain oxygen
Raw material	Known prepatterned polymers	Biomass, waste plastic, tire
Intended product	Carbon	Tars, Syngas, Carbon (Char)
Feed weight	Microgram to gram	Several kilograms/h
Primary application	Structure and device manufacturing	Waste to energy conversion

The mass balance equations of a continuous flow reactor are typically used in waste treatment plants. We know that pyrolysis results in (i) solid carbon, (ii) tar, and (iii) gaseous hydrocarbons.

Let us assume a fluidized bed reactor with the following parameters:

Input (feed): Mass flow rate = m_1; Mass fraction = x_1

SynGas: Mass flow rate = m_2; Mass fraction = x_2

Tar: Mass flow rate = m_3; Mass fraction = x_3

Large-scale continuous pyrolysis of waste materials

		Light Hydrocarbon (SynGas) Mass, m_2 Mass fraction, x_2
Waste Feed Mass, m_1 Mass fraction, $x_1 = 1$	**Fluidized Bed Reactor** $T = 550\ °C$ $P = 1\ bar$	**Tar (pyrolysis oil)** Mass, m_3 Mass fraction, x_3
	Residual (unprocessed) Waste Mass, m_5 Mass fraction, x_5	**Carbon (Char)** Mass, m_4 Mass fraction, x_4

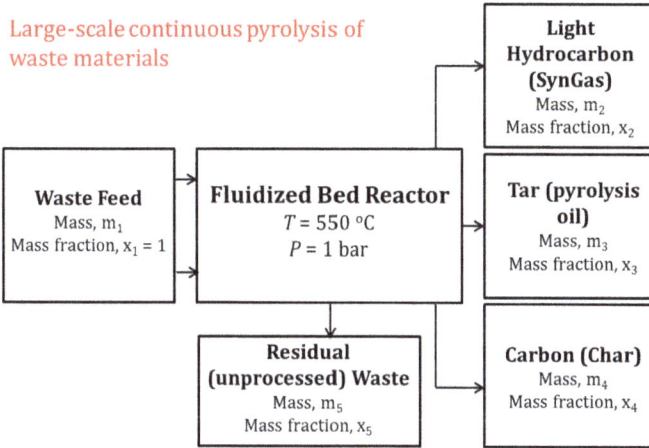

Figure 3.18: Mass-balance in a pyrolysis reactor designed for waste treatment.

Carbon: Mass flow rate = m_4; Mass fraction = x_5
Unprocessed feed: Mass flow rate = m_5; Mass fraction = x_5

In a reactor, such as the one shown in Figure 3.18, the mass balance can be calculated as follows:

(Rate of accumulation) = (Mass flow rate in) – (Mass flow rate out)

That is,

$$dm/dt = \dot{m}_1 - \dot{m}_2 - \dot{m}_3 - \dot{m}_4 - \dot{m}_5. \tag{3.1}$$

Now assume that the total feed rate is 50 kg/h. This feed can be, for example, agricultural waste. Prior to scale-up, laboratory tests are carried out on the same mixture in a proto-type reactor (100 % efficiency), from which the following mass fractions for each product are obtained: SynGas: 10 %, Tar: 70 %, and Carbon 20 %. When the process is transferred to a large-scale reactor (90 % efficiency), the mass balance equations are calculated as follows:

$$\dot{m}_2 = 0.9 \times 50\ kg/h \times 0.1 = 4.5\ kg/h,$$
$$\dot{m}_3 = 0.9 \times 50\ kg/h \times 0.7 = 31.5\ kg/h,$$
$$\dot{m}_4 = 0.9 \times 50\ kg/h \times 0.2 = 9.0\ kg/h,$$
$$\dot{m}_5 = (1 - 0.9) \times 50\ kg/h = 5.0\ kg/h.$$

Urban solid waste typically consists of both natural and synthetic polymers. Pyroly-sis is the only process that can be used for a safe degradation of these polymers, as most of them have a very high stability. The heat-treatment process, in the case of waste, is generally stopped at ≤650 °C, because the ultimate goal of the process is to obtain a high fraction of pyrolysis oil (tars) and syngas, rather than solid carbon. Since the precursor material being pyrolyzed is a mixture of various waste materials and one has no control over its exact composition, the carbon obtained from such a process may not have a good quality. Therefore, the state-of-the-art waste pyrolysis plants are designed to maximize the yield of syngas or pyrolysis oil, which can be used as fuels. Often, the process is a combination of combustion and pyrolysis (i. e., insufficient quantity of oxygen is present in the reactor). The small fraction of resulting carbon is low-grade, which implies that it contains a lot of impurities and features only very short range order with a poor electrical conductivity. This material is called "biochar," which is either discarded or is used in low-cost applications in bulk. There is a significant number of research articles on utilizing waste-derived carbon in high-end applications, such as electrode manufacture [106, 107, 185]. Waste materials, such as peanut shells, coconut shell, seeds of fruits, etc., have been utilized in their powder form for preparation of small-scale electrode. However, they are difficult to pattern in the micro/nano domain.

3.3.2 Carbonization and graphitization

During pyrolysis, hydrocarbon radicals of different types and sizes are generated with their highest concentration typically around 500–600 °C [207]. After 800 °C, a network of 2D carbon fragments, containing a large fraction of defects (non-six-membered rings, larger 2D and 3D voids, rough edges, etc.) as well as chemical impurities, start to develop. Further heat enables the annealing of these defects and an increase in the crystallite size (L_a). This is followed by a decrease in the stack thickness (L_c). As per the conventional nomenclature, the 700–et2000 °C temperature range is called the *carbonization* region, while between 2000–3000 °C, *graphitization* takes place. Graphitization, as the name itself suggests, is a process of converting the material into polycrystalline graphite. This is not possible for all polymer precursors, as some of them cannot be converted into graphite due to microstructural complexities. Further details are provided in Section 3.4.

3.3.3 Material characterization during heat-treatment

It is evident that it is extremely difficult to establish the exact composition of a carbonizing matrix during heat-treatment. There are some *in-situ* characterization techniques, detailed later in this chapter, that can provide information about the microstructural changes with the increase in process temperature. Typically the physicochemical changes during pyrolysis can be predicted via thermogravimetric analysis (TGA), which is a technique for analyzing the masses of temperature-induced fragments in a material. One can also investigate the structures of the byproducts, both in gas and liquid forms via gas or liquid chromatography combined with mass spectrometry. If the process is terminated during the pyrolysis stage itself, the solid residue of pyrolysis can be

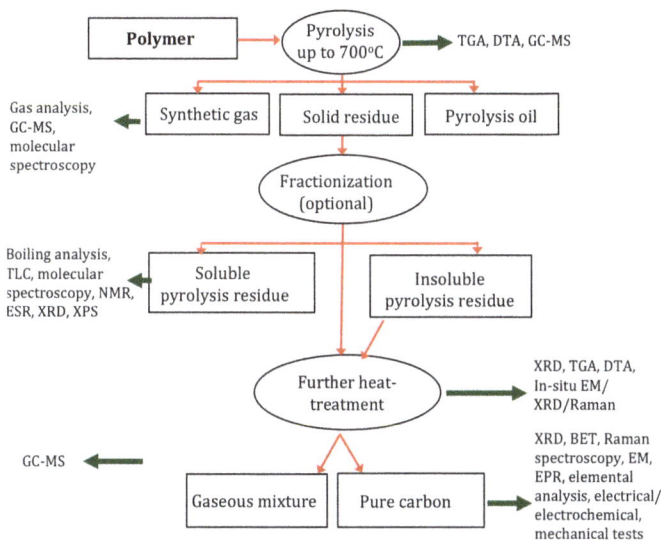

Figure 3.19: Characterization techniques used at various stages of heat-treatment of organic solids. Acronyms: GC-MS: Gas chromatography-mass spectrometry; TLC: thin-layer chromatography; NMR: nuclear magnetic resonance; EPR: electron paramagnetic resonance; XRD: X-ray diffraction; XPS: X-ray photoelectron spectroscopy; TGA: thermogravimetric analysis; DTA: differential thermal analysis; EM: electron microscopy.

evaluated via various standard chromatography and spectroscopy tests (e. g., thin-layer chromatography, electron spin/paramagnetic resonance, nuclear magnetic resonance spectroscopy, etc.). Once the temperatures are high enough to yield > 95 % purity of carbon, X-ray diffraction, Raman spectroscopy, electron microscopy, porosity analysis via gas adsorption, elemental analysis and surface functional group characterization (e. g., via X-ray photoelectron spectroscopy) can be carried out. General mechanical, thermal, electrical, and electrochemical characterization can be performed as per the targeted applications.

3.4 Polymer-derived carbon: Physicochemical aspects

When a polymer is converted into carbon via heat-treatment, it goes through chemical as well as physical changes. Both are responsible for determining the microstructure of the resulting carbon. It has been established over several years that the carbon materials from different polymeric sources exhibit different microstructure. Surface properties, porosity, and the overall weight fraction of these materials also vary. Some polymers yield graphite or graphite-like carbon, whereas others yield hard and brittle type carbons. In all cases, however, carbon-conversion leads to a dimensional shrinkage or porosity or both due to a net loss in the mass. In Figure 3.20, four possible scenarios on

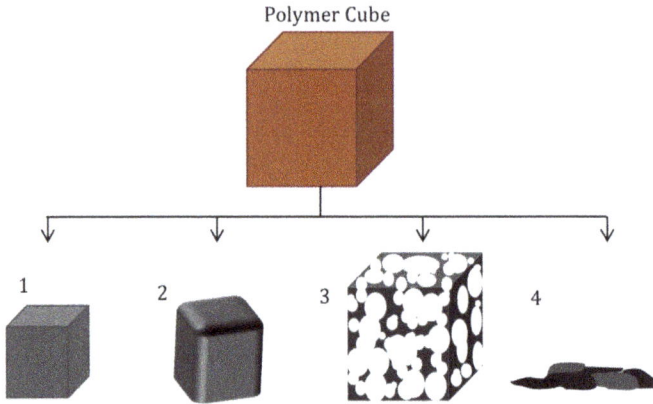

Figure 3.20: Possible carbon structures after heat-treatment of a polymer structure. 1. Shrinkage with no change in geometry, 2. Shrinkage with minor changes in geometry, 3. Porosity without any shrinkage, and 4. Extremely small fraction of carbon or complete loss of geometry.

carbonization of a polymer cube are illustrated. One can get a shrunk structure while maintaining its exact initial shape or with minor loss in shape, e. g., flattening of edges and corners. Mass loss can also lead to porosity if the carbon skeleton in the polymer remains rigid during heat-treatment. This leads to porosity, rather than shrinkage. Finally, if the carbon fraction is very little in a material, only little residue or flakes of carbon are obtained. The ability to train its shape becomes important when we are performing carbonization of microfabricated structures, because one needs to factor-in the shrinkage while designing the initial geometry.

3.4.1 Polymer-to-carbon conversion in nature

Polymer-to-carbon conversion is not essentially a synthetic process. In fact, all types of naturally occurring carbon forms on Earth are derived from solid organic materials. Different allotropes as well as many compounds of carbon exist in nature, at different depths in the Earth's core, or even at the same location in some cases. Graphite, diamond, coals, cokes, pitches, and other petrochemicals are all formed by a process similar to synthetic pyrolysis that entails removal of noncarbon atoms from organic precursors. Tectonic movements, extreme temperature and pressure, and other high-energy events that took place over millions of years, have led to the deposits of different various of carbon and hydrocarbon materials. The fact that we can find graphite near coal seams, or different types of petrochemicals at the same location indicates that not only external factors (T, P) determine the type of carbon. The chemistry of the precursor also matters. In fact, even within the type of carbon, such as coal, microstructural and chemical variations exist [69]. Some forms of deep carbon and their locations are shown in Figure 3.21.

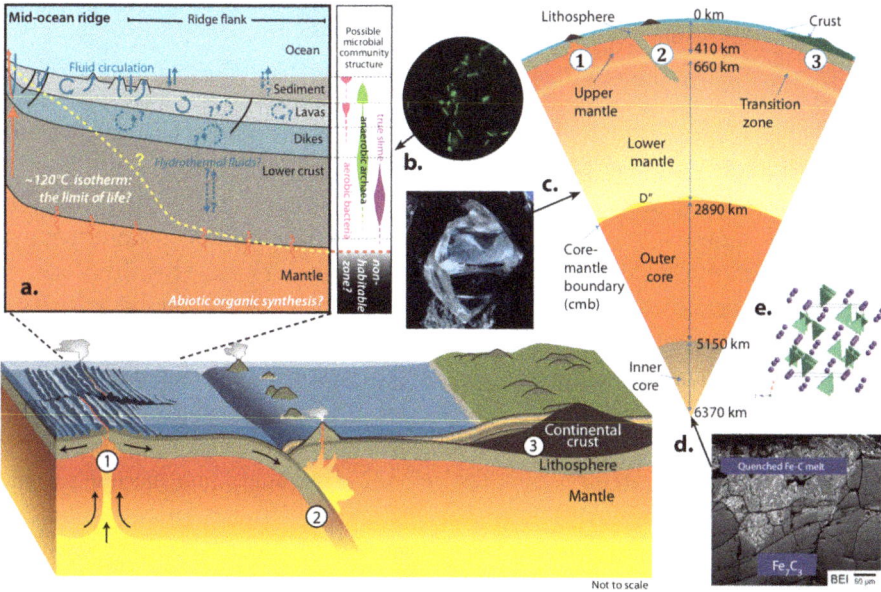

Figure 3.21: Earth's deep carbon cycle. (a) Microbial community on mid-ocean ridge flanks (b) An epifluorescence micrograph of an iron-reducing enrichment culture from a serpentinite-hosted habitat. (c) Diamonds and their inclusions in Earth's deep interior. (d) An optical image of iron carbide- a plausible constituent of Earth's solid inner core. (e) Proposed structure of a magnesium-iron carbon-bearing material similar to phase-II of magnesite at high pressures corresponding to depths greater than 1800 kilometers. Figure reproduced from [94].

Carbon allotropes present under the Earth's crust, petrochemicals, carbonates on the oceanic subsurface, and the carbonates expected to exist in the core of the Earth are collectively known as deep carbon materials [94]. The branch of science dedicated to studying them is known as the deep carbon science. Global programs, such as *Deep Carbon Observatory*, are supporting and encouraging research on understanding deep carbons to unravel mysteries pertaining to life and its evolution on the Earth.

Though deep carbon forms are also polymer-derived, they are not directly usable in technological applications. Natural graphite, for example, requires extensive ore processing (benefication) prior to its use. Certain types of coke (a solid petrochemical) can be used as a raw material for graphite manufacturing and pitches (highly viscous mixture of aromatic hydrocarbon generated in refineries and coal mines) are used for making carbon fibers. Carbon obtained from cokes and pitches is significantly different from each other. The modern-day carbon technology is inspired by these natural processes. The extreme conditions are mimicked and compressed into a workable time frame. The carbon materials deliberately prepared from well-defined precursors for the purpose of technological applications are known as synthetic carbons.

3.4.2 Early technological applications of synthetic carbon

In the early twentieth century, graphite and other forms of naturally available carbon became preferred substitutes for metals in various industrial applications. Notable examples include solid-state batteries, lightweight aircraft, automobile parts, inks for printing, and heating elements. Coals and charcoals were used for combustive heat generation, often aimed at steam production. Graphite, on the other hand, was recognized for its electrical and electrochemical properties. Graphite was also extensively used as moderator in early nuclear reactors (or nuclear piles), and as a high-temperature resistant manufacturing material in space vehicles (e. g., rocket nozzles, nose shields) [203]. This commercial demand for graphite was indeed the motivation behind the research on artificial graphite-like materials. The goal was to produce them in large quantities and with a relatively high purity. To achieve this, thermal treatment of various organic materials at high temperatures was attempted. The carbonization mechanism and properties of the resulting carbons were extensively studied in the early- and mid-20th century [243]. These carbon materials were designated "pyrolytic carbons" or "pyrolytic graphites," depending upon their microstructure.

It is believed that the inspiration for polymeric carbon came from wood and other cellulosic materials. Such materials were already in use in some applications, for example, as filaments in light bulbs. The major task, however, was to increase the carbon yield from the polymer, which could be achieved by lowering the amount of the CO_2 formed during the process, i. e., by reducing or completely eliminating oxygen from the furnace during the heat-treatment. For this purpose natural polymers as well as various hydrocarbons were carbonized via the pyrolysis process, and attempts were made to use the resulting carbon in technological applications. Here it is also important to note that the first synthetic polymers were studied in the early 19th century (around 1833), but their large-scale commercial production started only in the 20th century [116].

Phenol-formaldehyde resins, which are currently regarded as the most common precursors for pyrolytic carbon production, were first synthesized by Belgian scientist Leo Baekeland in the 1907 [116]. In 1920 Hermann Staudinger's work [215] shed light on the molecular structure and aggregation of polymers, which eventually led to the modern polymer science. Prior to synthetic polymers, hydrocarbon molecules from petroleum and other industrial byproducts were tested for obtaining a high-performance carbon, in addition to natural materials, such as cellulose and silk.

Before going into the details of the chemistry of the pyrolysis process and characterization of polymer-derived carbons, a brief history of some of the most popular applications of pyrolytic carbon is provided below. Many of these applications are still relevant today, however, the size of the devices and hence their manufacturing methods have changed over the years. This description also explains how social and commercial factors create the demand for a certain material, which directly or indirectly determines the direction of the research. Interestingly, the outcome of the research is often different from the original plan. Nevertheless, this leads to newer materials and applications, and provides the scientific advancements a whole new direction, which is indeed essential for the technological progress of a society.

Electrochemical cell and battery

The most remarkable application of carbon in the 18th and 19th centuries was electro-chemical cell and battery, which became the main source of portable electricity at that time. Carbon was an important component of these cells from their early development. The very first electrochemical battery, proposed by Italian scientist Volta did not utilize any carbon and was composed of zinc and copper electrodes. Nonetheless, the benefits of replacing the copper by charcoal in the Voltaic cell were already recognized by the French scientist Gautherot [180] in the early 19th century [180]. Owing to an extensive amount of research on cells and batteries throughout the 19th century, a large number of cell designs and materials were proposed and tested. Many of them utilized various forms of carbon, such as charcoal, graphite, platinized carbon, calcined carbon, carbon powder, and coke, from the blast furnace for cathode preparation. Some carbon-containing wet cells, as mentioned in the 1893 treatise on Voltaic cells by Benjamin Park [180] are as follows: Poggendorff cell (1842), Leuchtenberg cell (1845), Fabre de Lagrange Cell (1852), Walker Cell (1859), Tommasi Cell (1881), The Law cell (1882), Pabst cell (1884), Laborde Cell, R. Napoli Cell, and Crova and Delhaumucean Cell.

In 1868, French scientist Georges Leclanché designed a cell that comprised of the paste of manganese dioxide and carbon wrapped around a carbon-rod cathode (Figure 3.22A). Later versions of this cell laid the foundation of the modern day dry cell. However, the credit for inventing a truly dry zinc-carbon cell goes to Carl Gassner, who obtained a German patent on his design in 1886 [77]. He proposed a design consisting of two concentric cylinders as electrodes (see Figures 3.22 B and C), which were made of zinc (anode) and carbon-manganese (cathode) respectively. The space between the cylinders was filled with a mixture of zinc oxide, NH_4Cl (Salammoniac), plaster of Paris, zinc chloride, and water. The mixture was filled in liquid or semisolid form, which eventually

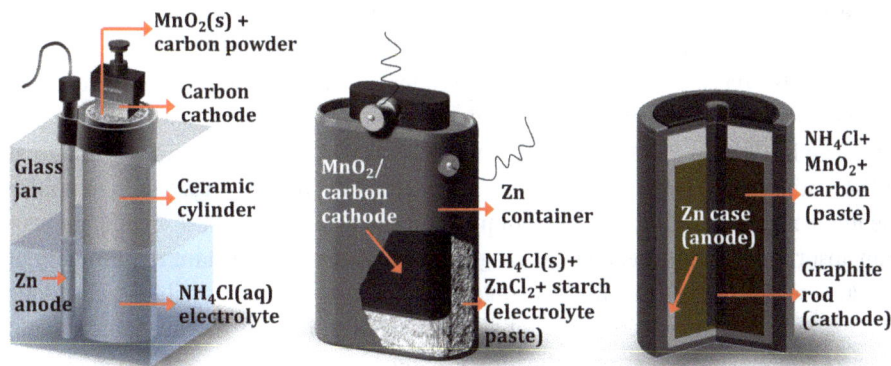

Figure 3.22: Early synthetic carbon-based batteries. Images are schematics of a Leclanché cell and two designs proposed by Carl Gassner.

solidified and became porous due to loosening of the plaster [180]. In successive years, multiple zinc-carbon cells inspired by Leclanché and Gassner's concepts were suggested. The present form of zinc-carbon cell is shown in Figure 3.22C. Most of them contained a carbon rod as electrode. Importantly, in these examples, the term "carbon" was used to indicate mainly wood-derived carbon (charcoal), or occasionally the residues from the blast furnaces. Graphite, despite its high purity and better understood electrical properties, was seldom used. From this, one can deduce that the material and manufacturing costs were too high for a commercial scale production using natural graphite.

Scale-up of chemical reactions

Today we take it for granted that chemicals are produced in large quantities in the reactors located at chemical plants, and that we simply need to purchase them. This was not the case 100 years ago. Reactions that involved harmful or corrosive chemicals (reactants, products, or byproducts), or required extreme processing conditions, such as high pressure and temperature, were only possible at laboratory scale. Even some well-studied chemicals, such as concentrated acids, had to be dealt with at a small scale. Their large volumes could easily get out of control and cause fatal accidents. Needless to say, construction materials of various types have always played an important role in the commercial chemical production. Certain polymer-derived carbons are very useful for making reaction pots and vessels due to their inertness.

Moderator in nuclear reactor

A moderator in a nuclear reactor is the material that is used to slow down the neutrons generated during the fission of U-235. Slower neutrons have a higher probability (cross-section) of being absorbed by other U-235 nuclei, and allow for the continuation of the fission chain reaction. The performance of a moderator improves if its atomic weight is close to the weight of a neutron. A low atomic weight material facilitates a larger energy dissipation on collision with neutrons, thus slowing them down more efficiently. The second requirement is that the moderator itself should not have a very high neutron absorption cross-section. Else, many neutrons might be lost during the process.

Graphite (via carbon atoms) and heavy water (via deuterium) facilitate a good collisional energy transfer with a modest neutron absorption. The early nuclear reactors, for example, *Chicago Pile-1* (operational in 1942) and *X-10 Graphite Reactor* (operational in 1943) utilized large amounts of graphite as moderator. The challenge was to minimize the impurities in graphite, since any impurity, e. g., boron, could cause significant neutron absorption. Therefore, high purity graphite became an important requirement in the 20th century. Scientists also investigated the properties and possible applications of the graphite after its neutron irradiation, since this material was available in bulk after its lifetime in the reactor. Unfortunately, some major nuclear accidents, including the Windscale fire (England, 1957) and the Chernobyl disaster (USSR, 1986), also involved graphite. In the event of fire, graphite may not be the safest material, as it can catch fire at high temperatures with ample oxygen. As a consequence to these accidents, some modifications were made to the graphite-moderated reactors that led to the current RBMK (*In Russian:* Reaktor Bolshoy Moshchnosty Kanalny) design. These reactors are cooled with pressurized water with individual fuel channels and using graphite as its moderator. At present, RBMKs primarily operate in the former USSR region.

Alternative moderator materials are standard (light) water, beryllium, and beryllium oxide. Although light water exhibits a relatively high neutron absorption cross-section, it is largely used in various commercial reactors across the globe due to its availability, low cost, and the fact that it can simultaneously serve as coolant.

First reports on polymer-derived carbon

As it can be inferred from the aforementioned examples, it became essential to produce artificial graphite or graphite-like materials in the 20th century. This was carried out by converting natural, and later synthetic polymers, into carbon. It was observed that carbon materials produced from different precursors, even under the same heat-treatment conditions, had measurable differences in terms structure and properties. Though some of them were similar to polycrystalline graphite, others exhibited randomly oriented small graphitic crystallites with a larger interlayer spacing and misaligned basal planes. Such carbons could not be converted into graphite, even at temperatures as high as 3000 °C. The differences in microstructure of polymer-derived carbons were also responsible for their properties, and hence, applications. As a result, they were extensively studied during the 19th and 20th centuries, and are still being studied due to the advent of new carbonizable polymers, which can also be micro- and nano-patterned. During their early development, the primary tool to study polymeric carbons was X-ray diffraction (XRD) [70, 243]. At present, Raman spectroscopy, neutron diffraction, magnetoresistance analysis, and TEM are also employed for their characterization.

Polymer-derived carbons are characterized by quantifying their proximity to graphite. This implies that regarding post-carbonization, typically the following questions are addressed: what is the fraction and size of the two-dimensional carbon fragments in the material? How well are these fragments stacked on top of each other? Are they organized similar to graphite and form small hexagonal crystals? What is the thickness of such stacks? What is the extent of defects and impurities in the given carbon material? etc. A comparison of polymeric carbon with graphite is natural, since the preparation of these materials was initially aimed at obtaining artificial graphite. In many texts, more graphite-like polymeric carbons are designated as "good" or "better" quality carbons. As such, there is no such thing as good or bad carbon. These conclusions simply depend on the intended application of the material. For example, to induce certain surface reactions, defects may be rather advantageous, which actually render the carbon "less" like graphite. Similarly, mechanical hardness could be a desirable property for the fabrication of load-bearing structures, but it may come with a lower electrical conductivity, thus making the material less useful for electrical devices.

In summary, one needs to choose the appropriate polymeric carbon suitable for the planned device. This often requires finding the optimum combination of properties with some trade-offs.

All organic materials, polymers or not, yield at least a small fraction of carbon when they are heated to elevated temperatures in the absence of oxygen. If the carbon cannot burn, i. e., cannot form oxides, it can withstand temperatures as high as 3000 °C, therefore below these temperatures one can always obtain some residual solid carbon. In this book our focus is device fabrication, so we are mainly concerned about the carbonization of polymers, more specifically, prepatterned polymers. Note that not all polymers yield sufficient quantities of carbon on pyrolysis. Their chemical structure plays an important role in not only deciding the weight fraction of the resulting carbon (known as the carbon yield), but also its microstructure and, in turn, properties. As a result, each polymer features a distinct carbonization mechanism. However, polymers belonging to the same chemical class do exhibit similarities in their carbonization mechanism, which helps us in deciding a few "rules of thumb" about carbonization. Based on this, a classification of polymer-derived carbon materials can also be established.

Carbonization of a polymer via heat-treatment takes place in four stages: (i) pyrolysis (up to approximately 600 °C), (ii) C–C bond formation and skeleton development (600–900 °C), (iii) increase in the 2D crystallite size or L_a (900–2200 °C), and (iv) increase in that stack thickness/graphitic content (L_c) (above 2200 °C). These stages are not entirely independent, for example, some C–C bond formation takes place at each stage, and some bond cleavage may also occur at temperatures above 900 °C. Semicrystalline polymers, which already feature graphite-like 3D arrangement of the graphenic sheets at stage (iii), can be further heated (generally up to 3000 °C) and converted into polycrystalline graphite. The additional stage (v) is known as graphitization.

Traditionally, the entire process involving all aforementioned stages was referred to as the "heat-treatment." In the contemporary literature, the term pyrolysis is used exchangeably with both heat-treatment and carbonization, since pyrolysis is the primary reason for the carbon-conversion of a polymer. Moreover, when the carbonizing material cools down to the room temperature, irrespective of the highest process temperature some C–C bond formation inevitably takes place. Hence, the step (i) is always accompanied with (ii) for all practical purposes. Consequently, polymeric carbon is occasionally also called "pyrolytic carbon."

3.4.3 Classification of polymer-derived carbon

Microstructure and properties of polymer-derived carbon depend on (i) chemical structure and crystallinity of the precursor polymer, (ii) heat-treatment environment (inert gas, reducing atmosphere etc.), (iii) process temperature, (iv) heating and cooling rate, (v) polymer processing technique, (vi) extent of cross-linking (in specific polymers), (vii) presence of trace metals or other elemental carbon materials in the polymer, and (viii)

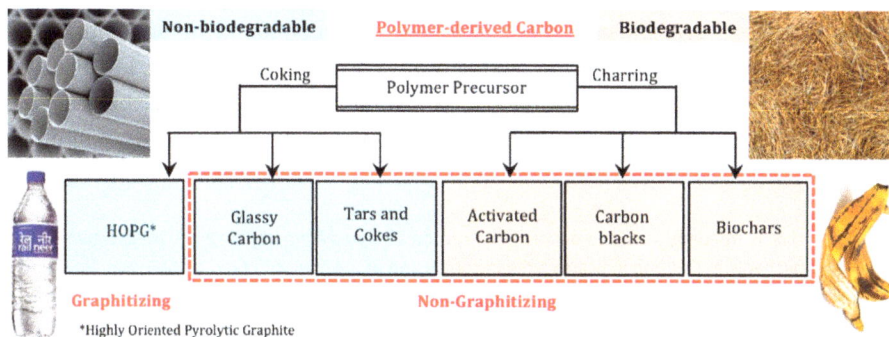

Figure 3.23: Classification of polymer-derived carbons based on precursor material and pyrolysis mechanism.

geometry and surface area of the sample. The most dominant factor, however, is the chemical composition of the initial polymer. Although each polymer has a characteristic carbonization mechanism, polymer-derived carbons can be classified based on the microstructure of the resulting carbon, which in turn reflects their carbonization mechanism. Such a classification is shown in Figure 3.23.

Coking and charring

If a polymer goes through a semisolid or rubbery state during its pyrolysis, this phenomenon is known as coking. Coking takes place if the glass-transition temperature of the pyrolyzing mixture falls just below the process temperature. Note that the composition of the pyrolyzing mixture constantly changes with the increase in process temperature. Its glass-transition temperature also continues to increase and is typically maintained very close to the process temperature. This means that while the material can become soft and liquid-like, it does not flow. However, it does become flat, because liquids and liquid-like materials try to reduce their surface energy. If glass-transition temperature of pyrolysis mixture is much lower than the process temperature, the material flows, and the given shape is completely destroyed. Many scenarios in between these two extremes can exist, where minor shape distortions are observed. Examples of polymers that undergo coking during pyrolysis are phenol-formaldehyde resins, furfuryl alcohols, and anthracene.

In the case of charring, a direct conversion of the rigid polymer structure into carbon is achieved with a more-or-less complete preservation of the skeleton of initial shape. Such a process typically results in porous carbon. Wood, cellulose, and many natural polymers often undergo charring. In the case of cellulose pyrolysis, an intermediate chemical, known as levoglucosan, is formed, which subsequently disintegrates and leads to the formation of tars (oil-like materials), volatile hydrocarbons, and solid carbon. The solid carbon backbone is replicated in the final char, and the oils and volatiles are collected and distilled if so desired. Cellulosic materials do experience a certain degree of softening in the 230–255 °C region, but it has not been directly correlated with either their glass transition or the melting point. Chars are predominantly nongraphitizing in terms of microstructure. Due to their porosity and surface chemistry, they can be easily activated via physical or chemical methods, and utilized in various applications as activated carbons. Pyrolysis of natural wood is more complex due to the presence of lignin and hemicellulose, which is not discussed here.

> ❗ The term *char* is frequently used to denote solid carbon obtained from any polymer, irrespective of its pyrolysis mechanism. *Charring* indicates a direct conversion of polymer into rigid carbon, without going through a prominent semisolid phase.

Graphitizing and nongraphitizing carbon

Graphitizing carbons are those polymer-derived carbons that can be converted into polycrystalline graphite by heat-treatment at higher temperatures, catalytic processes, pressure-induced graphitization, or any other method. It is known that graphite, by definition, features an ABABA-type layer arrangement with a separation of 3.354 Å along the *c*-axis between subsequent layers. R-graphite with ABCABC arrangement is also plausible, although less common in synthetic graphites. During initial pyrolysis stages, certain semicrystalline polymers, such as poly-vinyl chloride (PVC) or mesophase pitch, may convert into a carbonaceous material that resembles stacked graphene fragments or liquid crystal-like entities. Initially these fragments contain impurities and feature randomly rotated basal planes; their progressive ordering at higher temperatures leads towards graphite formation. These materials are known as graphitizing, as they can ultimately convert into graphite.

Non-graphitizing carbons, on the other hand, cannot be converted into crystalline graphite, irrespective of the heat-treatment temperature. They contain various structural defects (non-six-membered carbon rings as well as closed voids) and randomly oriented carbon fragments exhibiting a strong three-dimensional bonding. The fragments feature a Gaussian distribution in terms of size and shape, carbon-carbon bondlength and valence angles. They do feature some stacking, but the L_c is typically <10 nm. Within the stacks, the imperfect carbon fragments are rotated at arbitrary angles on top of each other, which leads to an interlayer separation slightly greater than 0.335 nm (generally below 0.344 nm). Such an arrangement, illustrated in Figure 3.24, is known as *turbostratic*, and the carbon material itself is called turbostratic carbon. Evidently, all non-graphitizing carbons are turbostratic. Additionally, the presence of defects causes these fragments to curl, fold, and form fullerene-like structures. One can occasionally also spot completely closed buckminsterfullerenes. The curved structures coexist with the larger, stacked carbon sheets. Non-graphitizing carbons exhibit a lower electrical conductivity, but an improved hardness compared to graphite. In graphite, the well-organized carbon fragments are allowed to slide over each other, since the bonding between consecutive layers is not strong. As the 3D-bonded fragments cannot slide in the case of non-graphitizing carbons, the material is rigid. These carbons are therefore also

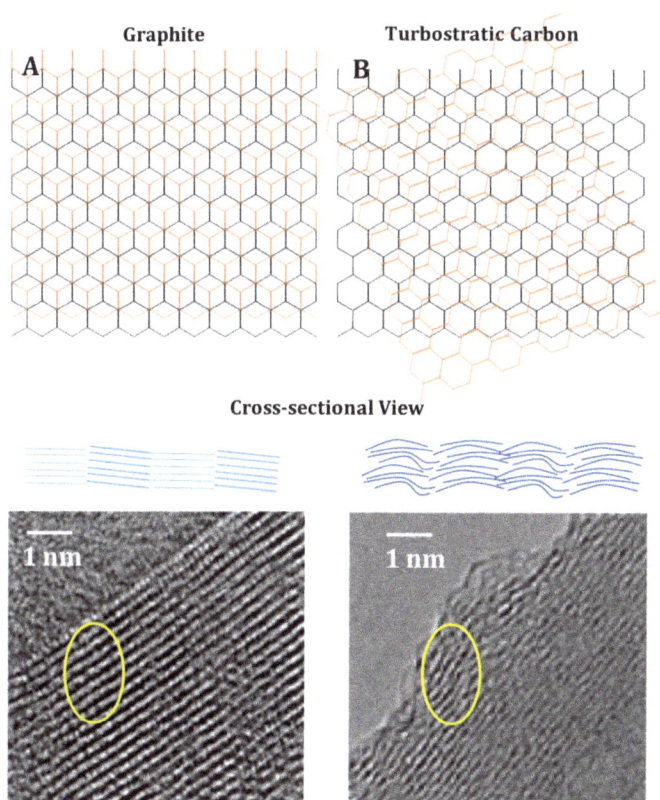

Figure 3.24: (A) Graphite-like and (B) turbostratic arrangements of sp^2 carbon sheets. TEM images in column A and B (bottom) display the layer spacing in graphite (0.335 nm) and in turbostratic carbon (0.336–0.344 nm).

called "hard" carbons. PF resins, cellulose, poly(vinylidene chloride), and certain polyimides yield non-graphitizing carbon on pyrolysis.

The first characterizations of graphitizing and non-graphitizing carbons was carried out by Rosalind Franklin in 1951 [70] (see Figure 3.25). This study provided the first experimental evidence of the microstructural differences between carbon obtained from different polymer precursors. The second microstructural model was suggested by Jenkins and Kawamura in 1971 [114]. Here, it was suggested that non-graphitizing carbons have an intertwined ribbon-like microstructure. This model was based on TEM investigations of glass-like carbon and was largely accepted. As the formation of ribbon-like structures (growth of sheets preferentially in one dimension) is unlikely, this model was contradicted by researchers. However, the authors indicated the presence of curved sheets via their ribbon analogy, which was later proven. The next model shown in the figure was suggested by PJF Harris around the year 2000. He proposed that rather than random crystallites or ribbons, these materials are made purely of curved carbon

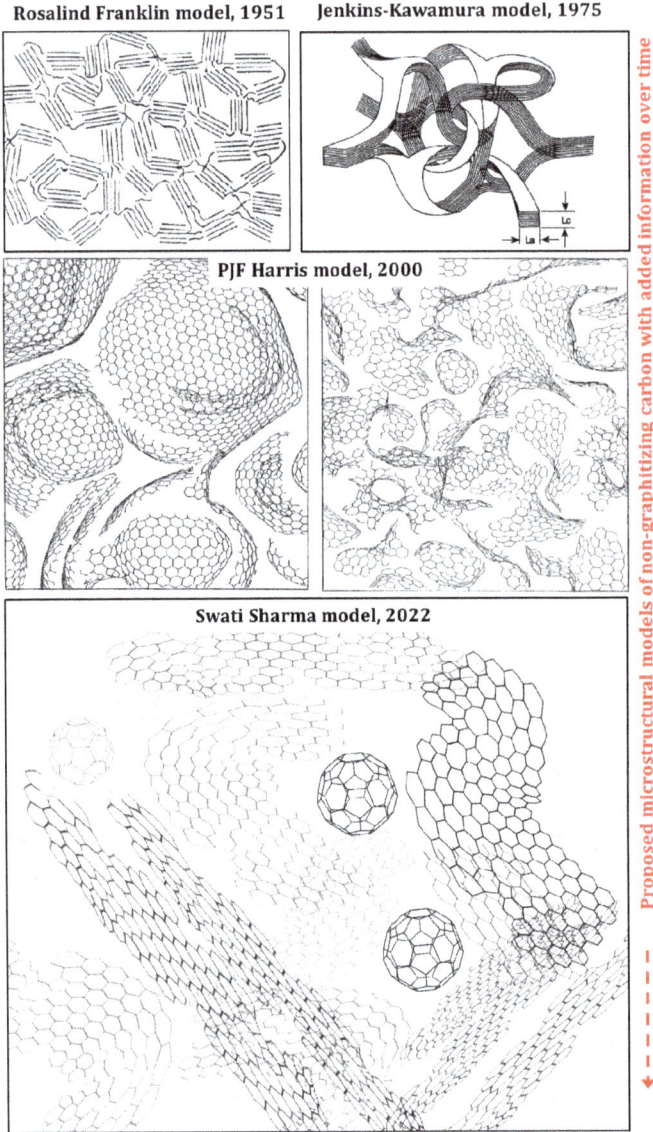

Figure 3.25: Various microstructural models of non-graphitizing carbon and their evolution from 1950 to 2022. Model 1: R Franklin, 1951. Copyright (scanned images) ©2017, Royal Society. Reproduced with permission from [70]. Model 2: GM Jenkins and K Kawamura, 1971 Copyright ©1971, Nature. Reproduced with permission from [114]. Model 3: Peter JF Harris, Copyright ©2004, Taylor and Francis.Reproduced with permission from [90]. Model 4: Swati Sharma, 2022. Copyright ©2021 Elsevier Ltd. Reproduced with permission from [211].

structure of varying sizes. He called such structures "fullerene-like" and also stated that completely closed fullerenes (e. g., buckminsterfullerene) are possibly present in non-graphitizing carbons. This theory was recently validated by Swati Sharma (author of this book) and coworkers [211], which leads to our last microstructural model. Sharma et al. showed that non-graphitizing carbon obtained from a phenol-formaldehyde resin as 900 °C contains up to 7 % buckminsterfullerenes along with carbon fragments having very low to very high curvatures [211]. These small and large curved carbon fragments coexist with the spherical fullerenes in the system. These fragments are partially stacked, thus leading to fractions of turbistratically arranged carbon in the form of small crystallites. This stacking increases with the increase in carbonization temperature, but the spherical fullerenes, clearly observable already around 800 °C, seem to be stable at higher temperatures [211, 210]. These fullerenes may lead to the closed porosity of non-graphitizing carbon, along with some voids, if present.

All microstructural models of non-graphitizing carbons are supported by some experimental investigation as well as material characteristics. For example, a non-graphitizing carbon known as glass-like carbon (described later) is impermeable to most gases and liquids, but it has a lower density than graphite. This is possible only if there are closed pores in the material. Since there are no easily detectable pores, presence of closed nanometer scale pores are highly likely. Similarly, the X-ray diffraction patterns of most non-graphitizing carbons suggest a wide range of bond-lengths (broad peaks), which is an indicator of some residual strain in the structure, which can easily lead to curved sheet formation.

Various theories exist on the growth mechanism of these carbon fragments. For example, it has been proposed that the polymer chains serve as the nucleation points for the resulting sheet-like formations, which later expand sideways, leading to ribbon or fibril-like geometries [188]. Another theory supports the formation of liquid crystals, which are believed to condense and yield graphitic (but not entirely graphite-like) carbon [188]. In a recent study conducted on a phenol-formaldehyde resin, it was confirmed that the fragments generated during heat-treatment of polymers feature a variety of shapes, sizes, and interaction patterns. They are highly mobile up to 800 °C, and continue to have some movement and rearrangement, even at higher process temperatures. These fragments seem to be constantly trying to attain a thermodynamically stable arrangement. They may separate from a large fragment or merge into a smaller one, thus resulting in a variety of shapes [210]. No specific pattern is followed during their formation. In this study, a small amount of a phenol-formaldehyde resin was placed on to a Joule-heated chip integrated with a transmission electron microscope. The *in situ* images taken during pyrolysis of the material are shown in Figure 3.26. One can also observe the formation of spherical or circular structures that can potentially be closed fullerenes. The presence and quantification of fullerenes in such non-graphitizing carbons was later studied and confirmed to be 5–7 % [211].

In addition to experimental work, there have been various theoretical investigations, for example, using the reactive molecular dynamics simulations, for understand-

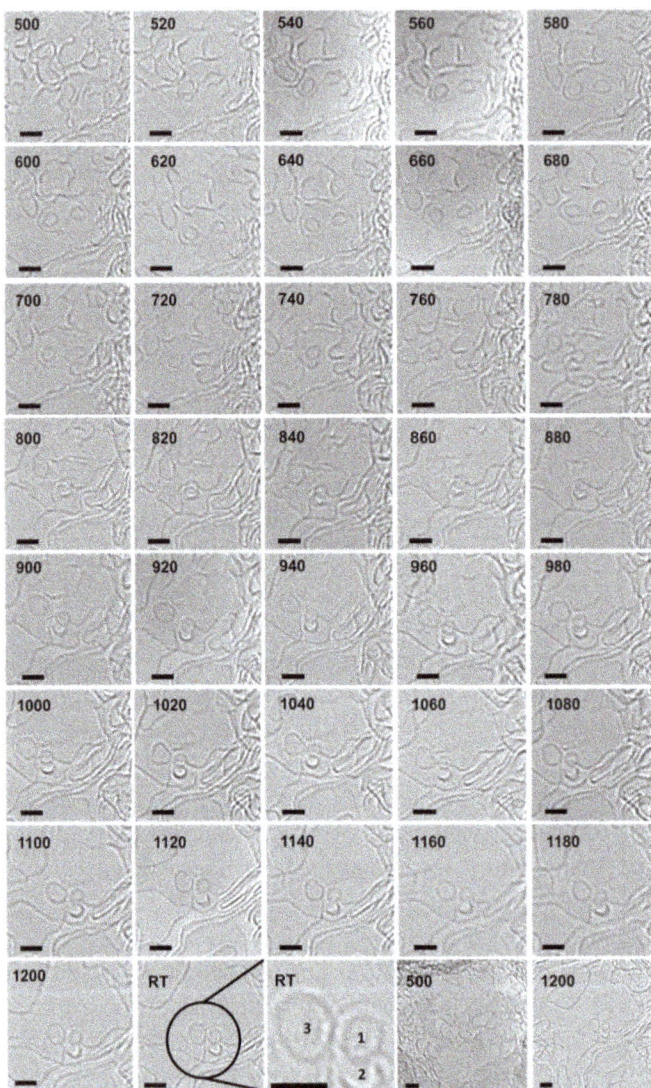

Figure 3.26: Series of transmission electron micrographs captured *in-situ* during the carbonization of a phenol-formaldehyde resin. All scale-bars are 1 nm. Heat-treatment temperature is mentioned on each micrograph in °C [210].

ing the mechanism of polymer pyrolysis and carbonization. Desai et al. [45] proposed that the primary fragments formed after the decomposition of a phenol-formaldehyde resin often contain atoms from different neighboring polymer chains, in addition to the parent molecule, rather than from one single polymer molecule. This finding is in line with the idea that there are no well-defined patterns or fragment nucleation sites during pyrolysis and C–C bond formation stages of the heat-treatment. Nonetheless, the fact that the polymer chemistry plays the most important role in determining the nature of the resulting carbon is something that has been proven again and again in both theoretical and experimental investigations. Therefore, polymer-derived carbons are often segregated, at least to a first approximation, based on the type of the precursor polymers. This is also the primary difference between carbons deposited from vapor phase (see Chapter 4 for details) and those derived from polymers via heat-treatment.

Only four representative models of non-graphitizing carbon are discussed in this section. Some other relevant microstructural models have been proposed in the context of glass-like carbon, porous carbons, and carbon fibers. Notable models, other than those discussed above, suggest the presence of curved carbon globules and fibril-like shapes formed by stacked carbon fragments [188].

3.4.3.1 Glass-like carbon

Non-graphitizing carbon with a high purity that has experienced at least some coking during pyrolysis is known as glass-like carbon. Its commonly used name include glassy carbon and vitreous carbon. This is a predominantly synthetic carbon, i. e., it is not found naturally. There may be some cokes that are similar to glass-like carbon in terms of microstructure, but their purity and other properties may be different. This is the most pertinent carbon when it comes to microfabrication, because many photolithography resists yield glassy carbon on heat-treatment. This material features a high electrical and thermal conductivity, thermal shock resistance, low thermal expansion coefficient, corrosion resistance, low porosity, mechanical hardness, and atomically smooth surface. Moreover, the fact that it can be produced using already molded polymers made it useful for manufacturing applications where graphite cannot be used. Some physicochemical properties of glassy carbon are listed in Table 3.2. Some initial industrial applications of glass-like carbon include fabrication of battery electrodes, high-temperature molding tools, lining for reactors containing harsh chemicals, laboratory crucibles, elements for high-temperature furnaces, bone and tooth implants, etc.

Table 3.2: Physicochemical properties of commercial glass-like carbon [203].

Property	Value	Special conditions, if any
Electrical resistivity	10–50 $\mu\Omega$ m	At room temperature
Young's modulus	20–40 GPa	–
Thermal expansion coefficient	$(2.0–3.4) \times 10^6 \, K^{-1}$	–
Poisson's ratio	0.15–0.17	–
Density	1.3–1.55 g cm^{-3}	–
Apparent porosity	0–12	–
Electrochemical stability window (potential limits)	(a) 0.9 to −1.1 V (b) 1.4 to −1.5 V (c) 0.5 to −1.6 V (d) 3.0 to −2.6 V	(b) in 1 M HCl (b) in Phosphate buffer, pH 6 (c) in 1 M NaOH (d) in 0.2 M LiClO4 in acetonitrile

> **!** Understanding glass-like carbon is key to understanding lithographically patterned carbon devices. Most resins that are currently used in lithographic techniques yield a carbon material similar to glass-like carbon. It is also important to note that while this material has properties almost the same as commercial glass-like carbons, whether or not it should be denoted with this name is still debated. Many carbon scientists believe that a material should only be called glass-like carbon if it is prepared above 2000 °C.

3.4.3.2 Pyrolytic graphite

Other than glass-like carbon, pyrolytic graphite and porous carbons are other forms of synthetic carbon materials that have been used in device applications. The highest quality pyrolytic graphite is called highly oriented pyrolytic graphite or HOPG, which is frequently used in X-ray diagnostics and spectroscopy equipment [142]. The constituting graphite crystallites in HOPG feature a very low mosaic spread angle (measure of variation in its angular spread) of typically <1°. HOPG is nowadays used as a precursor for few layer graphene that can be exfoliated from it [261]. The standard method of HOPG synthesis is via pyrolysis of gaseous hydrocarbons (CVD) followed by high-temperature annealing or hot-pressing for graphitization. Further details of HOPG synthesis are provided in Chapter 4. Pyrolytic graphites can also be obtained via heat-treatment of solid polymers, such as polyvinyl chloride (PVC). Such carbons do feature a graphite-like layer spacing, but they are largely polycrystalline and difficult to convert into HOPG. In the context of microfabrication, precursors such PVC are difficult to use, because they are not compatible with most of the state-of-the-art microfabrication tools. Moreover, carbonization of PVC entails the release of hydrochloric acid, which may corrode the chip material and support equipment, if not handled properly.

3.4.3.3 Porous and activated carbon

When polymers undergo direct charring (no semisolid state during pyrolysis), the resulting carbon material is porous and typically hydrophilic. These are non-graphitizing forms of carbon with a lower crystallinity and electrical conductivity compared to glass-like carbon. However, their surface area is much higher. In energy storage and certain sensing devices, the surface area and wettability of the electrode material is more important than a high electrical conductivity. As a result, the carbon material is sometimes made intentionally porous. A high porosity and accessibility of pores also makes these materials excellent adsorbers. One can perform further activation of a porous carbon via both physical and chemical processes that result in an increase in their porosity and/or incorporation of functional groups on to their surface. Porous carbons with activated surface are called activated carbons. Natural materials, such as wood, yield porous carbon materials with fractal-like geometries. Mathematically, fractals are repeating patterns that replicate themselves at smaller dimensions. By definition the process is infinite. But in real-life examples, such as tree branches, the replication stops at a certain physical dimension. Fractals can be expressed as follows:

$$Z_{n+1} = (Z_n)^2 + c, \tag{3.2}$$

where Z_n is a sequence of numbers and c is a constant. Figure 3.27 (A) shows an image of commercially available activated carbon in cylindrical shapes. Figure 3.27 (B) is a schematic representation of the pores of different dimensions (levels 1, 2, and 3) in carbon obtained from natural polymers, such as wood. The SEM image shown in Figure 3.27 (C) is from a tree-bark, which clearly shows that the initial structure of the precursor is well-maintained after carbonization.

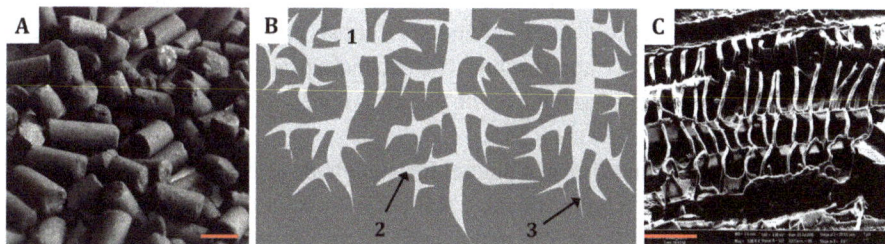

Figure 3.27: (A) Commercially available pellets of activated carbon (scale bar: 2 cm). (B) Fractal structure of pores in activated carbon (**1**: mesopore, **2**: macropore, **3**: micropore). (C) SEM image of a tree bark carbonized at 900 °C under nitrogen (scale bar: 10 μm). Note the retention of cellular morphology after carbonization.

Commonly, porosity is defined as follows:

$$\text{Porosity, } \Phi = \frac{\text{Volume of Pores}}{\text{Total Volume}} = \frac{V_P}{V_T}. \tag{3.3}$$

Porosity in activated carbon is often measured using indirect methods, such as gas absorption isotherms, dye and iodine absorption numbers, and mercury porosimetry. SEM, optical microscopy, and other material visualization techniques (such as MRI and X-ray tomography) are also used for porosity calculations. Although one cannot control the pore structure or the surface chemistry in the case of natural polymers, they are easily available from waste materials (for example, nut-shells, waste paper and cardboard, crop residues, etc.). Artificially produced cellulose, some synthetic polymers, and even impurity containing coals can also be used for activated carbon production.

All activated carbons are porous, but not all porous carbons are activated. Activation is an additional process, which can also be carried out for glass-like carbon, carbon fibers, etc., other than porous carbon. !

Activated carbon is the most common material for manufacturing water filtration units. Its surface can bind with a number of heavy metal ions and various bacteria and fungi. Depending on the surface chemistry, one can also adsorb chemicals such as antibiotic drug molecules dissolved in water [165, 256]. In addition to water filtration, activated carbon aids are utilized for the removal of a variety of pollutants from the air. Due to their electrical conductivity and high surface to volume ratio, activated carbons are also used in electrochemical applications.

3.4.3.4 Precursors for microfabrication of carbon shapes

Most of the polymers used as a precursor to glassy carbon are synthetic resins. Resins are organic materials with a high viscosity. There are several natural resins (e. g., pine-tree resin) that have been used for making natural rubbers and other structures. Most common precursors for glass-like carbon are phenol-formaldehyde resins and furan resins. Carbon derived from acrylate-based polymers, cellulose, and polyacrylonitrile (fibers) have also been used in device applications. The two main precursors and their chemical structure will be briefly discussed here. The pyrolysis mechanism of these polymers has already been described in Section 3.3.1.1.

The basic chemical structure of a phenol-formaldehyde unit is shown in Figure 3.28. These resins can be of two types: novalac and resol. Novalacs have a molar excess of phenol over formaldehyde, i. e., the molar ratio of phenol to formaldehyde is >1. In resols, phormaldehydes are in excess. Novalac type resins are liquid under standard pressure and temperature conditions, and their viscosity can be tuned by changing the fraction of phenol and formaldehyde. It is this type that is used in microfabrication. Such resins exhibit a thermosetting character after cross-linking, and their chemical structure facilitates an easy incorporation of epoxy side groups. Epoxy groups enable a chemical amplification after the photoacid generation for an enhanced crosslinking. Phenol-formaldehyde resins have the additional advantage of a high carbon content and relatively low oxygen in the primary (backbone) structure. They undergo coking process and yield high-quality glass-like carbon on heat-treatment.

Another common precursor to glass-like carbon is polyfurfuryl alcohol. This polymer is a type of furan resin obtained by polymerization of furfuryl alcohol. Furfuryl alcohols can be extracted from various biomaterials and are often produced from waste biomass. Figure 3.29 shows a furan ring, primary furfuryl alcohol, and a furan resin after its polymerization. These resins are not photosensitive. Microfabrication with furan resins is conducted using microscale molds, soft lithography, or other methods.

> ⚠ Development of photosensitive resins is a rapidly growing field. Various advanced 2D- and 3D-patternable resins are have lately become commercially available. However, many of them exhibit a low carbon yield, or cannot retain their shape after carbonization due to a high degree of coking. The chemical structure of many such resists or resins is patent-bound, which makes it difficult for scientists to predict their properties prior to carbonization. Some hit-and-trial is often required when a new resist is being used for making carbon patterns.

P:F<1
Access Formaldehyde

Resol
Solid

P

F

Nonvolac resin
Liquid

P:F>1
Crosslinked on additional acivation

○ Hydrogen   ● Carbon   ○ Oxygen

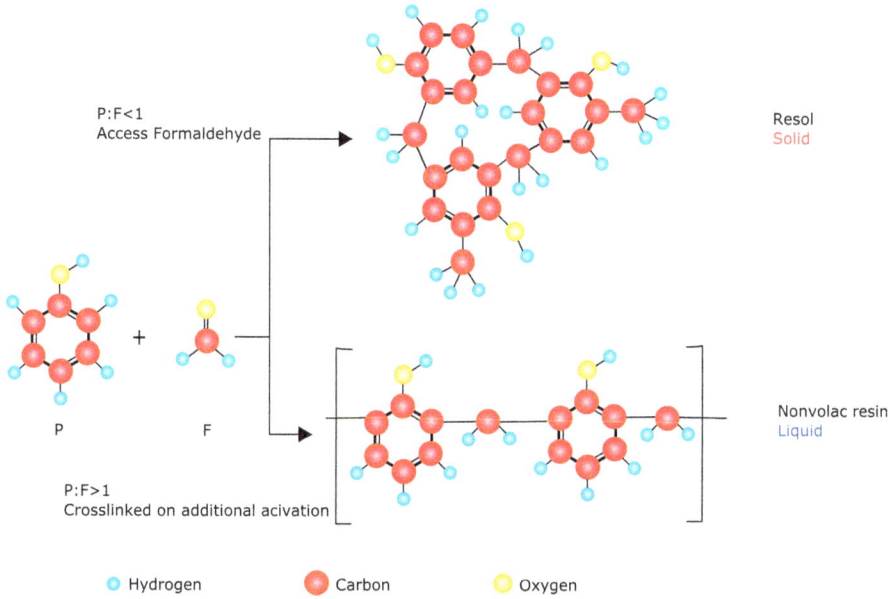

Figure 3.28: Chemical structure and constituents of a typical phenol-formaldehyde resin.

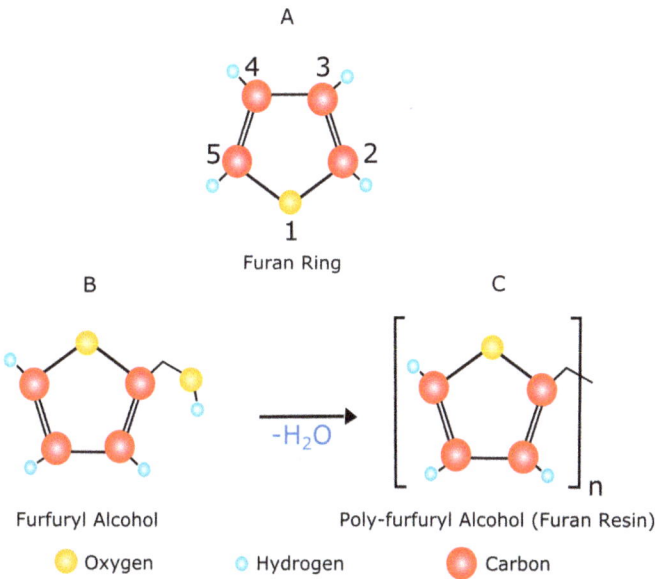

A

4   3

5   2

1

Furan Ring

B

C

Furfuryl Alcohol

−H$_2$O

Poly-furfuryl Alcohol (Furan Resin)

○ Oxygen   ○ Hydrogen   ● Carbon

Figure 3.29: (A) A furan ring, (B) furfuryl alcohol, and (C) furan resin after the polymerization of furfuryl alcohol, which involves loss of one water molecule per unit.

3.4.3.5 Pyrolysis shrinkage and size effect

Shrinkage of polymeric carbon structures is typically isometric, i. e., the structure shrinks uniformly in all directions. However, if one surface of the structure is attached to a substrate, this side may be unable to shrink compared to others. If the patterned structure is film-like (its spread in the xy-plane is much greater compared to its height), the shrinkage is only prominent in the z-direction. This means that the films become thinner after carbonization. However, if the structure has a significant height, a shrinkage in all three directions is observed. Though in large-scale structure fabrication shrinkage is considered a problem, for microfabrication it is seen as an advantage. This is because microscale structures can be made smaller by carbonization, which further improves their surface to volume ratio. This also enables the fabrication of much smaller structures than the capability of the microfabrication technique. A shrinkage allowance can be provided on the mask to avoid undesirable distortions. A detailed study of commonly used shapes in photolithographic carbon structure has been carried out by Natu et al. [167]. Authors showed that in addition to shrinkage, some sagging of the core (center) of the cylindrical structures is observed, particularly if the aspect ratio is >1. The exact % of dimensional shrinkage depends upon the chemical composition, extent of cross-linking, and presence of solvents, etc., in the polymer structure. A typical shrinkage value is 60–80 %, but some polymers can reduce by as much as 98 % [265].

In the case of the millimeter-scale part fabricated using glass-like carbon, the material must be heated to ≥2000 °C (typically above 2500 °C in commercial processes). This is done to ensure a good quality of material all the way to the core of the part. Good quality here refers to high electrical conductivity, uniformity in microstructure (of outer surface and core), uniformity in density, and controllable shrinkage. The intermediate materials generated during pyrolysis and carbonization processes are poor conductors of heat. If the part is too thick, a thermal gradient across the sample is created, hence, the temperature in the core may be lower than that on the outer surface. This can also result in differences in physical state of the core and the outer shell. Moreover, bubbles of volatile pyrolysis products (syngas) may find it difficult to escape the part. As a result, the part will display poor mechanical properties or nonuniform electrical conductivity. Nonetheless, it has been confirmed by several studies that in the case of micro-/nanoscale patterns, uniform properties of glass-like carbon can be achieved at temperatures as low as 900 °C [203]. This is attributed to their small sizes and often thin walls-/films-like geometries, which enable a faster cooling and plenty of surface for the release of volatile impurities due to their characteristic high surface-to-volume ratio.

3.5 Applications and example devices

Microfabricated carbon patterns have been used in applications that utilize their mechanical, electrochemical, or electrical properties. Biocompatibility of carbon is often used in combination with its other characteristics, which makes it suitable for biomedical devices. Most chip-based carbon device examples in the current literature are based on photolithographically patterned SU8 (a phenol-formaldehyde resin). Since this resist is available in different viscosities, patterns of dimensions up to 150 μm can be fabricated. Importantly, since the carbonization is only performed at 900 °C, the carbon derived from SU8 is occasionally not considered glass-like carbon. This is due to the fact that large-scale glass-like carbon is prepared at >2000 °C and is expected to contain impurities when the temperatures are lower. However, as discussed earlier, smaller patterns have a tendency to undergo carbonization relatively fast; one can potentially achieve a higher purity at 900 °C. Moreover, the properties of SU8-derived microfabricated carbon have been proven to be very similar to commercial glass-like carbon by many researchers. Details of several microfabricated carbon devices can be found in various review articles, for example, Sharma and Madou 2012 [206], Sharma 2018 [203], and Devi et al., 2021 [49]. Some representative examples from the prominent application areas are provided here.

3.5.1 Energy storage

From Table 3.2, it is evident that glass-like carbon has an excellent electrochemical stability window. This implies that the electrode can be used in wide range of voltage without itself participating in the reaction. Glass-like carbon is also corrosion-resistant to a great extent, hence, it is a very suitable material for electrochemical applications. With 2.5D microfabrication, one can increase the net surface area of the patterned structures. All these properties together render this material and process highly perfect for battery and supercapacitor electrode fabrication. Many examples in the literature can be found where microelectrodes made of glass-like carbon in the form of arrays of vertical pillars or interdigitated structures have been utilized in microbatteries and supercapacitors. In Figure 3.30, images of carbon electrodes for electrochemical applications and a device assembly are presented. Notably, the electrical conductivity value of glass-like carbon is lower than that of many metals, but it is high enough for common electrical and electrochemical applications.

In addition to batteries and supercapacitors, micropatterned carbon electrodes have also been employed as electrocatalysts for fuel cells. Fuel cells used for water splitting entail electrocatalysts for both hydrogen and oxygen evolution reactions. Different types of metal nanoparticles are used as catalysts, which are typically loaded on to commercially available glass-like carbon plates. If this carbon substrate can be microfabricated, the available surface area can increase manifolds. It has also been proposed that carbon itself can feature electrocatalytic activity with certain crystal modifications (e. g., by introducing a heteroatom like N, B in the crystal to create polarization).

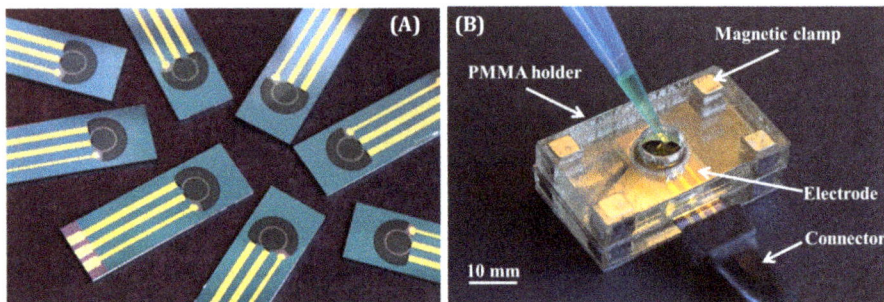

Figure 3.30: Digital photographs of (A) 2D carbon electrodes connected to metal contact pads, and (B) A device assembly with magnetic clamping containing the electrodes shown in (A). Copyright ©2017 Elsevier Ltd.Reproduced with permission from Hassan et al. [92].

3.5.2 Biosensing and cell culture

A number of biosensors are based on sensing certain chemicals released due to a stimulus provided to the biological tissue. This is often done via electrochemical route. A transducer is used for converting these so called biochemical signals into electrical signals. In the case of *ex-situ* biosensors, certain biomolecules are immobilized onto the surface of the electrode that can facilitate exchange of electrons. A higher surface of the electrode enables an enhanced mass-transport, hence, having smaller individual structures increases the sensitivity of such devices manifolds. The electrode material must be biocompatible for biosensing applications. Biocompatibity of a material can be of two types: *in-vitro* and *in-vivo*. *In-vitro* biocompatibility indicates that the material can be used as a substrate for cell culture and growth. It may or may not promote the growth of the cells, but it should not have any negative effect on their growth patterns or survival. *In-vivo* biocompatibility is evaluated when a structure made of a certain material is implanted in a human or animal body. The material for such applications should not only support cell culture and growth, it should not release any harmful chemicals or cause any physical or chemical damage while experiencing harsh biological environments over a long duration. If needed, it should also dissolve or disintegrated inside the body over time. Carbon's biocompatibility is more commonly used for *in-vitro* applications. There are examples of carbon-based short-term implants that are not expected to degrade by themselves. The inertness and stability of glass-like carbon prevents it from degrading or dissolving in most media.

Another variation of biosensing is the analysis of biochemicals, without having any biomolecules immobilized or attached to the electrode surface. Such detection methods can also be called chemical sensing. But if the chemical is a secretion from a cell or tissue or an indicator of certain biological phenomenon, biochemical is the generally used term. In this context dopamine sensing with carbon electrodes is important to mention. Glass-like carbon based electrodes have shown selective sensitivity towards dopamine

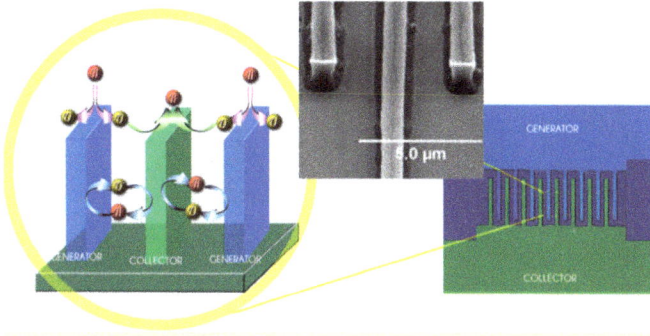

Figure 3.31: A schematic diagram along with a scanning electron micrograph showing carbon interdigi-tated electrode arrays used for dopamine sensing by redox amplification. Copyright @2014 Americal Chemical Society. Reproduced with permission from [124].

both with and without the presence of neural cells on their surface. A schematic diagram of interdigitated carbon microelectrode array used for dopamine detection is shown in Figure 3.31.

Carbon-based 2D and 3D patterns have also been utilized as a substrate for cell culture. When such substrates are 3D, they are called scaffolds. Carbon is an inert material that does not harm most cell types. It has been shown in multiple studies that it is a very good scaffold material for neural stem cells. Electrospun carbon fiber mats, 3D microfabricated pillars, and carbonized porous cryogels are some examples of carbon scaffolds for neural stem cell culture [161, 11, 73]. Figure 3.32 [73] shows a porous carbon scaffold obtained by pyrolysis of a cryogel.

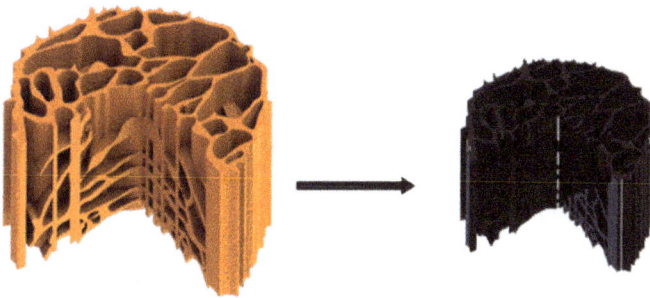

Figure 3.32: Carbonization of structured polymers into carbon. Structure shrinks due to a net mass loss [73].

3.5.3 Topography sensing

Atomic force microscopy is a commonly used tool for topography sensing. Such devices consist of silicon cantilever and a sensing element (tip) on top of it. These tips are made

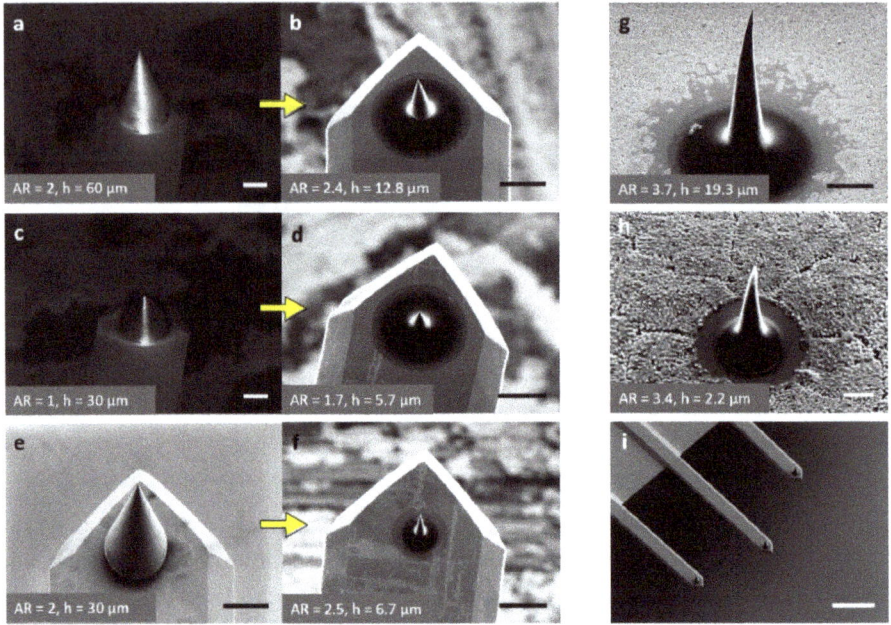

Figure 3.33: Carbon-based atomic force microscopy tips made by two-photon lithography of an acrylate based resist [265].

of a variety of materials, including diamond, silicon carbide, metals, and polymers. Polymeric carbon patterned by two-photon lithography process has also been used for tip-fabrication. The structures, shown in Figure 3.33, were patterned using an acrylate-based resist (or ink), which showed >98 % shrinkage, thus making the tips very small and sharp for better sensing. Conical tip-shaped structures fabricated using a phenol-formaldehyde resin have also been used for making carbon-based field-emission tips.

Chemical etching or laser-patterning of commercially available glass-like carbon sheets can also be used for obtaining surface roughness or microscale structures on the surface. This obviates the need for lithographic microfabrication. However, the structures may not feature smooth walls or uniform dimensions in such cases.

3.5.4 Other devices

In addition to aforementioned applications, microfabricated carbon devices have been used in microfluidics devices for the purpose of dielectrophoresis. Here, the carbon structures are used for a dual purpose of creating microfluidic channels as well as provide electrical stimulus to the flowing cells. The electrical conductivity of carbon comes in handy in biological devices, as many of them respond to electrical signals. For example, an additional feature can be provided to implantable neural electrodes with the possibility of neural stimulation. Such efforts have already been made, and they will be discussed in the chapter based on flexible device.

Some other examples of advanced, all-carbon devices include chemosensors and biosensors that utilize nanoscale electrospun carbon fibers integrated with carbon microelectrodes. Here, the lithographically patterned carbon simply forms the contact pads. The sensing element is a nanoscale fiber that can be further modified or decorated via vapor deposition or chemical routes. There have been reports of growing carbon nanotubes or fibers on the surface of polymeric carbon structures, but such structures have not been used for any specific applications.

3.6 Current limitations and future trends

Photolithography is a relatively inexpensive and commonplace technique. However, if the critical dimension is below 10 μm, the mask fabrication becomes very expensive. Mask is a one-time expenditure so for large-scale fabrication, photolithography remains cheaper compared to other techniques, even with a sophisticated chromium mask. The fact that it allows for a batch fabrication works to its advantage. Most of the commonly used photolithography resists have been optimized for carbon-device fabrication, which includes their design optimization according to expected shrinkage and design of the heat-treatment parameters that yield carbon with specifically engineered electrical and electrochemical properties. Owing to these reasons, photolithography seems to have an edge over other polymer patterning processes for carbon device fabrication.

Carbonized structures are much smaller than the critical dimensions possible with the initial polymer patterning process, thanks to carbonization shrinkage. It has been observed during longer post-exposure baking times at temperatures lower than the ones prescribed by suppliers. Carbonization parameters, such as heating rate, highest process temperature, and residence time at the highest temperature have also shown improvement in material properties. Further work on these aspects can potentially lead to carbon with much improved properties for specific applications.

Inducing desirable chemical functional groups onto the surface of carbon is yet another tool that can broaden its application areas. Though the material prepared by inert heat-treatment is inert, one can graft various functional groups onto its surface. Such methods have already been used for preparing bulk glass-like carbon electrodes for electrochemical applications [79, 236]. In microscale devices, at least some surface modification is almost always performed to ensure the wetting of the electrodes in a solution. In the case of biosensors, the biomolecule (e. g., enzyme) immobilization also requires surface pretreatment. In some reports, researchers have shown that addition of nanomaterials into resist prior to lithography can induce novel functionalities to the glass-like carbon structures, as well as improve their mechanical properties by tuning their crack propagation response. Such concepts pave the way for a large number of new hybrid polymeric carbon materials for micro- and nanodevices. The fabrication process-related challenges need more attention at this point.

Chip-based carbon devices can also be fabricated using bottom-up (additive) manufacturing techniques, such as printing and direct growth of nanomaterials onto a rigid substrate. In some instances, few-layer graphene has been transferred onto silicon or silicon oxide substrates for device applications. CNT and CNF forests can be grown onto rigid substrates in the presence of catalysts. Their tips can directly be used as sensing elements. Such assemblies have also been used in a variety of energy storage devices. Nanomaterials, however, are more commonly used in flexible electronics and nonsilicon substrates, which will be discussed in a separate chapter.

1. Which one of the following techniques is most suitable for the fabrication of insulating cylindrical pillars of height = 100 nm, diameter = 50 nm, and end-to-end gap = 50 nm: (a) photolithography, (b) nanoimprint lithography, (c) dip-pen lithography, or (d) soft-lithography?

2. What is a thermoplastic material? Can a thermopastic material convert into thermosetting on exposure to UV light?

3. Should the photoresist for nanoimprint lithography be more viscous compared to resists used in photolithography (negative tone)? Justify your answer.

4. Describe all steps of photolithography using a negative resist of your choice for the following structure: walls of length 250 µm, width 25 µm, and height 50 µm. Separation between two walls should be same as their width, and the total number of walls should be such that the overall structure is a square.

5. Which one of the following techniques is most suitable for the fabrication of insulating cylindrical pillars of height = 100 nm, diameter = 50 nm, and end-to-end gap = 50 nm: (a) photolithography, (b) nanoimprint lithography, (c) Dip-pen lithography or (d) soft-lithography?

6. What is a thermoplastic material? Why are thermosetting resins better than thermopastics for producing polymer-derived carbon shapes?

7. What is the role of change in glass transition temperature of a polymer in lithography? How does the chemical process differ in the case of thermally cross-likable and UV-cross-linkable polymers?

8. In your opinion, is pyrolysis of green plants/trees for the production of pyrolysis oils good for the environment?

9. What are the primary challenges in waste pyrolysis? Can urban solid waste be used as a precursor for high-quality carbon production?

10. In which cases would one use a batch reactor for large-scale waste pyrolysis?

11. Why does glass-like carbon have an atomically flat surface?

12. What is the reason behind porosity of carbons obtained from natural precursors (e. g., coconut shell)?

13. List at least five differences between graphitizing and non-graphitizing carbon.

14. In your opinion, what causes the active surface functional groups in natural polymer derived carbon?

15. What is the difference between combustion, decomposition, and pyrolysis?

16. Can porous carbon be used in micro-/nanodevices? How can it be patterned without using a binder?

17. Water filtration columns are often made of activated carbon rather than any other carbon form. Why?

18. What is the difference between physical and chemical activation of carbon?

19. Which characterization methods can be used during the heat-treatment of a polymer? Can this be done for microfabricated patterns?

20. A solid cube made of a resin needs to be converted into carbon at 1200 °C. Final (carbon) structure may have its edge length anywhere between 1 mm and 1 cm. If the cube is too big, it might contain some porosity in its core (as shown in the cross-section of the cube, Figure 3.34). Dimensional shrinkage of

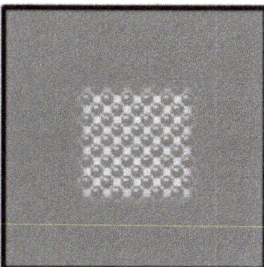

Figure 3.34: Cross-section of a carbonized cube containing pores in its core.

this resin is 50 %, but nothing is known about its pyrolysis mechanism and byproducts. Design a single heat-treatment experiment to determine the maximum size of the cube that can be made without any porosity. You will have sufficient supply of resin and any other material(s) and/or manufacturing process, if needed. After the heat-treatment, you are allowed to use up to two characterization techniques of any kind.

4 Carbon nanomaterial-based devices

Carbon has been used in devices pretty much since the need for solid-state energy storage was recognized by the scientific community. For almost two centuries, there have been constant and curious efforts to optimize the physicochemical properties of carbon, including the size, shape, and surface area of individual carbon grains or units, so as to improve the performance of the existing devices, as well as explore new application areas. In parallel, constant attempts are made to develop superior carbon materials that could meet these requirements. In this endeavor, scientists have encountered many new forms of carbon. The latest addition to this list is the entire class of carbon nanomaterials. These materials may not necessarily be new in nature, but their application in technology only started after their so-called discoveries that are relatively recent. Today, it is hard to talk about carbon devices without mentioning nanomaterials. In fact, since the 1990s, the momentum of carbon research has entirely shifted towards nanoscale materials and their applications.

Materials composed of well-defined and discrete nanoscale units are known as nanomaterials. Unlike any other element, carbon atoms tend to quickly bond with each other and adopt the most stable geometry possible with the number of available carbon atoms in the vicinity. If carbon atoms are present in abundance, bulk quantities of solid carbon, such as graphite, diamond, non-graphitizing carbons, etc., are generated. However, when the availability of carbon is limited or the atoms are distant from each other (e. g., in gas phase carbon) discrete nano-scale geometries are formed. Collections of these nanoscale units are designated as nanomaterials, although individual units may be directly used in certain applications. The shapes of nanoscale carbon geometries can be controlled and directed, for example, by confining their growth to a predefined region, (e. g., on the surface of nanoparticle). Other parameters, such as the composition of the gas phase carbon, which is in turn controlled by temperature, pressure, precursor, etc., also dictate the final product. Making these carbon nanomaterials and their patterning for devices are the focal topics of this chapter. Note that unlike bulk sold carbons, nanocarbons are generally produced via bottom-up manufacturing processes, starting at atomic or molecular length-scales. Consequently, the term *synthesis* or *manufacturing* are more commonly used rather than fabrication.

The organization of this chapter is as follows: first a brief overview of the most popular carbon nanomaterials is provided. Only elemental carbon forms are covered. This is followed by most commonly used synthesis processes for making carbon nanomaterials themselves. Once the material is prepared, the next step is to utilize it in a device. This entails a different set of techniques that fall within microfabrication, because the patterned shaped are typically microscale. Some manufacturing techniques that are based on direct use of nanoscale units in devices are described along with example towards the end.

https://doi.org/10.1515/9783110620634-004

> The term *structure* is typically used in material science to describe a crystal lattice or the geometries of that scale. In civil engineering, structures can be massive bridges and in mechanical engineering, automobile and aircraft parts can be referred to as structures. The term structure may allude to different length-scales in materials and manufacturing. In this chapter, nanoscale geometries formed by carbon are denoted nanostructures.

4.1 Introduction to carbon nanomaterials

Nanoscale 1, 2, and 3D carbon geometries commonly present in the form of a powder or suspension are known as carbon nanomaterials. Individual nanostructure is separable from its bulk powder form, and features its own unique set of properties, different from other carbon nanomaterials. Typically one nanomaterial contains only one type of carbon geometry, but it is not impossible to find different carbon nanostructures (e. g., CNT and fullerene) in the same material mixture. Carbon atoms adopt different structures so as to maximize their overall stability for the available number of atoms and hybridization states, as described below.

4.1.1 Formation and stability

Figures 4.1 and 4.2, drawn based on a review article by Shenderova et al. [212] elucidate the most stable carbon geometries or aggregates for a given number of carbon atoms having the possibility to bond together. The information and some data points in these plots are derived from computational studies, in addition to experimental results. It can be observed in Figure 4.1 that as soon as six carbon atoms become available for bonding, hexagonal carbon rings can be formed. The electron in the unhybridized p-orbital and its delocalization provides a special stability to carbon rings, which small linear carbon molecules, such as C_2, may not possess. Individual rings, however, are unstable, and they tend to club together with the increase in available atoms. Various shapes and sizes of such clusters of carbon rings can exist up to 40 carbon atoms. Some of them may also contain non-six-membered rings, and hence a curvature (bowl-like shapes). Buckminsterfullerene or C_{60}, a well-known football-shaped carbon molecule results from 60 carbon atoms. Above this number, other curved carbon structures, such as larger sphere or ellipsoid shaped fullerenes, as well as nanotubes, can develop. The transition between rings, curved fragments, and closed curved geometries is not very sharp. In fact, often these different carbon species can coexist in gas phase carbon. Chain-like carbon molecules, such as carbynes, may also be present in the system at the same time. There have been reports of smaller carbon molecules having 2 to 8 carbon atoms, which are believed to be highly unstable.

Fullerenes and nanotubes remain dominant carbon geometries all the way up to 1100 carbon atoms, as evident from Figure. Interestingly sp^3-hybridized carbon and

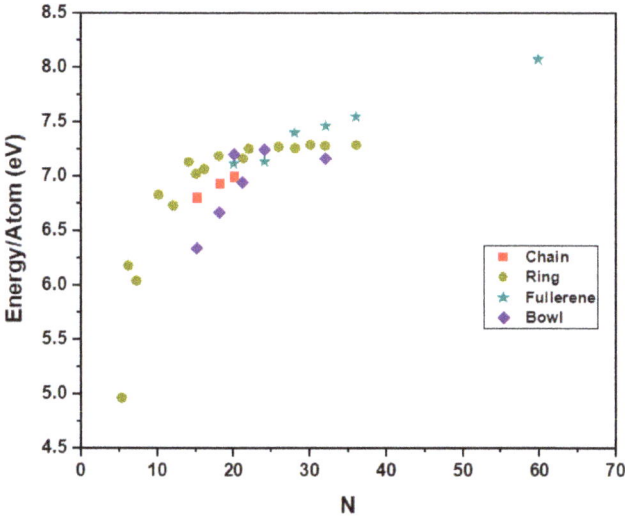

Figure 4.1: Stability of different carbon forms with respect to the total number of carbon atoms (N). Copyright ©1969 Plenum Publishing Corporation (article published in 2001 by Springer Nature). Adapted with permission from [157].

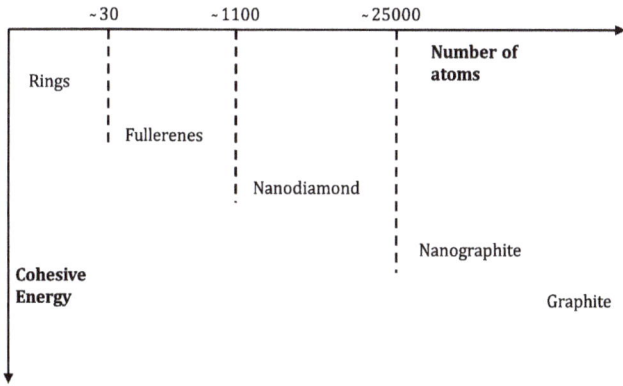

Figure 4.2: Stability of various carbon nanoforms with respect to number of carbon atoms and cohesive energy. Copyright ©2002 Taylor and Francis. Adapted with permission from [212].

nanoscale diamonds become a bit more stable than nanoscale sp^2 materials between 1100 and 25000 carbon atoms, and then graphite or other sp^2 carbons become the most stable carbon form again. In fact, a number of the computational studies that were referred for the purpose of this diagram were focused on understanding nanodiamonds and explaining their formation in high-energy events.

In this diagram, single or few layers of graphite (i. e., graphene) are not clearly shown. This is because as soon as there are sufficient rings present and there is a catalytic support, they tend to start depositing on the catalyst surface. This may occur even with relatively few carbon atoms, because other factors, such as the catalyst, promote carbon deposition. Depending upon the catalyst, pressure, and temperature, this process can be controlled, and single or few layers can be collected. This will be described in detail in Section 4.2.1.

The most common manufacturing processes for carbon nanomaterials are based on their collection from gas phase carbon. Gas phase carbon can be generated by (i) pyrolysis (cracking) of a light hydrocarbon at high temperatures, or (ii) evaporation of graphite (or other carbon) electrodes. Unlike most vapors, carbon species in the gas phase do not essentially represent "vapor phase carbon." To appreciate the complex composition of gas phase carbon, it is important to first understand the difference between vapor and smoke. Vapors are diffused in the system, and the chemicals present in them are generally of atomic or molecular dimensions. They can be condensed by changing pressure and/or temperature. Smoke, on the other hand, contains suspended solid particles of relatively large sizes. In both cases, the species present in the gaseous environment can potentially be reactive and may undergo rapid chemical or physical changes.

Now the question is: what is carbon in gas phase? Vapor or smoke? If we refer to the phase diagram of carbon (shown in Chapter 1), it is clear that vapor phase carbon can only be produced at temperatures >4000 K. Such conditions are difficult to achieve during carbon manufacturing, for example, in a CVD reactor. But hydrocarbon cracking does yield a gas phase carbon that is a combination of vapor and smoke. Carbon vapors have also been produced by electric arc generation across graphite electrodes. These vapors are known to contain carbon species of having 2–8 carbon atoms (C_2–C_8). In addition, some hexagonal carbon rings or small molecules containing a couple of carbon rings are also detectable. All of these are highly unstable forms of carbon and hence, they tend to readily convert into slightly larger fragments or formations within the gas phase. Chain-like carbon molecule carbyne, as well as small fullerenes have also been detected in gaseous carbon systems. All in all, it is very difficult to clearly distinguish between vapor and smoke at a given time, because of highly mobile carbon species that can potentially join to form larger fragments (collections of rings). Referring again to Figure 4.1 facilitates identifying regions of atomic concentration where more than one types of carbon species can coexist. In summary, gas phase carbon is a mixture of carbon species, ranging from atomic scale to tubular geometries containing thousands of atoms. Many of them are regarded as technologically significant carbon nanomaterials in their powder forms.

4.1.2 Classification

Carbon nanomaterials can be classified based on their dimensionality (i. e., whether they are 1, 2, or 3D), primary hybridization state (sp^2, $sp^{(2+n)}$ or sp^3), or their size and morphology. In dimensionality-based classification, fullerenes, quantum dots (solid spherical or near-spherical carbon structures <10 nm in diameter), and sometimes slightly larger carbon particles are considered zero-dimensional. Nanotubes and nanofibers that feature negligible diameters compared to their length are designated one-dimensional, and sheets or fragment-like shapes are considered two-dimensional. Diamond and graphite crystals, often with shape irregularities but confined within 100 nm equivalent diameter, are called three-dimensional. Though this kind of classification is more common due to its visual appeal and ease of explanation, it is becoming increasingly difficult to segregate newly discovered (or synthesized) advanced carbon materials based on their dimensions, as some of them may be a combination of more than one type of carbon nanostructure. Carbon peapods are a good example, which are composed of fullerene-like structures linearly arranged inside carbon tubes. Another issue with this kind of classification is the lack of well-defined limits for a structures be placed in a dimensional category or its size can be considered negligible.

Hybridization-based classification is more strongly correlated to material's properties, and it also takes the existence of non-six-membered rings into account. Fullerenes, for example, come under type materials, whereas defect-free graphene is a type carbon nanomaterial. This classification also connects carbon nanomaterials with their respective bulk counterparts (i. e., graphene and graphite), thus allowing researchers to find similarities and differences between nanoscale and bulk materials caused by their size, arrangement, and presence of defects. One challenge with this classification is the presence of (unintentional) defects, which can change its net hybridization state. An example is defect-containing graphene, which does not exhibit a pure hybridization.

The third type of classification is simply based on size. With the increase in size, the stability of the structure can change, which may lead to the formation of another material altogether. For example, a multiwalled carbon nanotube may collapse into a carbon fiber when the number of walls is above a certain threshold. Similarly, a large number of single graphene layers tend to be more stable as bulk turbostartic carbon, rather than discrete layers, and so on. Accordingly, one can place these materials in different size categories, e. g., <10 nm, 10–1000 nm, and >1000 nm, in terms of equivalent diameters.

It is evident that there is a large variety of carbon nanostructures, including carbynes, curved carbons, and carbon clusters of different shapes, which can exist at the nanoscale. Such geometries have been observed during experiments where high intensity electric arcs, plasma, or lasers cause a rapid sublimation of bulk carbon materials. The so-called carbon clusters are aggregates of generally <50 carbon atoms that can adopt random shapes, often with small carbon flakes containing small bunches of rings. The science of the formation and stability of carbon nanostructures is highly complex and will continue to develop for several more years, likely with many new material discoveries. In this chapter, we will focus only on those carbon nanomaterials that are commonly used for device manufacturing purposes. Some representative carbon nanomaterials are discussed in detail below.

4.1.3 Graphene

Introduction and brief history

Graphene is the single and defect-free layer composed of carbon atoms featuring pure sp^2 hybridization. Such layers are building blocks of graphite, where multiple layers are arranged in an ABABA fashion. Graphene and graphene-like layers are also found in turbostartic carbons and pyrolytic graphites (graphitizable carbon obtained via CVD of hydrocarbon gases). In these materials, the layers are not organized in any other specific order. Rather, they are randomly rotated on top of each other. Various carbon nanomaterials, such as CNTs and fullerenes, can also be considered analogues of graphene. But it is extremely important to understand that the term graphene is used only when we are talking about isolated layers, not those present in any other carbon material. By definition, these layers must be <10 in number and should not contain irregular defects.

It is known that in case of sp^2 hybridization there is one free electron present in the unhybridized p-orbital that forms a π-bond with its neighboring carbon atoms, while the remaining three electrons are bonded via σ-bonds. The electron present in unhybridized orbital is perpendicular to the plane consisting of carbon hexagons formed by the bonded electrons.

Figure 4.3 shows an optical microscope image of single-layer graphene. Graphene layers are optically detectable, because they reflect light that leads to an interference pattern. One can see it if the intensity of light falling on the graphene flake is different from that on the substrate. Typically, a 300 nm thick SiO_2 layer grown on Si substrates is used for optically observing graphene, as this substrate facilitates a good optical contrast [46].

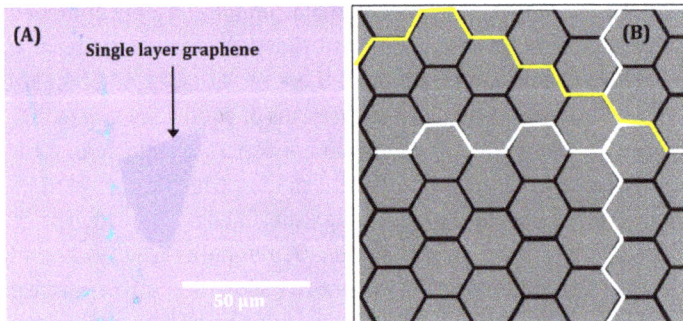

Figure 4.3: (A) An optical microscope image of a graphene flake on SiO_2 substrate. Image: courtesy Ayshi Mukerjee, Tata Institute of Fundamental Research, Mumbai, India. (B) A schematic representation of the distribution of carbon hexagons and possible electron propagation paths in graphene. Of course, many other combinations of electron paths are possible.

The anisotropic properties of graphite have always been of great interest to scientists. The electron present in the unhybridized *p*-orbital of sp^2 carbons, its delocalization, and interaction with the similar electron clouds present in top and bottom carbon layers strongly influence the electrical and thermal conductivity, diamagnetic susceptibility, and optical absorption of graphite. To understand these properties, one needs to first understand the behavior of electrons in a single and isolated graphite layer (i. e., graphene) followed by two-, and then multiple layers. In this context, graphene was already investigated as early as in 1947 by PR Wallace [246], who proposed a configuration for the electronic energy bands and Brillouin zones for graphite based on his analysis of a single layer. The lattice parameters and band theory deduced by Wallace still remains highly pertinent to the field and is used by material scientists for evaluating graphene's properties. In his later works, Wallace also investigated the effect of interstitial atoms in irradiated graphite. This material was available in large quantities from the early nuclear reactors that predominantly used graphite as moderator and reflector.

After Wallace's theoretical investigations, notable research, including experimental work on material synthesis, was carried out by a Boehm et al. in 1962 [25]. In their publication (in German language), the authors used the term "graphene" for the first time, which was added to the IUPAC database in 1994 [26]. Prefix "ene" is used for representing a variety of hydrocarbons (e. g., annulene, napthalene, etc). Taking "graph" from "graphite," this new material was designated graphene. Epitaxial growth of ultra-thin graphite films (now regarded as single- and few-layer graphene) was reported as early as in 1970s [175, 176], and CVD of hydrocarbon gases to obtain similar films was developed in the 1960s and 70s [176]. In many instances, the primary goal of these experiments may have been the deposition of thicker pyrolytic graphite films, rather than monolayers. These films were prepared under different conditions and also used for understanding the defect-annealing mechanism at high temperatures [245]. Evidently, graphene, as an independent material has been synthesized and studied for over 50 years. In the beginning of the 21st century, some new methods for graphene preparation were developed, which confirmed that the single layers are more stable than previously perceived. In this context, the most notable contribution was made by Geim and Novoselov, who isolated single graphene layers in a liquid suspension using an adhesive tape [173]. This experiment proved that one-atom thick 2D crystals, which were believed to be highly unstable in the absence of a solid substrate, could very well exist in solutions and on suspended membranes. The achievement was awarded with a Nobel in 2010. In addition to the aforementioned research groups, Mildred and Gene Desselhaus made significant contributions to understanding the electronic properties of graphene, particularly on the application of Raman spectroscopy to differentiate between single and bilayer graphene [53]. A chronological representation of the work on graphene, taken from Dreyer et al. [55], is shown in Figure 4.4.

It is the fundamental nature of carbon to bond with itself and form extensive sheets. If there are any carbon species present in vapor phase (smoke), and there is a heated transition metal surface nearby, formation of carbon sheets is almost inevitable. As a

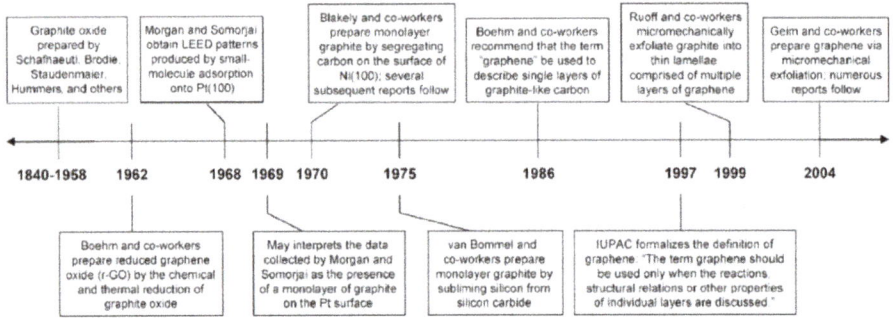

| Graphite oxide prepared by Schafhaeutl, Brodie, Staudenmaier, Hummers, and others | Morgan and Somorjai obtain LEED patterns produced by small-molecule adsorption onto Pt(100) | Blakely and co-workers prepare monolayer graphite by segregating carbon on the surface of Ni(100); several subsequent reports follow | Boehm and co-workers recommend that the term 'graphene' be used to describe single layers of graphite-like carbon | Ruoff and co-workers micromechanically exfoliate graphite into thin lamellae comprised of multiple layers of graphene | Geim and co-workers prepare graphene via micromechanical exfoliation; numerous reports follow |

1840-1958   1962   1968 1969 1970   1975   1986   1997 1999   2004

| Boehm and co-workers prepare reduced graphene oxide (r-GO) by the chemical and thermal reduction of graphite oxide | May interprets the data collected by Morgan and Somorjai as the presence of a monolayer of graphite on the Pt surface | van Bommel and co-workers prepare monolayer graphite by subliming silicon from silicon carbide | IUPAC formalizes the definition of graphene: "The term graphene should be used only when the reactions, structural relations or other properties of individual layers are discussed." |

Figure 4.4: History of the work done on graphene starting from 1840 up to 2005 [55]. Reprinted with permission. Copyright © 2010 WILEY-VCH Verlag GmbH and Co. KGaA, Weinheim.

result, few atom-thick graphene-like sheets can be observed near high-temperature furnaces, electron microscopy chambers, and various other equipment that utilize organic oils or gases for lubrication or heating, and contain metal or alloy surfaces or particles. The same goes for CNTs and other nanocarbon deposits, which will be discussed later in this chapter. Of course, a controlled growth of such "nanomaterials" is challenging. One of the biggest achievements of the last few decades is to acquire a good control over graphene and CNT growth processes for a relatively large-scale production. Development of sophisticated characterization tools, as well as the vast commercial interests in developing graphene as a next-generation electronic material have also played crucial roles in the tremendous attention received by this material.

4.1.3.1 Crystal structure and properties

A primitive unit cell of graphene consists of two nonequivalent atoms, as shown in Figure 4.5(A). The rhombus-shaped crystal unit cell feature two 60° and two 120° corner

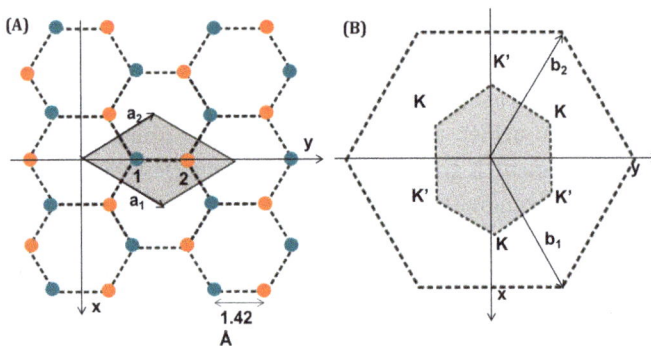

Figure 4.5: (A) Crystal structure showing the primitive unit cell of graphene, (B) The first Brillouin zone of graphene.

angles. Lattice vectors a_1 and a_2 are commonly taken such that there is a 60° angle between them. However, it is also acceptable to have the obtuse angle (120°) between them. Two neighboring atoms (shown in blue and orange colors in Figure 4.5(A)) are not equivalent in terms of crystal consideration, although they are chemically the same. This is because a blue atom cannot replace an orange atom by only translation operation. In other words, a single atom, or the entire hexagon, cannot form a Bravais lattice. One side of the hexagon or the C–C bond length in graphene is 1.42 Å. This is shorter compared to the 1.54 Å of diamond, and thus graphene is considered stronger than diamond. The first Brillouin zone of graphene is shown in Figure 4.5(B), which represents its primitive unit cell in the reciprocal space.

A Wigner–Seitz cell is a primitive cell that consists of one lattice point and the area surrounding it, which is formed by connecting all those points that are closer to this particular lattice point than to any other. One can take any lattice point and draw connecting lines to its nearest neighbours (atoms closest to it). Now perpendicular bisectors of these lines are drawn. The resulting enclosed area is the Wigner–Seitz cell. First Brillioun zone is nothing but the Wigner–Seitz cell in reciprocal space. For 3D crystals, bisector lines are replaced by planes and the area becomes volume.

Electrons in graphene behave differently compared to those in bulk carbon materials, owing to a quantum confinement (change in material's electronic properties when the sample thickness is <10 nm; diameter of a carbon atom is 0.14 nm). Theoretically, the electron mobility in (defect-free) graphene is >100000 cm^2/Vs at room temperature and 200000 cm^2/Vs (at 5 K). Its tensile strength is estimated to be 130 GPa and the Young's modulus as 500 GPa. By nature all carbon materials are light, and graphene being single-atom thick, is extremely light, with a theoretical surface area of 2000 m^2/g [135]. With all these properties, it is unsurprising that this material has drawn so much attention for numerous device applications. It is noteworthy that all these properties are for a single and defect-free layer, and they are rapidly compromised as the number of layers or fraction of defects increase.

Defects in graphene are mostly point defects caused by the presence of non-six-membered rings. Such defects are often found in pairs and can occasionally extend through the entire layer, causing a line defect in specific crystal directions. For example, if there is one five-membered ring in a graphene layer, it needs to be compensated with a seven-membered ring to maintain the overall hexagon-based crystal. It is energetically more favorable to have two such pairs of five- and seven-membered rings in a symmetrical fashion. Such defects, shown in Figure 4.6(A), are known as Stone–Wales defects; they are commonly observed in graphene [20]. Other types of defects include addatom (presence of an extra interstitial carbon atom), single and double vacancies $V1(5$–$9)$ of various types, as illustrated in 4.6(B-F). There may also be randomly shaped voids present in graphene crystal, which are often intentionally induced to improve the reactivity of the material [20, 228]. One needs to judge if the use of the term "graphene" is appropriate for such materials. A large number of non-six-membered rings in graphene cause

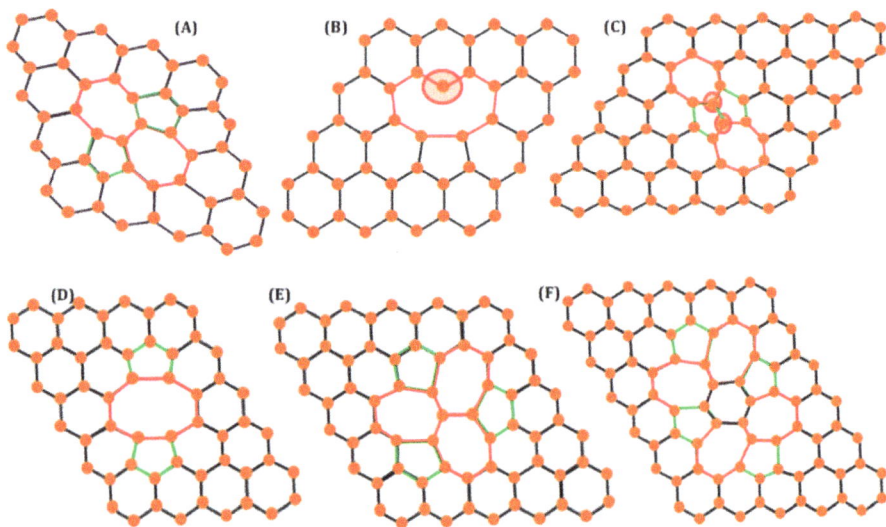

Figure 4.6: Defects in graphene, (A) stone-wales, (B) single vacancy $V1(5-9)$, (C) addatom, and (D-F) double vacancies $V1(5-9)$. Copyright © 2011, American Chemical Society. Adapted with permission from [20].

the sheets to attain curvature, which may ultimately collapse into spherical or tube-like structures. Even without any defects, graphene exhibits some curvature, which is due to its flexibility and characteristic vibrations of the carbon atoms. Most defects can be annealed out at elevated temperature or under high pressure. This principle of graphitization is indeed used for manufacturing highly oriented pyrolytic graphite (HOPG) at an industrial scale.

In addition to defects, graphene flakes may feature reactive edges due to lack of symmetry at their termination points. In principle, the edges of graphene should display either zig-zag or armchair-type configuration. This may not be the case in imperfect graphene, particularly if it is prepared by a top-down approach. Such edges contain a high fraction of dangling bonds, thus making the material more reactive under chemical or biochemical scenarios.

4.1.3.2 Nomenclature

The term graphene, as per the IUPAC definition, should only be used for a single layer, but in the recent literature, terms such as bilayer, few-layer, and multilayer graphene have become increasingly popular. Bianco et al. [22] suggested a terminology for the nomenclature pertaining to graphene in 2013. Accordingly, the following nomenclature should be used for denoting graphene: **only one layer:** graphene, **two (maximum three) layers:** bilayer; **2–5 layers:** few layer; **up to 10 layers:** multilayer graphene, **more than 10 layers:** turbostratic carbon (also see Figure 4.7). The term graphene should not be

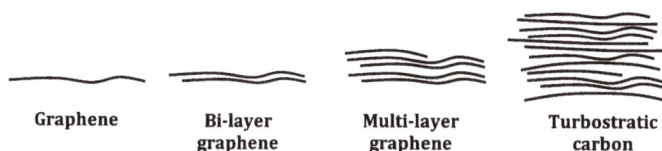

Figure 4.7: Illustration describing the recommended nomenclature for graphene according to the number of layers present.

used for any of its chemical derivatives, such as graphene oxide or reduced graphene oxide.

> Not just graphene, it is indeed many carbon materials, including graphite, carbon fiber, and non-graphitizing carbon, that suffer from variations in nomenclature. This is mostly due to the tremendous popularity of carbon among researchers, as well as industries. From time to time, a new carbon allotrope almost becomes a *fashion* and rapidly attracts a lot of attention as well as capital. Its production and application takes priority over its correct scientific representation. In the case of any confusion, IUPAC-recommended terminology can be useful.

Most common synthesis method for graphene is CVD. In addition, it can be produced by mechanical exfoliation (peeling) of graphite employing physical, chemical, or electro-chemical pathways or by reducing graphene oxide.

4.1.4 Carbon nanotube (CNT)

CNTs are hollow carbon filaments with diameters typically <50 nm composed of sp^2-hybridized carbon. The ends of CNTs may contain non-six-membered rings near their base or origin. The overall structure (that gives the impression of a rolled-up graphene) may experience some strain in the bonds due to a curved surface as compared to free graphene sheets. As a result, the net hybridization of CNTs is slightly shifted towards sp^3, i. e., it is $sp^{(2+n)}$ with a negligible-to-small value of n. CNTs are therefore considered a member of the fullerene family, although they are not spherical. Some of them may have capped ends with clear five-membered rings similar to a spherical fullerene. Notably, there have been reports on amorphous CNTs as well [198, 104] that do not feature any specific molecular arrangements.

Carbon nanotubes can be single or multiwalled. This means that the hollow cylindrical carbon structure can be one atom thick or feature multiple concentric cylinders. Single-walled CNTs feature either a zig-zag or armchair type symmetry, as shown in Figure 4.8 (pay attention to the terminating ends). Interestingly, this arrangement of carbon hexagons determines whether the tube will be semiconductor or metallic. For understanding this, it is essential to know the nomenclature used for denoting a certain type of CNT. The nomenclature is generally known as $n - m$-notation and is based on simply placing a coordinate according to the spatial the location of a hexagon. If one assumes

Zig-zag

Armchair

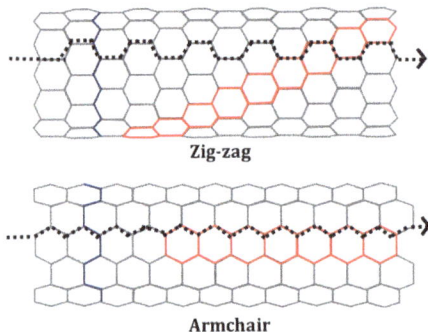

Figure 4.8: Illustration showing CNTs having zig-zag and armchair symmetries. The path traversed by an electron through the tube is shown by dotted lines.

that the CNT was unrolled to form a graphene, and considers the top-left corner hexagon to be the origin (coordinates: $(0, 0)$), subsequent hexagons in xy-plane can be numbered in sequence. This is illustrated in Figure.

Another important property used for structural representation of CNTs is their chirality. Any molecule of geometry that can be superimposed on its mirror image is called chiral. A single-walled CNT having (n,m) indices equal to $(n, 0)$ or $(0, m)$ represents a chiral geometry. These tubes are zig-zag-type. Armchair-type CNTs, on the other hand, have (n, n)- or (m, m)-type indices (i. e., $n = m$). When an electron travels through CNTs, it follows a zig-zag path in the armchair-type tube, and an armchair-shaped path in zig-zag-type tubes. As a result, armchair tubes are always metallic, whereas zig-zag tubes can be either semiconductor or metallic, depending upon the actual path followed by the electron. To better understand this concept, refer back to Figure 4.3(B). There is one horizontal and one vertical path indicated by white lines. The horizontal path requires the electron to cover slightly longer distances. When the electron has no choice but to take such a path, for example, because all paths are of the same type organized in parallel to each other, the CNT shows a semiconductor behavior. This can potentially also happen in graphene if the electron can somehow be confined to only horizontal travel (with respect to the geometry drawn in Figure 4.3(B)). But in practice, the electron can traverse through various combinations of zig-zag and armchair paths (one such path is shown by a yellow line); one never observes a semiconductor behaviour in graphene.

The nm-notations are used for calculating some important characteristics, including the circumference of the tube, the chiral vector, and the chiral angle. In Figure 4.10, you can see how these parameters are defined. If you recall, graphene lattice has two unit vectors: $\vec{a}_1$ and $\vec{a}_2$ (see Figure 4.5). If the same vectors are placed on an unrolled tube, the length of the tube in the respective directions of these vectors can be derived by multiplying these vectors by n and m, as shown in Figure 4.10.

The chiral vector $\vec{c}$ is then calculated as follows:

$$\vec{c} = n\vec{a}_1 + m\vec{a}_2.$$

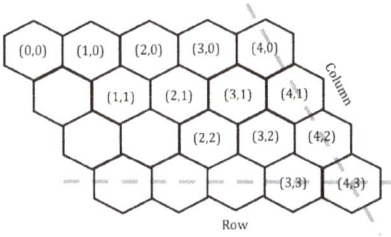

Figure 4.9: Representation of an un-rolled CNT and the methodology of *nm* indexing. Adapted from Frédéric Grosshans, Physics Stack Exchange Website [83].

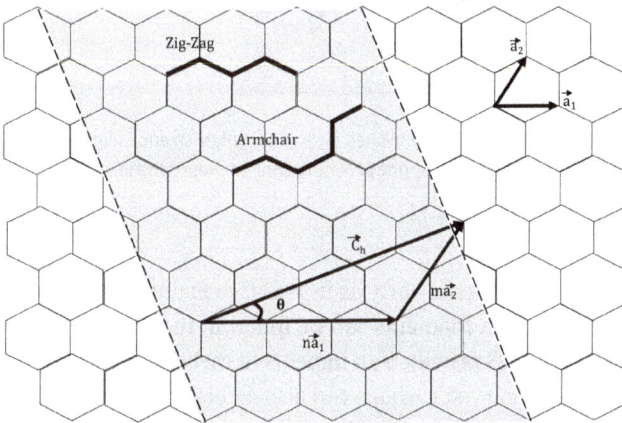

Figure 4.10: A pictorial representation of chiral vectors calculations in an unrolled CNT. https://www. sciencedirect.com/science/article/abs/pii/S1386947714000368, Copyright @2014 Elsevier BV. Reproduced with permission from [27].

The chiral angle θ is the angle between $\vec{c}$ and $\vec{a}_1$, which ranges between 0–30°. The relationship between *nm* indices and θ can be easily derived from the diagram as follows:

$$\sin \theta = \frac{\sqrt{3}m}{2\sqrt{n^2 + m^2 + nm}}.$$

Armchair and zig-zag CNTs have chiral angles of 30° and 0°, respectively.

Problem: Calculate the radius of a (9, 14) CNT if the lengths of lattice vectors $\vec{a}_1 = \vec{a}_1 = 2.46$ Angstrom.
Solution: $(n, m) = (9, 14)$
$\vec{c} = n\vec{a}_1 + m\vec{a}_2$
$\vec{c} = (9 \times 2.46) + (14 \times 2.46) = 56.58$
$\vec{R} = \vec{c}/2\pi = 58.58/2\pi = 9.01$ Angstrom

4.1.4.1 History

CNTs are often present near blast furnaces or other high temperature metallurgic processes involving ferromagnetic metals, particularly Fe. The first TEM images of such

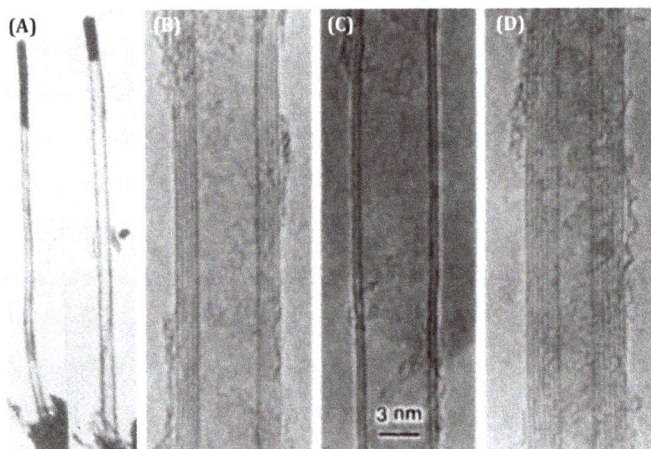

Figure 4.11: Transmission electron micrographs of CNTs published by (A) Radushkevich and Lukyanovich in 1952 [91, 197], and (B-D) Iijima in 1991. Copyright ©1991, Springer Nature Limited. Reproduced with permission from [108].

tubes were published by Russian scientists as early as in 1952 (see Figure 4.11). Following this, some reports on hollow carbon filaments can be found in the literature. The discovery of spherical fullerenes in 1985 provided an impetus to carbon nanomaterials research and suggested that non-sp^2-curved carbons can also be highly stable. In 1990, Sumio Iijima purified single-walled CNTs from the solid carbon deposits obtained from arc evaporation of graphite electrodes in an experimental set-up very similar to that used in fullerene synthesis. This work is regarded as the starting point of the use of CNTs in technological applications. In some of the previous research work from Iijima's group, multiwalled CNTs were also prepared in a similar fashion.

In addition to Iijima, P. M. Ajayan [4, 56], Mildred and Gene Dresselhaus [54], and several other scientists have contributed significantly to synthesis, characterization, and applications of CNTs and their work can be found in various books and articles.

> **!** CNTs and vapor-grown carbon fibers are sometimes collectively called carbon filaments. During CVD, partially hollow or entirely solid nanoscale carbon fibers are often produced along with CNTs.

4.1.5 Carbon nanofiber

Solid carbon filaments having diameters smaller than 100 nm are known as carbon nanofibers. They can be grown via (i) CVD in a similar fashion as CNTs, or (ii) electrospinning of polymers followed by carbonization. Among these, CVD grown fibers (commonly known as vapor-grown carbon fibers or VGCF) are considered nanomaterials. Electrospun fiber may also feature diameters <100 nm, but usually they are typically utilized in the form of mats having an average diameter >100 nm. The details of fiber-based devices and their manufacturing process, etc., are provided in Chapter 5. Here, a brief introduction to VGCFs is covered, mainly for the purpose of comparing them with CNTs.

Multiwalled CNTs tend to collapse into solid cylindrical structures during their CVD. As a rule of thumb, if the catalyst particles are >10 nm, there is a higher probability of obtaining carbon fiber than tube. Some intermediate arrangements of cylinders prior to completely converting into a solid fiber are also commonly observed during CNT growth, particularly through CVD. Similar microstructure may also be present while preparing carbon fibers via top-down processes, such as electrospinning. TEM micrographs of these tube-fiber intermediates are shown in Figure 4.12. Fibers obtained via CVD are more graphitic compared to electrospun fibers, as they are closer to CNTs in terms of microstructure. In both VGCFs and multiwalled CNTs, the layers (or coaxial cylinders) are randomly rotated (turbostratic) with respect to each other. In other words, there is no 3D organization of the hexagons, even if they feature no point defects.

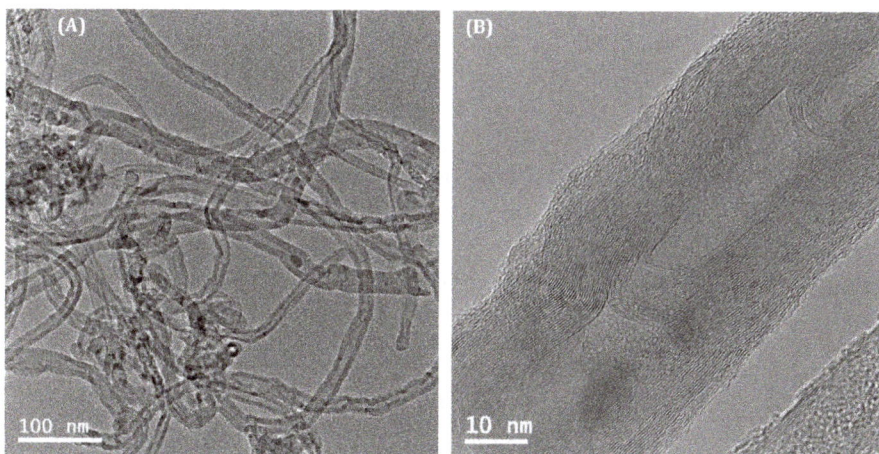

Figure 4.12: TEM micrographs showing partially hollow carbon fibers that show a transition between tubular and solid fiber geometries. Image in (B) is reprinted from [202].

4.1.6 Buckminsterfullerene

Bukminsterfullerens are football-shaped C_{60} molecules, which are semiconductor in nature. They are mainly synthesized via arc evaporation of graphite electrodes or some chemical synthesis pathways. Although the term fullerene can be used for a large variety of curved carbons, bukminsterfullerenes (that may also include C_{70}) are the strongest representatives in terms of being stable, despite a high curvature. In Chapter 1, the angle of pyramidalization has been described in detail, which can be referred to for understanding the molecular geometry of spherical fullerenes. Buckminsterfullerenes are

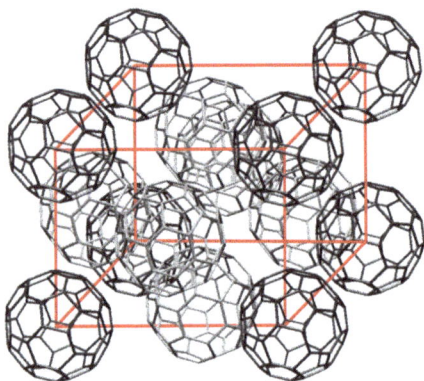

Figure 4.13: Representation of a face-centered cubic crystal formed by buckminsterfullerenes.

semiconductors, and they are extensively used in organic photovoltaics. Typical size of a C_{60} molecule is between 0.7–0.9 nm. In bulk, they tend to form FCC crystals, where individual C_{60}s serve as atoms. This makes their chemical separation and purification simple, as they can be obtained in a relatively pure crystalline form. Figure 4.13 contains a schematic representation of such a fullerene crystal and a small angle neutron scattering (SANS) spectrum of crystalline fullerenes in dry solid state.

Notably, fullerenes are not always perfectly spherical. Several variations including distorted spheres, ellipsoids, and some open-cage strongly curved structures are also a part of the fullerene family. These curved carbon geometries are also found in non-graphitizing carbons, as described in Chapter 3.

4.1.7 Carbon black

Carbon black is a synthetic form of carbon composed of aggregates of nanoparticles (see Figure 4.14 A–C). These aggregates are not spherical, they rather resemble randomly shaped bunches of grapes. Each aggregate or cluster has its longest dimension typically around 1 μm. Each constituting nanoparticle features a near-spherical geometry with the most commonly observed diameter range of 10–12 nm. These nanoparticles contain >98 % carbon [213] and exhibit a turbostratic arrangement of carbon sheets with a relatively small crystallites (Figure 4.14 D, E) [28] in their outer shells. Their core part is made-up of highly disordered carbon. Owing to a large number of defects, predominantly turbostratic nature, and the geometrical configuration of each aggregate, carbon black displays an extremely low density in its bulk form and has earned the nickname "fluffy carbon" in the manufacturing industry.

Carbon black has been known since the 1930s, and it is indeed one of the most extensively produced carbon materials worldwide. Its main application is as a vulcanizing agent for rubber used in the typing materials manufacturing industry. In addition,

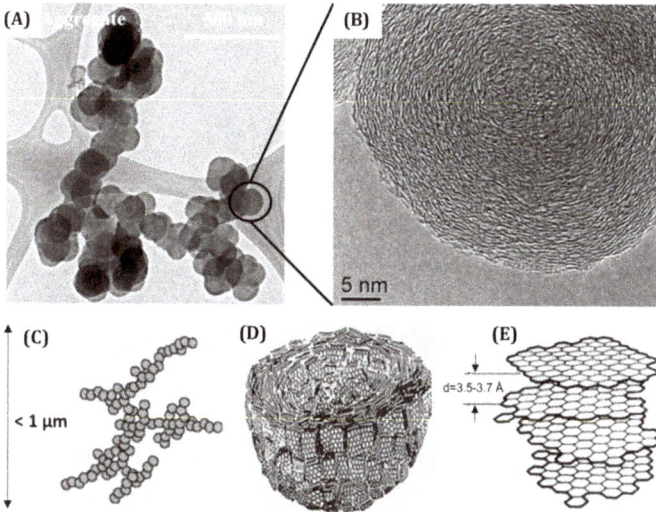

Figure 4.14: (A) Transmission electron micrograph of a carbon black aggregate composed of nanoscale particles. (B) A Transmission electron micrograph of individual nanoparticle unit of carbon black. (C) Arrangement of nanoscale carbon aggregates in a carbon black particle. (D,E) Proposed microstructural model showing a turbostratic carbon sheet arrangement in the building blocks of carbon black, proposed by Heidenreich et al. [97]. (A) and (B) are reprinted from [213]. (C) Copyright @ 1999 Elsevier Science Ltd. Reproduced with permission from [28]. (D, E) Reproduced with permission of the International Union of Crystallography from [97].

it has also been employed as reinforcement in composites and as a pigment in a variety of printable inks. In 2015, the International Organization for Standardizations (ISO), in its Technical Specification 80004-1 of 2015, declared carbon black a *nanostructured material*. Today carbon black is gaining a rapid popularity in device fabrication, because (i) it is an already established ink-pigment, and (ii) it is inexpensive and readily available compared to other carbon nanomaterials. The electrical conductivity of carbon black powder is poor, hence, it is often mixed with graphite particles in printable electrode applications.

4.1.7.1 Carbon black and soot

Carbon black is occasionally confused with soot (candle soot, diesel soot etc). Soot is a collection of partially burned particles resulting from an incomplete combustion of hydrocarbons. Though it may feature some similarities with carbon black, soot contains a significant fraction of impurities, including tars, ash, and even inorganic residues [196]. They often form random clusters of variable sizes and are produced in a less controlled fashion. Carbon black, on the other hand, is made up of well-defined nanoparticles in pretty much the same diameter range. A schematic diagram shown in Figure 4.15 can be helpful in understanding the difference between the two. It is important to note that

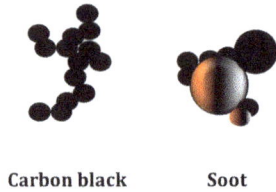

Carbon black   **Soot**

Figure 4.15: Drawings of carbon black and soot particles. The brown areas in soot represent organic impurities, such as tars.

there may be some fraction of carbon black in soots, which can be observed in TEM, for example. But it is extremely difficult to separate them from the rest of the particles containing impurities. One can anneal the soot at high temperatures to remove the tars to a great extent, and can then use it for various applications. The reactive species and radicals of organic compounds present in soots are harmful for the environment, as well as for human health. One should therefore take utmost care while handling soot in the laboratory.

Carbon black can be graphitized at >2500 °C. Under graphitization conditions, the outer shells of carbon black particles release defects and gain order. The fundamental process of graphitization involves increase in L_a followed by decrease in L_c. Consequently, the loosely bonded carbon atoms present in the disordered core get pulled towards the shell, where they contribute to forming larger 2D sheets. A decrease in d-spacing between these sheets leads to further shrinkage of the shell, thus causing voids or empty space in the core of the particle. It is clear that such hollow graphitized carbon black nanoparticles are extremely light and display a very high surface area. They are still found in their characteristic grape-bunch-like aggregates. Different grades of graphitized carbon blacks (GCBs) are commercially available and have been widely used as adsorbents in gas chromatography, environmental sciences, as well as energy storage applications [52, 32, 31].

The aforementioned materials are the most common carbon nanomaterials used in device fabrication. There are, however, plenty of other advanced materials and their hybrids, which are being rapidly developed for specific applications. Some examples include nanodiamonds, carbon nanoparticles, carbon quantum dots, graphite whiskers, fullerene-CNT peapods, and chemical analogues of all these. All carbon nanomaterials can be classified into film, fiber, or particle-type when it comes to their patterning at the microscale. Even if they don't have these exact shapes, they can more or less be processed employing the techniques compatible with these representative materials. Device fabrication with carbon nanomaterial has two stages: (i) synthesis of the nanomaterial itself and (ii) its patterning into larger device-compatible shapes. Of course, device characterization and testing follow. Below some of the most common nanomaterial synthesis are described first. Subsequently, the processes used in putting these nanomaterials together to achieve device-friendly geometries is detailed.

4.2 Synthesis of carbon nanomaterials

4.2.1 Chemical vapor deposition (CVD)

Vapor deposition is the process of collecting a solid material on a substrate directly from its vapor phase or smoke. If the desired vapor is generated via a chemical reaction, the process is known as chemical vapor deposition (CVD). If physical methods, such as low pressure, high temperature of electron beam assisted material removal is used for converting the source into its gas phase, the process is designated physical vapor deposition (PVD). In most cases, the material deposition is in the form of thin films, but depending upon the size and shape of the substrate (that often also serves as a catalyst), other geometries, such as fibers and tubes, can also be obtained. Here it is more appropriate to use the term growth rather than deposition. CVD has been used for several decades for the purpose of depositing thin films and industrial coatings of ceramics, diamond, corrosion resistant materials, silicon, etc. A general schematic diagram of the CVD process is shown in Figure 4.16. Further details are as follows:

Basic principle: Products of a chemical reaction, occurring at a certain combination of pressure and temperature, are confined to a closed chamber. One primary product from this vapor/smoke state is collected onto a solid support that facilitates a faster deposition compared to other locations in the chamber due to its catalytic activity and/or higher temperature with respect to its surroundings. Substrates may play additional roles in guiding the microstructural evolution and morphology of deposited material.

Materials used: Various carbon allotropes, including from both diamond and graphite families, ceramics, silicon, metals and alloys, etc. Substrates: transition metals as flat pieces or nanoscale particles, graphitic carbon materials, any manufactured part in the case of wear resistant coating (e. g., of diamond-like carbon).

Process steps: If the reactant is in gas phase, it is passed through a (typically) tubular furnace or CVD chamber at a low pressure, with or without a carrier gas. If the reactant

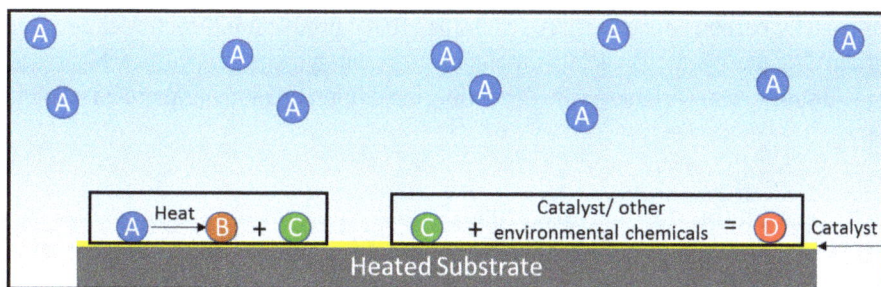

Figure 4.16: Schematic representation of a chemical vapor deposition process. Reactant *A* thermally decomposes to form *B* and *C*. One of these products deposits onto the catalytic substrate and takes the ultimate product form, *D*.

is solid or liquid, it can be placed in a separate chamber connected to the main CVD chamber and converted into vapors at low pressure or high temperature. In the case of CVD of carbon, a mixture of hydrocarbon (primary reactant or feed gas), H_2, and N_2 is flown into the chamber with their flow rates controlled via mass flow controllers. The chamber is essentially maintained oxygen free. In the case of vapor deposited diamond and diamond-like carbon, CVD is assisted by plasma. A substrate at high temperature is used for collecting material in solid form. Substrates used in carbon nanomaterial CVD are mostly transition metals and their alloys that additionally serve as catalysts, thus promoting material deposition.

Optimization parameters: Chamber temperature and pressure, substrate temperature, substrate shape and size, substrate-feed gas contact time, reactor geometry, initial concentration of the reactants, flow rate of all gases.

As stated earlier, CVD is one of the most commonly used techniques for manufacturing carbon nanomaterials. It is suitable for relatively large-scale production quantities and has therefore gained a considerable commercial attention. One of the reasons for its popularity in carbon technology is that the basic know-how has already been available for 5–6 decades for the purpose of manufacturing pyrolytic graphite and HOPG. In fact, the history of pyrolysis of gaseous hydrocarbons dates back to early 19th century when Dalton and Henry decomposed methane and ethylene using an electric arc [101]. Following this, other hydrocarbons, including propylene, acetylene, mixture of organic gases, and even light liquids (such as alcohols), have been studied for their carbon deposition ability. With the advancements in CVD equipment, as well as nanotechnology in general, better parameter control, deposition of extremely small quantities and modifications in catalyst morphology has become possible. Today, removal of single layers from HOPG has become a well-known graphene production method. Interestingly, it is the graphene that is converted into HOPG via thermal/stress annealing in the first place. This graphane-HOPG-graphene cycle is illustrated in Figure 4.17. Of course, the number of layers for HOPG manufacture is not single. Multiple turbostratic and potentially defect-containing layers are deposited prior to their thermally assisted crystallization. Nowadays a number of manufacturers are developing CVD equipment specially designed for carbon nanomaterials, rather than HOPG production. To better understand carbon CVD, one needs to look into the chemical thermodynamics of hydrocarbon cracking and subsequent carbon species formation, in addition to the general process parameters.

4.2.1.1 Mechanism of CVD of carbon films
The core chemical reaction responsible for carbon generation from a hydrocarbon source (e. g., methane) is pyrolysis. As explained in Chapter 3, pyrolysis is defined as the thermochemical decomposition of a hydrocarbon in the absence of oxygen. When the hydrocarbon is a simple gaseous molecule, one can evaluate its decomposition conditions based on thermodynamic stability and enthalpy of formation. In Figure 4.18, free

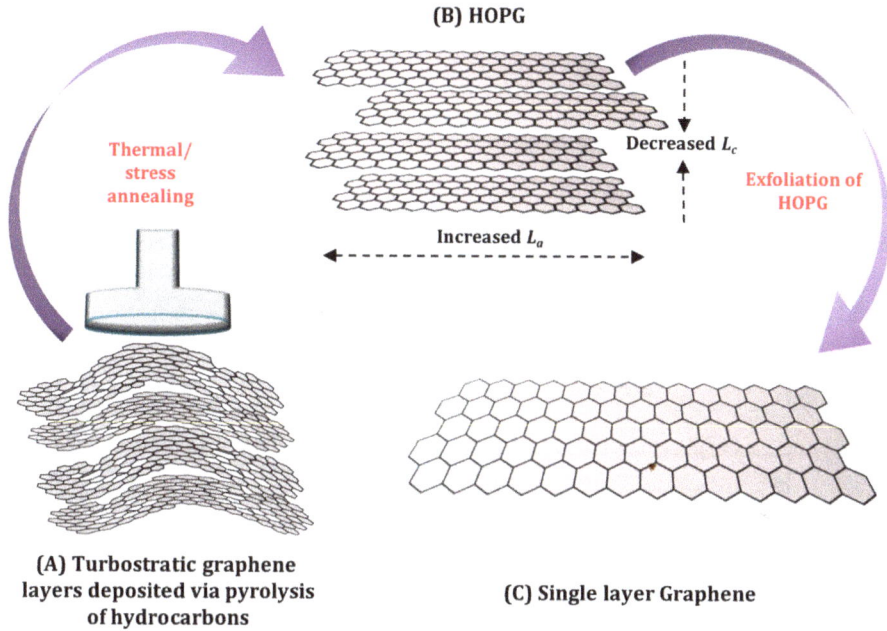

Figure 4.17: Illustration showing how multiple graphene layers are converted in HOPG, which is back-converted in graphene via exfoiliation. Image reprinted from [49].

enthalpies of formation of various hydrocarbon compounds from their constituting elements are plotted as a function of temperature. If the enthalpy value is negative, it shows that the compound is stable. Once it crosses the zero line and the enthalpy becomes positive, its formation becomes less likely. This means that its decomposition becomes more likely at/ above the corresponding temperature value.

Let us take the example of methane. It can be observed in Figure 4.18 that at 500 °C methane's standard enthalpy of formation becomes positive. Due to this thermodynamic instability, it can be dissociated into carbon and hydrogen above this temperature. However, this dissociation never reaches 100 % efficiency. Rather, an equilibrium state is achieved where methane, hydrogen, solid carbon species, any other hydrocarbons formed during the process, and the carrier gas will all have finite partial pressures. At this equilibrium, a quantity known as "carbon solubility in gas phase" becomes important. This is defined as the total amount of all carbon present in the system at the time of equilibrium. Naturally, this property is sensitive to temperature and pressure. When the solubility of carbon is minimum, it is the most favorable deposition condition as the carbon species are prone to precipitation. Similar to a solid dissolved in liquid, minimum solubility is when precipitation is maximum. An example of gas phase carbon solubility curves (indicated by C/H) at different pressures and increasing temperature is shown in Figure 4.19. The minima of each curve shows the optimum pressure condition at that temperature.

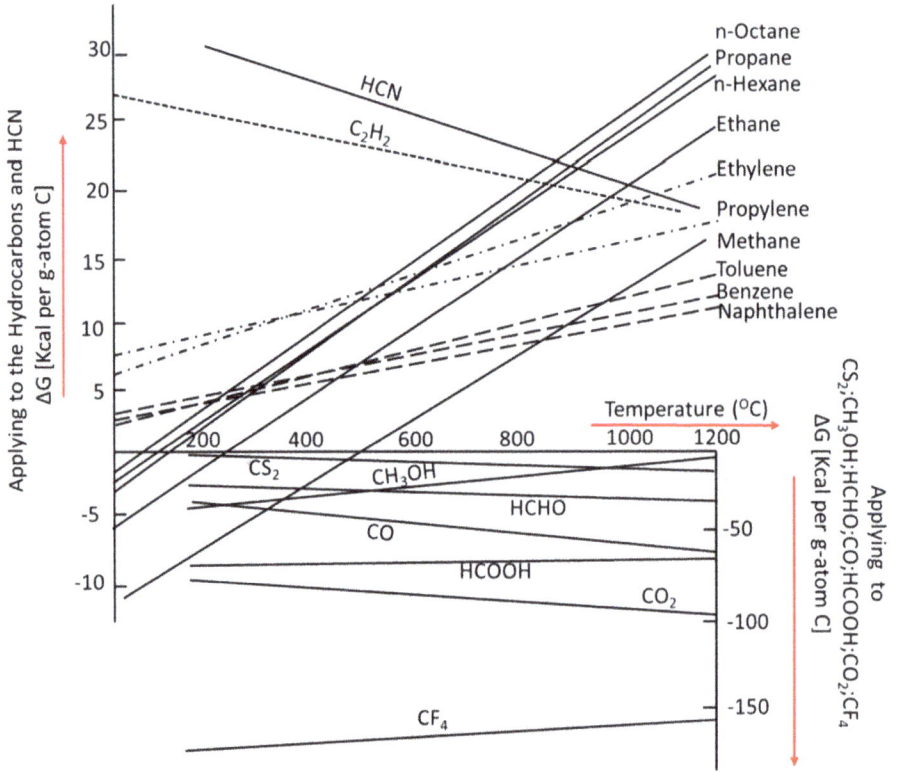

Figure 4.18: Free enthalpy of formation for various hydrocarbons and other compounds that can release during pyrolysis. All values refer to 1 g-atom of carbon. Adapted from Fitzer et al. [66].

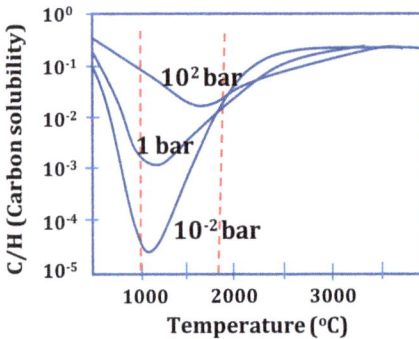

Figure 4.19: Curves showing carbon solubility in gas phase during pyrolysis of a hydrocarbon gas. Copyright ©1967 Elsevier Ltd. Adapted with permission from [145].

If you go back to Figure 4.18 and analyze it for, say, 1100 °C, we see that methane has become unstable (positive free energy), but other hydrocarbons such as toluene, benzene, and naphthalene are now less positive, and hence more stable than methane. At this temperature, the possibility exists that some carbon species present in the gaseous environment will form these hydrocarbons. It is also clear from this diagram that even very small quantities of oxygen in the system will rapidly enable the formation of CO_2 and CO, as these compounds are very stable at higher temperatures. One can also evaluate the difficulties involved in decomposing alcohols and acids.

In physical sense, one can summarize the entire process taking place inside CVD chamber as follows: **Step 1:** hydrocarbon dissociation at temperatures above its thermodynamic stability that leads to the formation of gas phase carbon (vapor + smoke); **step 2:** migration of carbon species towards heated substrate, which occurs at the temperature and pressure that facilitate minimum carbon solubility in surrounding gas mixture; **step 3:** migration of carbon species on the substrate surface and removal of H, if any; **step 4:** formation of islands of carbon film, typically at several nucleation points, and **step 5:** film growth and removal of byproducts. These steps can also be seen in Figure 4.20. Note that the carbon species formed after step 1 (dissociation) may still contain some H, which may release during or after step 2.

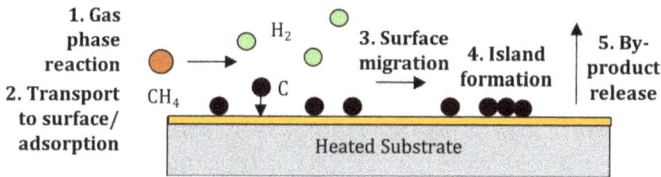

Figure 4.20: Schematic diagram showing steps involved in carbon deposition from chemical vapors produced by pyrolysis of hydrocarbon gases.

In addition to optimizing the chemical reaction conditions, one also needs to optimize the substrate. Interestingly, though most reaction parameters during CVD of graphene, nanotubes, and nanofibers remain more or less the same, it is the substrate that makes all the difference and determines the morphology of the resulting carbon material. Some substrates enable penetration of carbon inside their crystal, whereas some keep it restricted to their surface or near-surface regions. Based on this, graphene growth is mainly carried out on Cu and Ni substrates. Alloys of these two metals, as well as various other materials have been utilized as substrate material. They do facilitate graphene growth, but the overall material quality and yield is relatively poor. Nonetheless, this is an active research area, and several efforts on developing new substrates for graphene growth have been made in the recent past. Several review articles and books are available for further reading.

In the case of the Cu and Ni catalysts for graphene film growth, the substrate is taken in the shape of a flat piece. The carbon species collect or adsorb on its surface at appropriate temperature, and then migrate to form hexagonal rings. It is reported that while carbon atoms on Cu remain surface-bound [218], in the case of Ni, they are able to feature minor penetration. The substrate temperature also plays an important role here. Typically, at the temperatures and pressure present inside the CVD chamber, the top surface of the substrate becomes soft. This may enable some carbon atoms/species to penetrate through it. When we are interested in making 2D structures, such as graphene, it is generally better to use the substrates that keep the carbon atoms on the surface and allow for their faster surface migration.

4.2.1.2 CVD for CNTs and VGCF growth

The fundamental principles of pyrolysis of a gaseous hydrocarbon remain the same whether the deposited material is a film, tube, or fiber. The primary difference here is the chemical structure, size, and shape of the catalyst material. It has already been stated that carbon filaments, hollow or partially hollow, are commonly observed near blast furnaces. Iron, indeed is one of the best catalyst materials for obtaining CNTs. Other than that, magnetic metals, such as Ni and Co, also serve as catalysts.

Unlike graphene, in the case of CNTs, carbon species penetrate inside the catalyst particles. Depending upon the catalyst size, tubes can grow from either top or bottom of the particle. There are two widely accepted mechanisms for CNT growth: (i) vapor-liquid solid (VLS) and (ii) vapor solid solid (VSS) mechanism (see Figure 4.21). In VLS, the HC pyrolysis is believed to take place post-adsorption on the catalyst surface to yield atomic carbon. This atomic carbon then forms a solid solution with the catalyst, yielding an intermediate carbide phase. Finally, carbon is precipitated from this carbide, which leads to tube growth. In VSS, on the other hand, pyrolysis reaction takes place in the gas phase, and the generated carbon species are adsorbed/diffused into the catalyst without the formation of carbide. Carbon is then precipitated and tubes are grown.

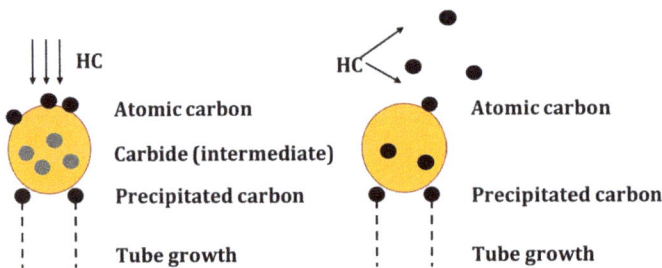

Figure 4.21: Two accepted mechanisms of CNT growth during CVD. The first one suggests a possibility of an intermediate carbide formation, whereas the second doesn't. Tubes can grow from both top or bottom of catalyst in either case.

4.2.1.3 Vapor deposition of diamond films

Plasma enhanced (or plasma assisted) chemical and physical vapor deposition are the names to given to processes where material deposition from the gas phase is promoted by the presence of plasma in the chamber. These processes are utilized to make polycrystalline diamond films or coatings. Diamond coatings are extensively used on industrial tools for rendering them wear resistant, as they protect the tool material from any diffusion or corrosion. In addition, such synthetic diamonds have been used in batteries, reaction vessels, crucibles, radiation detectors in photonic devices, optical coatings (lenses, laser protection etc.), heat sink, sensors, field effect transistors, etc. Although, the film itself cannot be designated a *nanomaterial*, the individual diamond crystals or particles present in such films are often in nanoscale. Moreover, the vapor deposition of diamond is also used for obtaining nanodiamonds (nanoscale diamond particles), which have their use in sensor fabrication, reinforcement in composites, and biomedical engineering [195]. In principle, vapor deposition of diamond can be performed without the plasma, but that would require temperatures as high as 2000 °C, because diamond formation entails an additional intermediate structure involving atomic H, as discussed here.

There are two types of diamond films: (a) CVD (also called VD) diamond and (b) diamond-like carbon (DLC). CVD diamond, as the name suggests, is prepared via a chemical reaction (pyrolysis), whereas DLC is obtained through physical vapor deposition (PVD) in the presence of plasma. DLC typically contains a higher fraction of sp^2-hybridized carbon atoms along with some H impurity. CVD diamond, on the other hand, is relatively pure sp^3 carbon. Consequently, CVD diamond features a polycrystalline characteristic, whereas DLC films are more amorphous in nature. In both CVD and PVD, a low pressure is essential. Notably, PVD also utilized gas phase carbon as the material source. The only difference is in the method of obtaining this gaseous carbon. In CVD, a hydrocarbon gas (e. g., methane) is used as carbon precursor similar to graphene CVD. In PVD, carbon species are generated by arc evaporation of graphite electrodes.

The mechanism of diamond formation from gas phase carbon is very interesting. In principle, CVD yields graphitic carbon, and hence used for graphene, CNT synthesis. Since graphite is the most thermodynamically stable carbon form, it is obvious that its formation is preferred during CVD. Then how should one get diamond from essentially the same precursor inside a CVD chamber? Unlike graphite, diamond exhibits a 3D geometry. Its crystal structure manifests two interpenetrating face-centered cubic (FCC) lattices, as shown in Figure 4.22(B). Growth of this geometry in a process that is suitable for layer-by-layer deposition is not very likely. But this becomes plausible when some atomic H is present in the system, and it forms an intermediate bridge-like structure by attaching itself to the growing carbon film, as illustrated in Figure 4.22(B). These H bridges prevent the carbon structure from collapsing and flattening. Once the next carbon layer is deposited, H is replaced with C atoms. But of course, it is hard to achieve

(A)

(B)

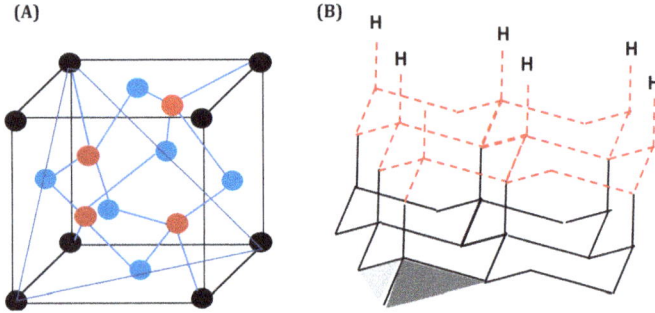

Figure 4.22: (A) A diamond crystal (two interpenetration face-centered cubic cells) showing {111} planes. All atoms are chemically equivalent. (B) Formation of H-bridge structure during PECVD growth of diamond.

100 % diamond this way. Some graphitic carbon and/or H is always present in the material.

It is known that H_2 prefers to be in a diatomic molecule form. Atomic H generation requires high energy (ΔH = 434.1 KJ/mol) and is endothermic. This process can only take place at 2000 °C. Importantly, nanoscale diamonds are not stable above 1000 °C as it converts into graphite. This means that if the substrate is heated to 2000 °C for CVD diamond or DLC deposition, this will ultimately yield only graphite. To obtain atomic H at lower temperatures, high intensity plasma is used in the chamber. Plasma promotes atomic H generation, as well as activation of carbon species, which facilitate H-bridge formation.

4.2.1.4 CVD reactor

Most commonly used CVD reactors are tubular furnaces that contain an oxygen-tight gas inlet and an outlet. Flow rates of different gases flowing inside this reactor are controlled via sophisticated mass flow controllers. The reactor is connected to vacuum pumps, various temperature sensors and other essential electronics. One can also have additional connected chamber and vaporizer if a liquid or solid precursor is to be used.

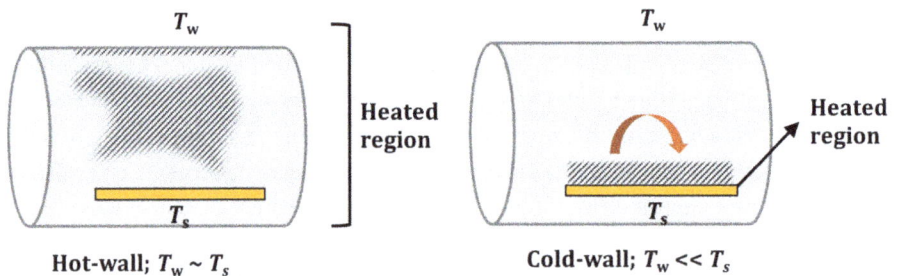

Figure 4.23: A schematic showing a hot- and cold-walled CVD reactors. T_w and T_s indicate wall and substrate temperatures, respectively.

There are two main types of CVD reactor design: hot-walled and cold-walled. In the case of hot-walled reactors, the entire CVD chamber/tube is heated, whereas in the case of cold-walled reactors, only the substrate is at the hydrocarbon cracking temperature. In microfabrication processes, cold-walled reactors are more common, because they enable direct collection of graphene films or tube/fiber growth on to desired substrates. One can also modify the catalyst or pattern the substrate to facilitate a prepatterning of the materials. Bulk production of fibers and tubes is sometimes done with hot-walled reactors. The catalyst particles are suspended along with the carrier gases through the heated chamber and tubes or fibers are allowed to grow in their suspended state. Some product deposits onto the walls, but there may also be remaining products, which can be collected using filters after the process. Evidently, hot-walled reactors are more suitable for particle-type catalyst rather than plates. They can provide larger production quantities, whereas cold-walled reactors yield a better quality and control over fabrication process.

4.2.1.5 Boundary layer

The flow of gases inside a CVD chamber is maintained laminar, i. e., the Reynolds number is always <2000. In the case of cold-walled reactors, the gas velocities very close to the substrate are nearly zero. As one moves away from the substrate, the velocity of each gas layer increases and finally matches with that of the remaining system. This velocity profile causes the formation of what is known as a boundary layer close to the substrate (see Figure 4.24).

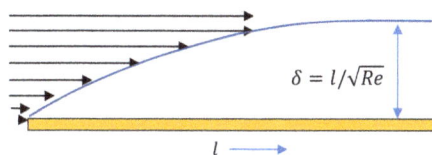

$$\delta = l/\sqrt{Re}$$

Figure 4.24: Formation of boundary layer close to the substrate during CVD in a cold-walled tubular reactor.

This influences the mass transport of the carbon species and hinders its diffusion towards the substrate. The boundary layer is thinner when the pressure is low and the diffusion coefficients are large. The rate of a CVD process is limited either by the reaction kinetics or by the diffusion of carbon species into the substrate. Reaction kinetics limited processes are more commonly observed in low pressure-low temperature systems, as they that do not facilitate a fast reaction. Diffusion-limited processes, on the other hand, occur in high-temperature systems with a thick boundary layer. The boundary layer decreases the rate of the diffusion of the generated species into catalyst. Desorption of by-products also becomes slow. Flow rate, concentration and residence time of gases can increase the deposition rate.

Boundary layer, δ, can be expressed as follows:

$$\delta = \frac{l}{\sqrt{\mathrm{Re}}},$$

where l is the length along the substrate, and Re is the Reynolds number defined as follows:

$$\mathrm{Re} = \frac{\rho \, v \, D}{\mu},$$

where ρ is material density, v is velocity of gases, and D is the tube (reactor) diameter.

Among these parameters, primarily the tube diameter, substrate length and velocity of gases can be controlled by the user. One can optimize these for a given reactor to avoid the boundary layer formation. Boundary layer can also be avoided by tilting the substrate plate, or making other design changes that physically disable the boundary layer formation.

Comparison of CVD for different carbon materials

As stated earlier, various carbon nanomaterials, including graphene, CNTs, and some CNFs, can be synthesized in large quantities using the CVD process. Though the basic process principles remain the same, specific parameters lead to the manufacture of different carbon materials. In Table 4.1, a comparison of CVD parameters to obtain different carbon nanomaterials is provided.

4.2.2 Exfoliation of graphite

Exfoliation of single or few layers of graphene from commercially available graphite is another method of producing graphene. Exfoliation can indeed be considered one of the first methods for graphene production. The German scientist Schafhaeutl prepared individual graphene flakes employing this method as early as in 1840 [55, 260]. He utilized the intercalation method, which relies on graphite layer separation caused by inserting small molecules (e. g., metals) between the basal layers, which weaken the binding forces between them. Such layers were then exfoliated using acids to obtain separated layers. In this process, oxide of graphene is formed and subsequently reduced. Similar intercalation-based experiments were performed by Brodie in 1859. This material was given the name *graphon* [55, 260].

Despite these contributions, for several decades 2D materials, such as graphene, were considered unstable in the absence of a solid support, and despite their astonishing properties, their applications were not explored. In 2004 Geim and Novoselov [173] showed that graphene layers removed using adhesive tapes were stable while suspended in a solvent. Following this work, a large number of efforts were made to chemi-

Table 4.1: Chemical Vapor Deposition parameters used for carbon nanotube and graphene.

Parameter	Carbon nanotube	Graphene
Catalyst material	(a) Transition (d-block) metals, mainly Fe, Co, Ni (b) Organometallocenes that release metal nanoparticles *in situ*	(a) Transition metals and alloys mainly Ni, Cu, Co, Pt, Ag, Ru, Ir, Pd (b) Liquid metals: d- and p-block (Ga, In)
Catalyst morphology	Nanoparticles (particle size determines tube diameter, typically >50 nm particles yield solid fibers rather than hollow tubes)	Films (single crystal or polycrystalline)
Process temperature	600-900 °C: MWCNT 900–1200 °C: SWCNT	Typically >1200 °C at atmospheric pressure
Mechanism	Growth from tip or base of catalyst nanoparticle	Adsorption/desorption or surface migration of carbon species
Precursor	Most C_xH_y for MWCNT, Methane/CO for SWCNT	Most C_xH_y, mainly methane

Information is based on widely used protocols as on date. CVD research is progressing rapidly and parameter optimization is constantly ongoing. Primary sources: For CNT: Kumar and Ando [136]; Graphene: Kalita and Tanemura [121].

cally disintegrate graphite into individual layers. It has been reported that such exfoliated graphene can spread over several micrometers, which is generally difficult to achieve via CVD. Highly oriented pyrolytic graphite (HOPG) is the most suitable precursor for this purpose as it is prepared by high-temperature annealing, and therefore contains extremely few defects. However, intercalated graphite, graphene oxide, and some graphitizing carbons have been used as precursors.

Common methods for separating carbon layers include ultrasonication, use of shear forces in fluidic systems, heat-treatment, high electric fields, ball milling, electrochemical exfoliation of graphite electrodes, etc. Some of these processes are physical and simply based on applying mechanical forces. In some cases chemical, electrochemical or a combination of these with physical separation is utilized. Electrochemical peeling pathways include intercalation of salts inside graphite layers to increase the layer separation (see Figure 4.25), which can then be more conveniently removed. Several chemical methods that are based on reduction of graphene oxide are also combined with mechanical exfoliation for graphene production. Of course, removing one layer of HOPG using an adhesive tape is also one possible way of making graphene.

The graphene yield in exfoliation processes is usually very low, as it is very hard to obtain the single layers (that essentially are graphene). A lot of precursor material, often expensive high-quality graphite, is wasted. Moreover, mechanical forces on the material can lead to defect generation. Further improvement in exfoliation methods is

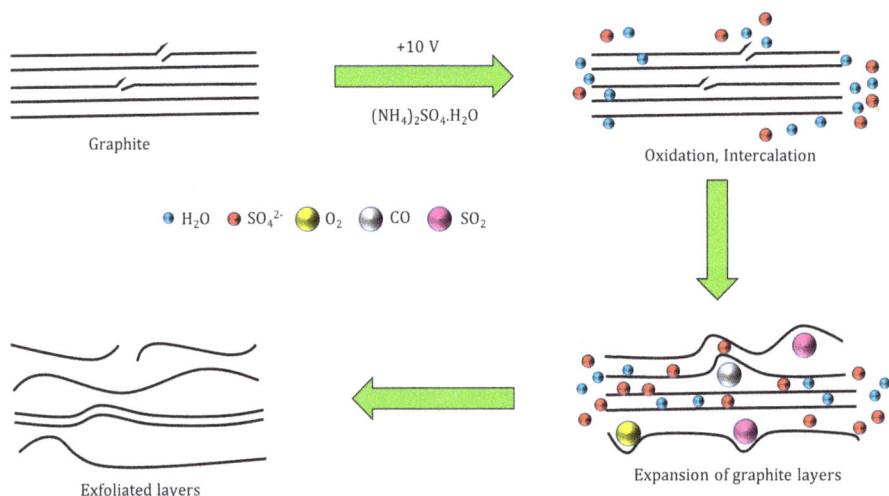

Figure 4.25: Schematic diagram showing intercalation of various small molecules from salts into graphite prior to its exfoliation. Copyright ©2014 American Chemical Society. Reproduced with permission from [183].

an ongoing research area, and it is expected that more efficient methods will soon be developed.

4.2.3 Epitaxial growth

Growth of a crystalline material on the surface of another material having a similar crystal structure is known as epitaxial growth. The first couple of layers are strongly influenced by the structurally similar substrate. This effect is also known as *templating* effect. Silicon carbide (SiC, also called carborundum), a wide-bandgap (2.3–3.3 eV) semiconductor, has a crystal characterized by close-packed stacks of double layers of Si and C atoms. In fact, SiC has been associated with the graphite-like materials since the very first patent filed by Acheson on the synthetic production of graphite in 1896 [2].

The fundamental principles of this method are as follows: at high temperatures, silicon phase starts to sublimate and the remaining carbon tends to form layers of hexagonal network similar to graphene. The sublimation rate of silicon is faster than that of carbon. Hence, the evaporated silicon imparts a higher vapor pressure, which promotes graphene growth. Graphene grows on both silicon and carbon faces of the SiC crystal. On the silicon face it is generally easier to control the number of layers. The process is carried out in standard high-temperature furnaces, typically between 1350–2000 °C under Ar environment. It is clear that no specific carbon precursor is needed, since the graphene layers are generated due to removal of silicon from SiC. To promote graphene growth on the silicon face, hydrogen etching on the carbon face can also be included.

Epitaxial growth of graphene is highly suitable for device fabrication, as it obviates the need for pattern transfer or any wet process. SiC substrates are commercially available, and it has been shown that large-area graphene with single-to-feew layers can be epitaxially grown on to them. Further details on the crystal structure of SiC and various graphene growth conditions can be found in the review article by Yazdi et al. [260].

4.2.4 Arc evaporation and laser irradiation of graphite electrodes

Arc discharge, laser irradiation and similar high-energy processes are a means to vaporize carbon from the surface of graphite electrodes. The composition of carbon vapor is complex, as also discussed in the context of CVD. In addition to atomic, chain, ring- or few-ring-type species, these vapors contain fullerenes and nanotubes of different sizes and shapes. By varying the gaseous environment, chamber pressure, current and voltage, power supply type, electrode geometry, temperature, catalysts (if any), and laser fluence (in case of laser irradiation), one can change the fraction of different carbon species in this so-called carbon vapor. Solid deposits can be then collected and purified to separate the desired nanomaterial from it.

Spherical fullerenes and CNTs were collected from such carbon vapors in two famous experiments: Kroto et al.'s laser ablation of graphite disc and Iijima's arc discharge on graphite electrode. Kroto et al. [132] carried out through Nd:YAG laser irradiation in a He environment and produced vapors from a graphite source and extracted a small fraction of C_{60} and C_{70} from it. Following their work, Krätschmer et al. [131] purified relatively larger (1 %) quantities of buckminsterfullerenes from the carbon vapors produced by an arc discharge between two graphite electrodes in He environment at 100 Torr. In a very similar set-up filled with Ar at 100 Torr, Iijima [108] synthesized CNTs on the negative side of the electrode. He also observed spherical fullerenes in his carbon product mixture, as expected. Various modifications have since been made to further optimize these processes and increase the yield of the desired carbon nanomaterial. Here, a standard set-up and procedure of arc discharge used nowadays is described.

In arc discharge, a plasma (or plume) is generated between two carbon electrodes due to the electrical breakdown of a gas or liquid filled in between. Among these two electrodes, the cathode is almost always pure graphite, whereas the anode may be composed of graphite or other carbon materials. In certain cases, catalyst particles can also be packed along with the carbon source in the anode. Electrodes, mounted either horizontally or vertically, are brought into contact with each other once the power supply is switched on. This contact leads to an arc (electric spark containing plasma) generation. The electrodes are then slightly pulled back and maintained at a distance of 1–2 mm to attain a stable arc. The high temperatures (>4000 K) on the anode caused by resistive

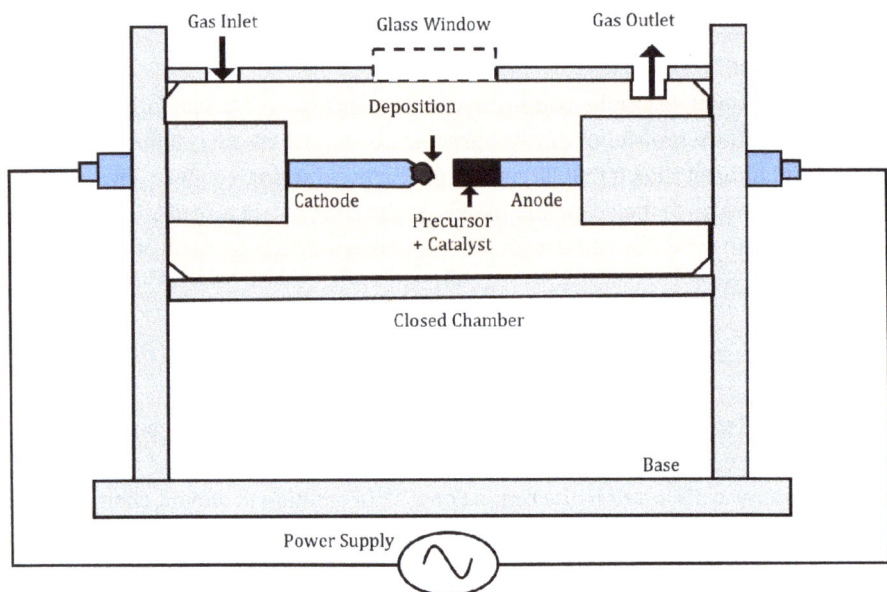

Figure 4.26: Illustration of arc discharge set-up for the production of carbon nanotubes. Copyright ©2014 Elsevier B.V. Adapted with permission from [15].

heating, as well as the presence of high-intensity plasma induce carbon surface evaporation. Once the density of carbon species in this evaporated mass reaches a critical value, this mass starts to move towards the cathode. The cathode has a tapered or pointed shape (see Figure 4.26), which results in a temperature gradient. It can also be often water-cooled for lowering its temperature. The carbon mass deposits on it as a result of cooling and can be collected after the power is switched off, typically in a few minutes. This carbon mass contains CNTs, fullerenes, soot, and other carbon structures that can be separated after further purification. High-performance liquid chromatography (HPLC) is the most common purification process for fullerenes.

Laser ablation of graphite or other carbon materials is simply carried out by irradiating the sample with a high intensity laser. Different types and intensities of lasers have been used for this purpose. Other than graphite, polycyclic aromatic hydrocarbons can be used as carbon source, specially for spherical fullerene synthesis. Laser irradiation is more commonly applied for making fullerenes, but CNTs can also be produced this way (with a relatively low yield). In addition to the aforementioned techniques, spherical fullerenes can be synthesized via various chemical routes. For further details, review article by Mojica et al. [163] and various books may be referred.

4.2.5 Manufacturing process of carbon black

Carbon black is one the few carbon materials that is synthesized by a combination of pyrolysis and combustion (thermo-oxidative decomposition) rather than pyrolysis alone. This implies that some oxygen is also supplied during its preparation at elevated temperatures from light organic precursors in vapor form. There are different carbon black manufacturing methods based on the type of precursor, configuration of the furnace, and fraction of supplied air. The standard process temperature is 1300–1500 °C [213]. The hydrocarbon fuel (hydrocarbon gas, alcohols, light oils) is sprayed inside a heated chamber with a controlled supply of nitrogen and air. Nonvaporized hydrocarbons may also be used in some thermal-oxidative processes, where they are dissociated into smaller fragments prior to their carbonization. Once the materials is converted into carbon black aggregates, it is present in the chamber as a colloid (smoke-like appearance). This is subsequently quenched in cooling chambers, followed by collection through filters. This collection of carbon black is sold as fluffy carbon, which may be further processed if required for vulcanization, for example. A typical process flow for carbon black manufacturing is represented in Figure 4.27.

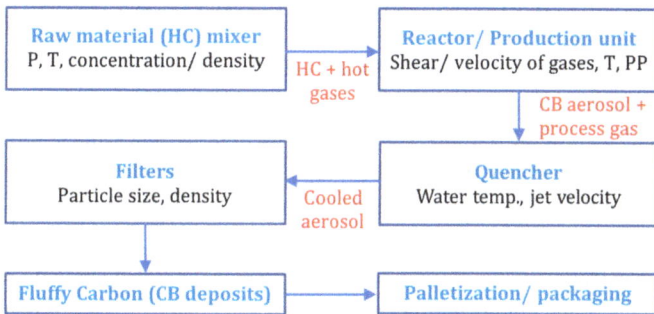

Figure 4.27: Typical process flow of carbon black manufacturing. In each box, the name of the equipment or unit is stated on top, and the bottom line indicates optimization parameters. **CB:** Carbon black, **HC:** Hydrocarbon, **PP:** Partial pressure.

There are a few variations of the carbon black manufacturing processes: channel black, furnace black, lampblack, Degussa (gas) black, thermal decomposition, thermal black and acetylene black. Figure 4.28 gives a comparison of these methods in terms of process parameters. Both *Channel* and *Degussa* black processes were used during the early production of carbon black in the USA and Germany, respectively. Later, the other variants were developed, some of which (e. g., acetylene black) are primarily thermal decomposition processes with little to no oxygen during synthesis.

Production system type	Process	Precursor
Thermal-oxidative decomposition		
Closed system (turbulent flame)	Furnace black	Aromatic oil, natural gas
Closed system (turbulent flame)	Lampblack	Aromatic oil
Open system (diffusion flame)	Degussa (gas) black	Coal tar distillates
Open system (diffusion flame)	Channel black	Natural gas
Thermal decomposition		
Discontinuous	Thermal black	Natural gas/ oils
Continuous	Acetylene black	Acetylene

Figure 4.28: Variants of carbon black manufacturing process and their process parameters.

4.3 Patterning of carbon nanomaterials for devices

4.3.1 Printing

Once the carbon nanomaterial is synthesized, the next step towards device fabrication is its patterning onto a suitable substrate. These materials are often in a dry powder or liquid suspension form after their preparation. For dry powders, printing is one of the most commonly used patterning technique.

Printing is an ancient technology that has been used for wide variety of applications spanning across science, technology, and arts. Carbon materials, such as lampblack, have been considered some of the oldest pigments for making black inks. Subsequently, various dyes and coloring agents were printed either in a serial or batch type process. The first use of printing for electronic device fabrication was around late 1980s for electrical conductors, such as molten solder and metal nanoparticle colloids. These materials were used as electrodes, heaters, and certain passive microscale structures in the integrated chips and other electronic packaging. This was also the time when microprocessor-controlled printers became popular and commercial printer manufacturing industry experienced a rapid growth. This was a simple and inexpensive method with excellent design choices, hence it promptly gained attention as an alternative microfabrication tool.

By early 2000s, the use of printing expanded to fabricating patterns for photonic devices, MEMS, sensor-microelectrodes, and various other integrated chip-based devices [74]. Printing was an obvious choice for carbon nanomaterials, as there were very few other techniques that would enable their low-temperature micropatterning. The know-how developed for metal nanoparticle printing could be easily translated to carbon, because, as mentioned earlier, certain carbon materials were already in use as black ink pigments [213]. Unlike metal nanoparticles carbon nanomaterials did not require any sintering or oxidation protection, even at higher printing temperatures compared to metals. As a result, 2D printing became the most extensively used microfabrication process for carbon nanomaterials, particularly for flexible device manufacturing. Notably, one can directly grow or deposit carbon nanomaterials onto a substrate via directed CVD (described later), but this entails temperatures >800 °C, which are unsuitable for polymeric (flexible) substrates.

Printing is an additive technique. It can be carried out in serial, batch, or continuous fashion. The jet of ink can come out of a nozzle, or the ink can be squeezed through prepatterned stencils. In all cases, the ink is the most crucial component of printing, and its viscosity needs to be optimized such that it remains functional while maintaining a good printability (i. e., without nozzle blocking or smudging or spreading outside of the desired patterns). The ink should also solidify reasonably fast without developing internal stresses or cracks. Certain properties, which are common to most carbon nanomaterials, render their ink preparation challenging. For example, their chemical inertness, extremely light weight, accumulation of electrostatic charges on to the surface, poor solubility in common solvents, and agglomeration to form clusters. Below we will discuss some standard preparation methods employed for carbon material ink preparation.

4.3.1.1 Ink preparation

There are two types of inks used in 2D printing: water-based and organic solvent-based. Essential characteristics of a good ink include optimum viscosity, low surface tension, appropriate boiling point, and uniform distribution with desirable density of the additive (carbon material) in solvent or carrier. Often various types of surfactants, rheological additives, and stabilizers are added in the mixture during ink preparation to improve ink's printability, as well as performance. Carbon-based inks are conductive, and they are indeed used in devices as electrode materials. It is important, therefore, to ensure that the electrical properties of the carbon material is not compromised during ink preparation. This can happen due to excessive use of noncarbon components in the ink, as well as due to agglomeration, especially in the case of nanomaterials. Carbon nanomaterials feature a high specific surface rate, inherent Van der Waals' forces (e. g., between layers of multilayer graphene), carbon nanomaterials becoming prone to agglomeration and subsequent precipitation when their concentration in the ink is low. A good ink should have a uniform dispersion along with a reasonably high concentration of the carbon material.

There are two primary components of a printable ink: (i) functional material (carbon) and (ii) a solvent or carrier in liquid phase. The solvent can be organic (including polymers and wax) or water. Different types of solvents have their own advantages. For example, light organic solvents, such as alcohols, evaporate fast and exhibit a low surface energy, thus enabling the fabrication of finer features without nozzle clogging. Polymeric carriers, such as n-methylpyrrolidone (NMP), dimethylformamide (DMF), polyvinylpyrrolidone (PVP), chloroform, etc., serve as good dispersion agents, as their surface energies are closer to those of carbon nanomaterials. Water's biggest advantage is its nontoxicity. Water-based inks typically require chemical functionalization of carbon surface for inducing hydrophilicity.

In addition to the primary carrier, surfactants are used for facilitating an improved dispersion, especially in the case of organic solvents. These compounds are amphiphilic (i. e., they contain both hydrophilic and hydrophobic groups), and their primary function is to reduce the interfacial tension between the carbon surface and solvent to enable a stable dispersion. It has been reported that surfactants can attach themselves to the carbon surface noncovalently via Van der Waals forces, hydrogen bonding, static electricity, and $\pi - \pi$ interaction, etc. Examples of surfactants used with carbon nanomaterials include sodium dodecyl sulfate, dodecyl-benzene sodium sulfonate, Tween-60, Tween-80, etyltrimethyl-ammonium bromide, polyethylene oxide (10) nonylphenyl ether, Triton X-100, and polystyrene-block-polyacrylic acid (PS-b-PAA) [237]. Several proprietary surfactants are available in the market with specific functionalities. Carbon nanomaterials, such as nanotubes, may also be directly dispersed in water when coated with surfactants, exhibiting a high hydrophyle-lyphophyle balance [237].

To maintain the functionality of the carbon nanomaterial, their percolation or particle–particle contact in the ink is extremely important. Some carbon nanomaterials, such as nano-graphite (discrete, nanoscale crystals of graphite in a powder form) and multilayer graphene display anisotropic properties due to their layered arrangement. It is well-known that the electron transfer along the basal planes in such carbons is much faster compared to its perpendicular direction. In such cases, the particle-particle interaction and the size of each particle plays a crucial role in maintaining the conductive properties of the overall printable dispersion system.

If we take the example of CNTs or VGCFs, these materials are one-dimensional, and they tend to get tangled with each another. This is both good and bad. Good because entangled tubes are in good electrical contact with each other, and bad because this inhibits their uniform dispersion in the ink. If the CNTs have a low aspect ratio, they may not get entangled, but some isolated CNTs may be electrically disconnected with the rest of the material, and thus unable to contribute to the system (see Figure 4.29). To minimize this, one can print multiple layers of the ink (on top of each other). It has been reported that the sheet resistance of CNT inks drastically reduces when the second layer is printed. With subsequent layers this effect becomes less prominent, ultimately approaching the bulk conductivity of the CNTs [231].

A poor interface may cause precipitation or result in a significant change in the physicochemical properties of the functional material on drying, which is undesirable. To overcome some of these challenges, usually a surface treatment is carried out prior to

Figure 4.29: Schematic diagram showing printed carbon nanotubes and fibers with (a) significant entanglement and (b) relatively tangle-free distribution.

ink preparation. Although in this book the focus is on devices and fabrication of micro-/ nanopatterns, chemical surface modification and interface engineering of carbon is highly pertinent to carbon-based composite manufacturing. It is a vast research area in itself, and here we may only be able to cover the aspects relevant to carbon-based printed devices.

4.3.1.2 Surface treatment

It is intuitive that for preparing inks in water, a hydrophilic surface would provide a cleaner interface between the components of the ink. In the case of organic solvent-based inks, it becomes a bit more complex to optimize the surface as it is often solvents-specific. In general, most as-synthesized carbon materials are hydrophobic, and they feature low surface energies. They are also extremely light weight and have a tendency to agglomerate due to strong interparticle Van-der Waal forces. This results in an incompatibility with polar solvents, and some surface functionalization is almost always performed. These functionalization methods vary for different carbon nanomaterials, but there are some common principles and processes.

The first step in most of the carbon surface treatment processes, whether they are for device applications or for composite material preparation, entails oxidation. Oxidation can be performed through dry plasma or wet chemical processes. Among wet processes, interaction with strong acid and other oxidizing agents (e. g. HNO_3, H_2SO_4, $KMnO_4$, H_2O_2) lead to the attachment of oxygen-containing polar groups (e. g. –OH, –COOH, $-NH_2$) on to carbon surface [168]. Ozonation of carbon materials is another processes that may be carried out in solution (although rarely) or in gas-solid phase. Ozone removes disordered carbon and also creates oxygen-containing functional groups. One can also use ozone or oxygen plasma for the same purpose. These reactive plasmas are also useful for creating surface roughness (in the case of materials such as glass-like carbon), as they exhibit different etch rates for graphitic and disordered carbon [205]. In addition, they generate reactive oxygen containing functional groups. There is, however, one difficulty in using plasma treatment—it is not suitable for powders. If it is a carbon fiber mat, or there is a graphene film attached to a substrate, or the sample is a microfabricated pattern, plasma can be used. But powders cannot be uniformly exposed, and they also present challenges in equipment operation (for example, they may get sucked into the vacuum tube). Hence, for materials such as CNTs, VGCFs, and carbon black, wet chemical treatment is preferred.

The oxidized materials can be directly used for ink preparation (especially if it is an aqueous ink), or can be further modified with more complex chemicals. Some post-oxidation functionalization processes include silanation, esterification, alkylation, arylation, etc. It has been reported that some functionalization processes may disturb the $\pi - \pi$ conjugation system, create defects, or even change the hybridization states of surface carbon atoms [168, 205]. This can influence electrical, electrochemical, or mechanical properties of the material. Hence, the correct treatment must be selected for a given material.

There are five common types of 2D-printing techniques that are used for carbon materials: (i) screen-printing, (ii) ink-jet printing, (iii) gravure printing, (iv) flexography, and (v) microcontact flexography. Other than carbon materials, these printing techniques are used for metals, ceramics, and other materials for device manufacturing purposes. Among these, the three primary techniques, inkjet, screen, and gravure are described below. Typical ink viscosities are 1–100 Pa s for screen printing (thick pastes), 0.1–1 Pa s for gravure printing (medium viscosity), and 1–10 m Pa s for inkjet and aerosol jet printing (low viscosity) [113].

A compilation of different carbon inks, prepared using both nano- and large-scale carbon materials in powder form, is provided in Table 4.2.

4.3.1.3 Inkjet printing

Basic principle: A functional material is mixed with appropriate binders to form an ink. Electrical forces are applied (via passing current or changing voltage) to the nozzle as per the design requirements. This leads to an ejection of droplets in continuous or discrete fashion. Ink is partially absorbed on the substrate and the solvent is evaporated.

Materials used: Carbon materials (graphene, CNTs, VGCFs, carbon black, graphite powder, etc.), metal nanoparticles, ceramic suspensions, dyes/colorants, metallo–organic and light-emitting materials, solders, and other materials available in powder form. For substrates, silicon wafers, flat glass, woven cloth, polymer sheets, paper, etc., have been used. Papers with a range of coatings compatible with inkjet printing are commercially available.

Process steps: Inkjet printing can be divided into continuous and drop-on-demand type printing. In the case of continuous printing, a stream of ink droplets is constantly ejected from the nozzle, which are deflected by voltage plates. Whether the droplet will be deposited onto the substrate or recycled (if no pattern needs to be printed at that instance of time) is controlled by the applied voltage. In drop-on-demand inkjet printers, on the other hand, a droplet of ink only ejects from the nozzle when the printing on the substrate takes place. There is no recycling of waste droplets involved in the process. Based on the printing mechanism, drop-on-demand inkjet printers can be further divided into two categories: thermal and piezoelectric. Thermal inkjet printers (also called bubble jet printers) have a thin film resistor in their nozzle that heats up when a current is passed through it. Consequently, the ink vaporizes, and a bubble is generated. The increased pressure pushes the ink droplets out of the nozzle. Piezoelectric inkjet printers have a piezoelectric transducer in their nozzle that deforms when a voltage is applied, forcing the ink droplets out of the nozzle. A schematic representation of thermal and piezoelectric type drop-on-demand printers is shown in Figure 4.30.

Primary factors that influence the jet formation and printing efficiency are pressure at the nozzle inlet, viscous losses in the nozzle due to constriction of flow, and surface

Table 4.2: Carbon inks for printing prepared using combinations of various carbon materials, solvents and additives.
Acronyms: CNT: Carbon nanotube, MWNT-f-OH: multiwalled (carbon) nanotubes functionalized with dihydroxy phenyl, PVP: polyvinylpyrrolidone, PEG: polyethylene glycol, DMF: dimethylformamide, GNP: Graphene nanoplatelets.

S. No.	Carbon material used	Primary solvent	Binder	Additive/Surfactant	Application	Ref.
1	CNTs	Water	Acrylic-styrene resin	Poly-vinyl-pyrrolidone	Electrocatalysis and enzyme biosensing	[75]
2	MWNT-f-OH/ carbon black/ graphite	Water	Aqueous acrylic resin		Electroluminescent devices, flexible wearable electronics, printed capacitive sensors	[148]
3	Graphene powder and carbon black	Dibasic esters	Vinyl chloride-acetate co-polymer	BYK-066N (antifoaming agent)	Printed flexible electronics	[149]
4	Graphite/ carbon black	Ethanol/Diethylene glycol butyl ether/4-hydroxy-4-methyl-2-pentanone	Ethyl cellulose/PVP		Conductive carbon ink for screen printing	[93]
5	Graphite powder	Acetic acid	Chitosan	Glycerol	Electrochemical sensing and biosensing	[34]
6	Carbon soot (from vehicles)	DMF	Calcium carbonate (thickener)	Polyurethane resin powder	Flexographic printing	[80]
7	Graphene nanoparticles	Ethylene glycol	PVP		Flexible printed electronics	[96]
8	Graphene/CNTs	Terpinol/ ethanol	Ethyl cellulose		Flexible organic integrated circuits	[255]
9	Graphene/ activated carbon	DI water	Carboxy methyl methylcellulose	2,3-butanediols/ diethylene glycol	High performance planar micro supercapacitor	[35]

tension acting on various free surfaces. In addition to plenty of experimental research on optimization inkjet printing, several computation studies have also been focused on understanding the behavior of individual droplets and jets. One of the most common numerical analysis methods was suggested by Fromm [72], which is based on utilizing a stream-function to express the flow through the nozzle. The streamline function adopts

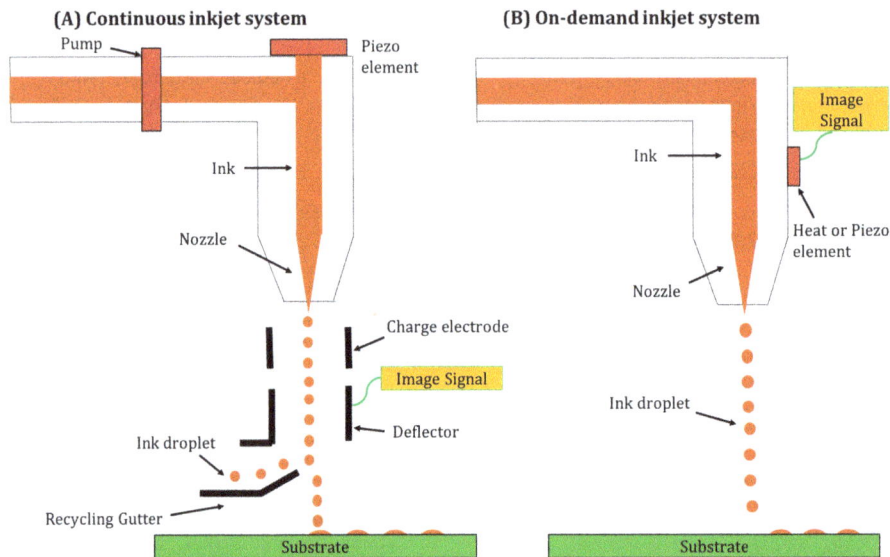

Figure 4.30: Schematic diagram showing (a) continuous and (b) drop-on-demand inkjet printing methods.

a value according to the flow rate, which is in turn governed by the process parameters (pressure, chamber-nozzle interface, viscosity, etc).

Advantages and challenges: Inkjet printing does not entail any prefabricated templates or screens, which facilitates a rapid printing process of a range of designs at a relatively low cost. As this is a serial process, post-printing steps, such as removal of leftover ink, are not required. Several types of high-precision printers are available commercially, so the technology is already widespread and can be used for device manufacturing with minor modifications. Advanced inkjet printers allow for multiple ink cartridges that can simultaneously deposit different materials, thus enabling the use of additives in the ink in a controlled fashion. One can print multilayer patterns with a good precision with this method, which helps in increasing the height and surface area of the overall structure. In general, an inkjet printer ink should have a low viscosity, good dispersion of nanomaterial without any floculation, and a low surface tension. Achieving optimum material properties and good percolation while ensuring its printability is the biggest challenge in inkjet, as well as other forms of printing.

4.3.1.4 Screen printing

Basic principle: Functional ink is squeezed through a pattern-containing woven mesh on to a substrate. This may be followed by curing or other forms of ink post-processing.

Materials used: Functional materials (often a combination of one or more carbon materials), a polymer binder (or *organic vehicle*), and inorganic additives are used for ink preparation. Particularly in the case of microscale screen printing, glass frit or an oxide admixture is employed as the inorganic binding agent along with the other two components. Substrate materials can range from flexible (e. g., cloth for the woven mesh (or screen)) is made of synthetic fibers, such as nylon or polyester.

Process steps: (i) Creating the design or stencil (microfabrication techniques may be used for microscale stencil fabrication); (ii) placing the stencil on a flat substrate on top of the printing board; (iii) spreading the ink using squeeze, which presses through the pattern and transfers on to the substrate; (iv) curing or post processing of ink. A schematic diagram of the screen-printing set-up is shown in Figure 4.31.

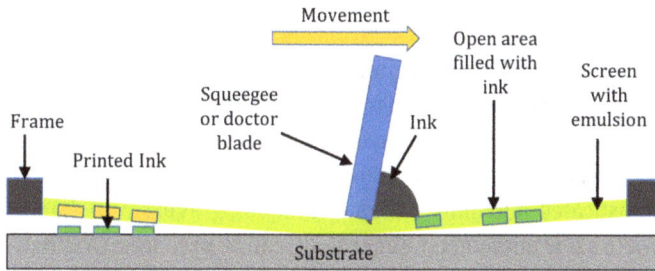

Figure 4.31: Schematic representation of typical screen printing process.

Optimization parameters: Screen-printing is a thick-film technology. Obtaining a smooth film surface as well as uniform chemical properties and stability of the ink material is achieved via post-processing techniques, such as thermal curing. Preprocessing of the substrate is done by using chemical modification. In terms of screen design, the mesh count needs to be optimized. A higher mesh count enables printing of finer features, whereas a low mesh count facilitates a better flow of the ink without clogging. Protection of the mesh from dust and other contaminants can be carried out by decreasing and application of photographic emulsions on it.

Use of traditional screen-printing is sometimes unsuitable for direct application in microscale capillaries or membranes owing to a high surface tension and a higher probability of trapping air bubbles. To avoid this, the inks formulations are made less viscous for microprinting. In addition, the widespread application of screen printing in microelectrode and MEMS fabrication has led to the development of certain variants of this process, collectively known as microprinting techniques. In microprinting, the ink or paste is deposited via a microinjector and/or a micropipette, often positioned using a micromanipulator. Dip-pen lithography and soft-lithography (a type of printing process where a PDMS mold having the negative pattern filled with the printing ink, followed by removal of PDMS) may also be regarded as a printing process due to their additive nature.

Advantages and challenges: Compared to inkjet printing, screen printing offers the obvious advantage of batch manufacturing. The inks used in screen printing can have a higher viscosity, and thus a good density of the functional material. Screen printing has always been dominated by carbon materials, and therefore plenty of optimized carbon ink/paste preparation processes are already in place. Ink curing and post-processing for obtaining a smooth surface is often a challenge. Moreover, though this technique is very suitable for patterning multiple layers of more or less the same shape on top of each other, multimaterial printing with varying shapes is difficult. For example, if a metal interconnect is to be printed along with the carbon electrode, it is not easy to align it, especially in the micro regime.

4.3.1.5 Gravure printing

Gravure printing is a roll-to-sheet or roll-to-roll printing technique. It is more than a century old technique and is still widely used for mass or continuous printing at industry scale. It has been reported that a critical dimension of 5 µm can be achieved by gravure printing [82]. The stencil here is of a cylindrical shape, which is called the gravure roll or just gravure. With sophisticated microfabrication techniques, gravure rolls, having single cells (grooves) down to a few micrometer dimensions, can be prepared. Originally developed as a magazine and poster printing techniques, gravure is rapidly gaining popularity in micro-device printing.

Basic principle: Ink is captured inside the grooves of an engraved pattern on a cylindrical surface, followed by its transfer to another (generally flat) substrate by rolling the cylinder.

Materials used: Gravure cylinder is generally made of metals. One can also attach prepatterned plates or chips, e. g., made of silicon, onto the cylinder surface. Techniques used for direct engraving of the cylinder include mechanical engraving using a stylus, which in turn performs electrochemical etching of the material. Another engraving technique is based on laser ablation, which offers the advantage of fully automated computer-to-plate fabrication [99]. Photolithographically patterned Si chips may be directly pasted, or used for transferring the patterns onto a polymer or metal film, which is subsequently wrapped around a cylinder to be used as the gravure roll. Substrate materials can range from paper and polymers to glass, ceramics, and metals. A flat substrate is used in most cases but roll-to-roll printing is also possible if a cylindrical substrate is used.

Process steps: A cylindrical stencil with the desired patterns engraved into it, or wrapped around it, is prepared. It is passed through an ink container or reservoir for ink suction into the engraved cavities (known as cells). Excess ink (from the surface other than cells and from the convex meniscus formed on the cells) is removed applying a doctor blade. The cylindrical roll is then pressed to make a contact with the substrate. Due to a capillary action, the ink patterns are transferred on the substrate material.

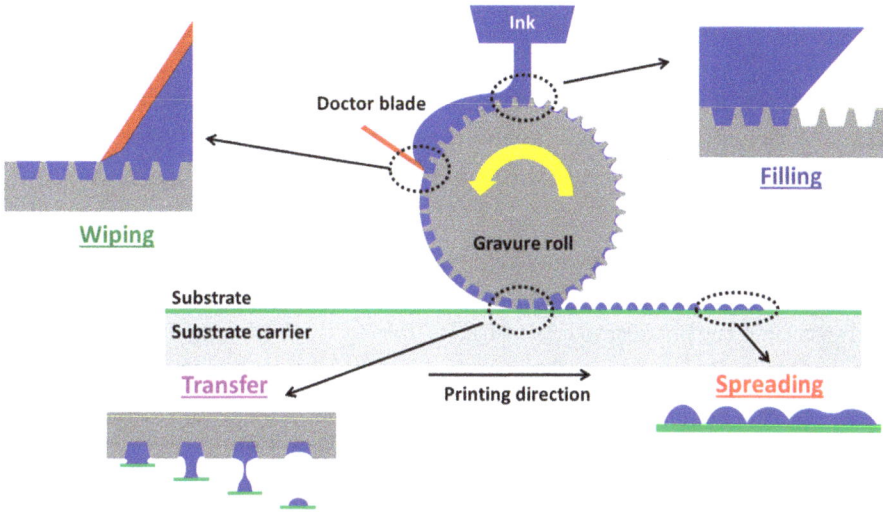

Figure 4.32: Illustration of gravure printing process. Copyright ©2016 IOP Publishing Ltd. Reproduced with permission from [82].

Finally, the ink is spread and/or absorbed on the substrate. A schematic representation of gravure printing is presented in Figure 4.32.

Advantages and challenges: The advantage of gravure printing is the fact that it utilizes moderate ink viscosities and allows for a good concentration of the functional material in the ink, even for 5–10 µm feature sizes.

The use of gravure printing in microfabrication is mainly limited by the cost and engraving techniques in the microscale. The resolution of the engraved patterns is typically controlled by the radius of the stylus in the case of electrochemical engraving. Laser ablation may create heat-affected zone, thus introducing surface roughness. If the feature size is very small, surface finish and polish becomes very difficult, as it may destroy the cells. During the printing process, the roughness caused by the doctor blade is yet another reason for surface roughness and tool wear. To protect the roll surface from wear, it can be coated with, for example, boron nitride particles in a nickel matrix. Another issue is the incomplete removal of extra ink by the doctor blade due to homogeneous lubrication residue and ink drag-out from cells. This may be addressed by finding the most suitable geometry for the doctor blade for a certain ink type.

A complete filling of the cells without air trapping may limit the designs. Long and narrow channels or shapes containing straight deep walls, sharp corners, etc., are generally difficult to pattern using this technique. For printing such patterns, one need to optimize the ink viscosity. A dimensionless quantity known as **capillary number**, (Ca), is used to describe the relative magnitude of surface tension and viscous forces during printing.

$$Ca = \frac{\text{Print speed} \times \text{Viscosity}}{\text{Surface tension}}$$

Evidently, the print speed needs to be optimized for each ink composition, which is in turn governed by the required concentration of the functional material for specific applications.

4.3.1.6 Flexography and microcontact flexography

Flexography (also known as flexo) is a direct convex printing method that can be used for both absorbent and nonabsorbent substrates. There are multiple (usually 4) metal cylinders that are in contact with each other for ink transfer. Printing plates, made of elastomers or photopolymers patterned onto flexible substrates, contain the design. These so-called flexo-plates are then wrapped around the cylinder, which is in contact with the substrate material to be printed (generally called web). A schematic diagram of flexography is shown in Figure 4.33.

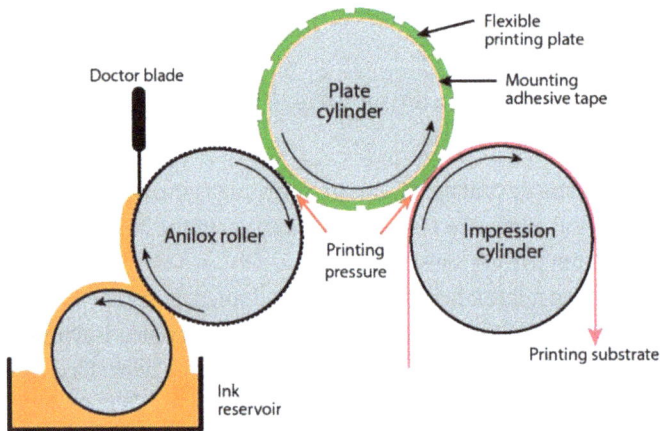

Figure 4.33: Illustration of flexography process. Image courtesy: Dean Valdec, University of North Croatia. Originally published in [238].

Flexography inks typically have a low viscosity, but their composition can vary from water-based, organic solvent-based, to curable resins. This technique is most popular for printing large-scale food wrappers, etc., that require safe and quick drying inks.

Basic principle: Same as other printing techniques.

Materials used: Flexographic printing plate can be made of elastomers (e. g., natural or synthetic rubber, PDMS) or photopolymers by laser-photochemical machining, direct engraving, and in the case of microscale printing, lithographic techniques. Cylinders are generally made of metals or plastics.

Process steps: Ink is transferred from one to another cylinder through micro-scale grooves or cells. Any extra ink is removed by a doctor blade in order to have only a controlled quantity of ink on the second cylinder's surface. The third cylinder contains patterned plates, and is in contact with the film of the material to be printed (web). The fourth and last cylinder, called the impression cylinder is for providing support and pressure when the web passes in-between itself and the plate-containing cylinder.

Flexography is useful in carbon nanomaterial printing as the printing pressures are low, which prevent any mechanical damages, especially when multiple layers of the material are printed. Some microfabrication techniques are indeed already used for the fabrication of flexo plates. In this respect, some concepts are similar to soft-lithography, where elastomers are patterned using a master mold. Further details on plate materials and design optimization may be found in the review article by Liu and Guthrie [150].

4.3.2 Spray coating

Carbon nanomaterial inks and their suspensions in organic solvents can be directly spray-coated onto large areas. Sprayed CNT coatings are perfectly black and one of the best absorbers of light. Spray coating is a traditional industrial coating process that is used for a variety of materials, including metals and ceramics. Its common variants are thermal spraying and plasma spraying. One can either already premix the components of the ink, or add them during the spray injections. In the case of carbon nanomaterials, typically the ink is already prepared, given some surface functionalization, or additives may also be required.

Among carbon nanomaterials, CNTs are the most preferred material for spray coating. The inks for this purpose can be prepared in a relatively simple fashion. For example, CNTs suspended in an aqueous solution using a dispersion agent can serve as spray coating ink. Films deposited this way are generally thick, but they suffer from lack of uniformity. The technique, however, is inexpensive and suitable for large area coverage.

4.3.3 Direct growth

There are a few methods where these nanomaterials are directly used in devices. For example, CNTs and CNFs can be grown on to a patterned substrate as individual units or in bulk if the substrate also serves as a catalyst. Another option is to prepattern the electrodes structures, for example, by metal sputtering on a silicon substrate, and then grow the nanomaterial on top of it. The most crucial aspect here is the thermal stability and compatibility of all prepatterned components at high temperatures, which are inevitable for nanomaterial growth/deposition. Generally metal electrode pads are used for direct nanomaterial growth. Noble metals, such as gold can deform at temperature lower than their bulk melting points when they are in a thin film form, especially if the film contains any defects or voids. Other than electrode deformation, the adhesion of different components due to thermal expansion, followed by shrinkage at cooling, needs to be factored in. If single or few CNTs or CNFs are being grown on to prepatterned substrates, it is important to control the direction of their growth, especially if the goal is to have an electrical contact on both ends of the filament.

4.3.4 Pattern transfer

Pattern transfer implies that a material or pattern is separately fabricated, and then transferred on to a set of prefabricated electrodes, other circuitry, and device elements. After that the device can be electrically insulated or packaged. This method is often used for transferring films such as graphene. Generally, a cross-linkable polymer (e. g., PDMS) is poured onto the substrate where graphene layers (single, bi-, few- etc.) are already deposited via CVD. The polymer is then allowed to cross-link and peeled off. Graphene comes with it, which can be released onto the desired device geometry. Notably, the adhesion to PDMS plays an important role here. Graphene (or other nanomaterials) should only adhere to the top surface of PDMS and should have weak enough adhesion for a good release without much mechanical damage. Cross-linking parameters can be accordingly optimized. The new surface where the material needs to be transferred may contain adhesion promoters, so as to have a better affinity to the transferred material compared to PDMS. Though pattern is very commonly used in microfabrication, there are low fabrication yields due to crack formation or incomplete transfer. In addition, contamination with polymers and solvents is a major issue that can require various post-processing steps.

4.4 Applications of carbon nanomaterials

Carbon nanomaterials have been used for a wide range of applications covering almost everything, from advanced electronics and sensors to structural materials in large-scale manufacturing. Extensive literature is available for each nanomaterial, which sheds more light on material-specific properties and applications. Below, some primary applications along with suggested further reading are detailed.

Graphene: Owing to its electrical and electronic properties, graphene has been used in various chemo- and biosensors, transistors, transparent electrodes and displays, protective coatings, functional textile, energy storage devices, Hall effect sensors, and as a reinforcement material in composites, which are in turn used for manufacturing biomedical implants, lightweight sports gear, lightweight aircrafts and spacecraft, and numerous other fields [10, 38, 235]. Composite of graphene in a concrete matrix is used as a structural material in civil engineering. Application of graphene/asphalt composite has been reported in road construction. These applications utilize graphene in the form of large sheets, nanoparticles, nanoflakes, ribbons, as well as GO and rGO. Devices may be in both direct and printed (mixed with a carrier) form.

CNTs: As discussed earlier, certain SWCNTs are semiconductors that broadens their applications beyond electrical, electronic, and structural components. CNTs have been used in micro- and nano-electronics, sensors, flat panel displays, energy and hydrogen storage devices, and as a reinforcement in composite materials. CNTs are used as single units, forests, films, ropes, and bundles. Applications that set CNTs apart from other carbon nanomaterials include, but are not limited to, field emission devices, electron guns, AFM tips, microwave amplifiers, gas-discharge tubes, electrochemical actuators, and nanochannels [4, 60]. Chemically functionalized and doped CNTs (e. g. with N) have their own specific applications, particularly in biomedical and environmental applications.

Fullerene: Since many fullerenes have slightly different applications owing to their hollow spherical shape and the variation in their conductivity (semiconductor to conductor, depending upon structural configuration). As they are chemically synthesized, it is possible to trap chemical moieties, such as metal ions or small molecules, inside them by modifying their precursors. For example, Gadolinium trapped inside fullerenes leads to magnetic properties that render the material useful in biomedical imaging. Applications of fullerenes encompass gas absorbers, rectifiers, drug delivery systems, biosensors, energy storage devices, and organic photovoltaics [117, 71, 29]. Fullerenes are excellent electron acceptors having a low reorganization energy after the electron transfer. They display an electron mobility of $0.1 \, \text{cm}^2/\text{Vs}$, which renders them very suitable for organic electronics and photovoltaic applications. Fullerenes have almost become a gold standard in both solution- and vacuum-deposited organic photovoltaics. Compared to other organic materials, fullerenes lead to more stable devices with a much improved power efficiency [29].

In Chapter 2, a brief description of conversion of fullerenes into open nanoscale curved carbon material is provided. Noteworthy also those lines is that KOH activation fullerenes in ammonia has been reported to yield a new type of porous carbon material (N-containing curved-carbon units in bulk), which is useful for energy storage purposes [219]. One could envision the use of fullerenes in designing new carbon materials with very high surface area and specialized functionalities.

Vapor-grown carbon fibers: VGCFs have similar applications as those of CNTs, but they are most commonly used as a reinforcement in composites [89, 8]. As they are compatible with printing (unlike electrospun carbon fibers, see Chapter 5), they can be more conveniently used in devices. They are most commonly employed in biosensing, filtration, and environmental pollution control devices, tactile senors, energy storage devices, etc. [59, 250, 127]. As they are often produced along with CNTs, in many electronic and structural applications they are mixed with CNTs.

Carbon black: Carbon black's primary application is reinforcement of rubber in the tire manufacturing industry; it is also carbon black, the most common support material when it comes to making carbon-based electrodes for electrochemical devices, such as batteries and supercapacitors. This material is added to increase the conductivity of the carbon system [24, 125]. Graphitized carbon blacks have a much higher electrical conductivity and an extremely light weight. This material is a well-known adsorber in gas chromatography systems [31]. Recently, carbon black has been used for nanoparticle trapping aimed at achieving conjugated microporous polymers, which are emerging materials for cutting-edge energy storage device fabrication [125]. Carbon black is indeed considered a promising electrode material for Na-ion battery [9].

4.5 Future prospects

The performance of microfabricated carbon nanomaterial-based devices can be tremendously improved if one can use (i) single nanoscale units (e. g., one CNT) instead of their bulk forms as the functional element, and (ii) nanomaterial without any additives. The solutions have been found, but most of them are still limited to laboratory scale owing to challenges involved in terms of reproducibility.

Carbon nanomaterials are highly stable. But during their synthesis from organic precursors, they are very sensitive to every single process parameter. It is not straightforward to obtain the exact same dimensions, size distribution, and structural uniformity among different batches, even if they are prepared under macroscopically the same synthesis conditions. High-precision furnaces, mass flow controllers, and high purity gases are essential for carbon manufacturing. Despite ensuring all this, batch-to-batch variations are common in carbon nanomaterials, including in those available commercially. This may lead to deviations in device performance when they are fabricated using bulk nanomaterial (i. e., a collection rather than individual nanounits). Microfabrication techniques, such as printing, require binding agents, dispersion agents, polymeric carriers, or other additives. This additionally results in a compromise on the properties of an individual nanounit. This implies that a lot of nanostructures in the bulk material do not get an exposure to the surroundings, and hence they cannot contribute to device functionalities.

Very promising initial research supported by good marketing strategies created a high demand of carbon nanomaterials worldwide, which attracted a lot of monitory investments. Unfortunately, this demand could not be sustained for too long, and the material quantities needed in practice were much less compared to what was predicted. Unlike other chemicals, carbon nanomaterials do not easily expire due to an excellent shelf-life, which implies that researchers don't have to frequently buy these chemicals. As a result, some carbon nanomaterial producers started fearing losses. To sustain the carbon device technology and the supporting industry, it became essential to directly use the bulk or powder forms in the devices rather than individual nanounits. As a result, the research direction shifted towards developing processing and manufacturing methods for bulk powders of nanomaterials that would increase the viability of commercial production of these nanomaterials. Though there are still several ongoing efforts aimed at designing single nanounit-based devices, more investigations are needed to simplify the isolation of single nanounits from the powders, and also to increase the purity of the materials obtained via CVD.

4.5.1 Health and environmental safety of carbon nanomaterials

Handling carbon nanomaterials in the bulk (powder) form is challenging due to their extremely light weight and a lack of safety data on the potential health hazards. For example, during CVD, it is not possible to control the size and shape of each graphene flake. Since individual graphene flakes are extremely small and often feature reactive edges, their inhalation and interaction with various types of cells in the human body may cause potential health hazards [61]. Additionally, the disposal of these materials into the environment, whether it is direct or is mediated through the human body, may lead to their mixing into the soil. Finally, graphene will enter the food chain and will become extremely difficult to eliminate. A diagram of the carbon nanomaterial cycle, taken from Fadeel et al. [61] is shown in Figure 4.34. Although these materials are generally unreactive, their prolonged exposure can certainly raise human and animal health, as well as environmental concerns.

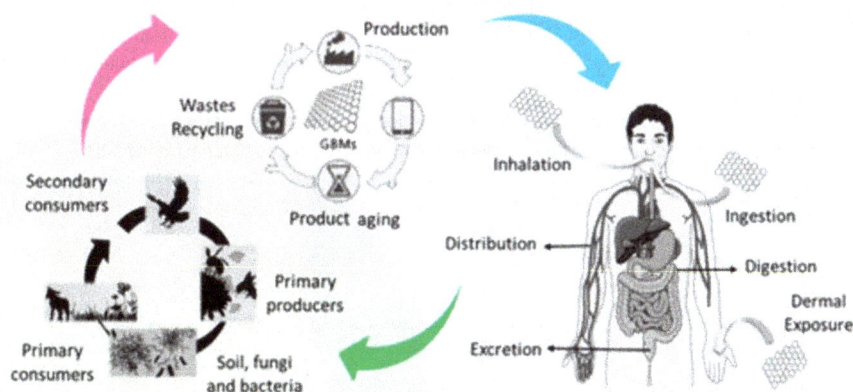

Figure 4.34: A schematic diagram showing potential human and animal health hazards caused by exposure to carbon nanomaterials. Copyright ©2022 American Chemical Society. Reproduced from [61].

1. What are the primary differences (in terms of mechanism) between pyrolysis of a light gaseous hydrocarbon and a solid organic polymer?
2. Segregate these carbon nanomaterials based on their hybridization states (refer to Chapter 2, classification of allotropes): buckminsterfullerene, vapor grown carbon fiber, carbon black, multilayer graphene, CNT.
3. The L_c value of a graphitic carbon sample having, on an average, four graphene-like layers in each stack is 10.38 Å. What is the average d-spacing between the 002 planes?
4. Calculate the length of lattice vector a ($= a_1 = a_2$) for a graphene unit cell.
5. What are the important optimization parameters while establishing a new CVD process, for example, while trying out a new catalyst or precursor gas?
6. Can carbon nanomaterial-based devices be fabricated on to rigid chips? Explain with one example process for fabricating such a device.
7. Which synthesis method, in your opinion, is most suitable for obtaining a single-layer graphene (refer to Section 4.2), and why?
8. Compare graphite and graphene in terms of structural defects.
9. What is the difference between a nanoparticle and a quantum dot?
10. What are the different types of a CVD reactor? Which type of reactor should be preferred for single-layer graphene growth?
11. Mention all steps in graphene CVD using Cu catalyst? Will there be any difference of Ni catalyst is used?
12. Explain the difference between few-layer graphene and turbostratic carbon.
13. Explain the difference between diamond-like carbon and vapor deposited diamond.
14. Explain the roles of (a) hydrogen and (b) plasma in diamond growth. How does diamond CVD differ from graphene CVD?
15. Explain the reasons for the boundary layer formation during CVD.
16. What is colloidal carbon? Explain in the context of carbon black.
17. What is the difference between thermal decomposition and thermo-oxidative decomposition? What can be the influence of oxygen's presence during the decomposition of a gaseous organic precursor?
18. What are the main ingredients of an ink prepared using a carbon nanomaterial for printing purposes?
19. What is the difference between inkjet and screen printing? Which of the two is more suitable for large-area patterning?
20. What is pattern transfer, and how is it relevant to device fabrication using carbon nanomaterials.
21. Which general precautions must be taken while handling carbon nanomaterials in the lab?

5 Carbon fiber based devices

5.1 Introduction

Carbon fibers are one-dimensional architectures having an aspect ratio >100, composed mainly of sp^2-hybridized carbon. Depending upon the fabrication method, some carbon fibers can have their aspect ratio (length/diameter) in the range of several thousand, whereas others may have it just close to 100. Their diameters vary between tens of nanometers up to a few micrometers. Typically, fibers with a diameter <100 nm are called carbon nanofibers (CNF), and those featuring micrometer scale diameters, or a mixture of fibers whose diameters range from nano-to-micro meter, are simply designated carbon fibers (CFs). The extent of sp^2 hybridization and the stacking of carbon fragments also varies.

Before discussing carbon fiber based devices, it is important to understand that >98 % carbon fiber production is for large-scale manufacturing applications, e. g., for making body parts of automobile and aircrafts. In these applications, carbon fibers-based composites are utilized for their mechanical properties, rather than electrical or electrochemical. This composite material, known as carbon fiber reinforced plastic (or Polymer) (CFRP) is rapidly replacing metals and traditional polymers and are indeed considered the next-generation manufacturing solution. Its production has seen an accelerated growth worldwide in the last 20 years. Modern CFRP manufacturing plants have hundreds of kilogram of daily production capacities. Carbon fibers used in CFRPs for manufacturing applications have their diameters in the micrometer scale and are essentially uniform, i. e., only very little diameter variation among fibers is acceptable. Fibers used in devices, on the other hand, are often nanoscale with a wider diameter tolerance. Their electrical, electrochemical, and thermal properties are also used in combination with mechanical. In Figure 5.1, a car interior part made of CFRP and a flexible device having nonwoven carbon fiber mat as its functional element are shown. One can observe that there is not only a difference in the dimensional scale of these two materials; they also possess significant differences in terms of physicochemical characteristics.

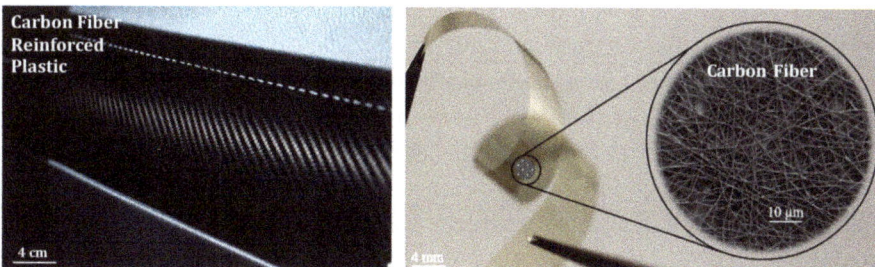

Figure 5.1: Images of carbon fiber reinforced plastic (CFRP) used in large-scale manufacturing; and of electrospun nonwoven carbon fiber mat directly used in device fabrication. Image of CFRP, curtesy: Hudson fiberglass. Image of carbon fiber mat adapted with permission from [84].

https://doi.org/10.1515/9783110620634-005

As stated above, for manufacturing purposes, carbon fibers are not used in their pristine form. They are used as a reinforcement in CFRP, a composite material. Composites are defined as materials having at least two chemically and physically distinct phases: reinforcement and matrix. Reinforcement is the phase that brings tensile strength and modulus to the material. Matrix is the base in which the reinforcement is either dispersed or placed in a layered manner. Matrix can be a polymer (that is, plasticized), metal, alloy, ceramic, concrete, or even another form of carbon. Accordingly, the resulting material is denoted as CFRP (polymer matrix), CFRM (metal or alloy matrix), CFR-ceramic (ceramic matrix), CFR-concrete (concrete or cement matrix), and CFRC (carbon matrix). A notation CF/matrix is also used, where CF indicates reinforcement, and the second term part after "/" denotes the matrix. In principle, other than a woven fiber mat, the reinforcement can also be dispersed fibers, tubes, or particles; but woven fiber mats arranged on top of each other with a well-defined rotation are most preferred reinforcement due to their superior mechanical properties.

Besides carbon fibers, glass and polymer fiber mats are also used in commercial composite manufacturing. One must note that CFRP is sometimes denoted as just carbon fiber for industrial purposes. In addition to this being a technically appropriate representation of the material, it has become a convention. One must always be careful while using this nomenclature and understand that the properties of bare carbon fibers may be quite different from those of CFRPs. Other factors, for example, whether the fibers are spun or vapor grown, woven or nonwoven, uniform (in terms of diameter) or nonuniform, have undergone any surface treatment or not, must also be clarified prior to comparing their properties with any given standard.

> **!** CFRP are composed of woven carbon fiber mats in a layered (laminated) fashion, surrounded by resin as matrix. This resin is hardened after this composite is provided the desired shape, e. g., by molding. Laminated composites have anisotropic (directional) mechanical properties, which are very different from those of individual carbon fibers, carbon fiber bundles, or nonwoven mats that are used for small device applications. While searching for information about carbon fibers on the internet, one must ensure that the given properties are those of pristine carbon fibers, not of CFRPs.

Similar to other carbon materials, carbon fibers are derived from hydrocarbon precursors via top-down or bottom-up manufacturing processes. Common top-down processes are based on spinning of a liquid/semi-solid high molecular weight precursors, such as polymers and petroleum pitches. Bottom-up synthesis is via chemical vapor deposition (CVD) by pyrolyzing hydrocarbon gases, similar to that of carbon nanotubes (CNTs), but with slightly larger catalyst particles. They can be grown via seeding (substrate-bound) catalyst or floating catalyst (catalyst particles floating inside the CVD chamber) similar to CNTs.

The term spinning is derived from yarn spinning used in the textile industry, where natural or synthetic fibrous materials are converted into long threads and yarns for fabric weaving purposes. The raw material for spinning should feature a property known as *viscoelasticity*, which will be discussed in detail in this chapter. Spun fibers are converted into carbon via heat-treatment, which more or less follows the general carbonization chemistry (pyrolysis → carbonization → graphitization). Depending upon the heat-treatment temperature and the type of precursor, the resulting carbon fibers can be turbostratic or graphitic. Their ultimate tensile strength (UTS) and elastic (Young's) modulus in turn depend upon the arrangement of the carbon sheets. Of course, due to a high surface-to-volume ratio, fibers have a slightly different carbon formation and microstructural evolution mechanism compared to other bulk forms of carbon.

CVD-grown carbon fibers are commonly referred to as vapor-grown carbon fibers or VGCFs. Basic principles of CVD and the relationship between CNT and VGCF CVD is described in Chapter 4. Similar to CNTs Fe, Co, Ni, and some other transition metals and alloys can be used as catalysts for VGCF growth. Both substrate bound and floating catalyst type growth can be used in cold or hot-walled CVD reactors. In Figure 5.2, SEM images of carbon fibers grown via CVD and spinning are compared. Evidently, they have different morphologies. VGCFs usually grow as forests (dense volume on a substrate), whereas spun fibers are collected as fabric-like mats. They may also be wound to form reels, and then woven or braided if the ultimate goal is to use them in a composite. In this chapter, the focus will be on nonwoven mats, bundles, and individual fibers that are device-compatible. Composite manufacturing will not be discussed in much detail.

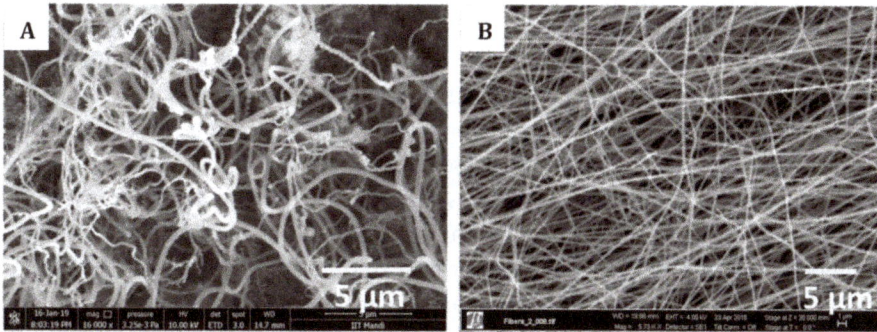

Figure 5.2: (A) Vapor grown carbon fiber (VGCF) on a substrate and (B) carbon fiber mat prepared by electrospinning of polyacrylonitrile followed by carbonization. Image (A) courtesy: Piyush Awasthi and Viswanath Balakrishnan, Indian Institute of Technology Mandi.

Carbon fibers are used as electrode materials in energy storage, sensor, and bioelectronic devices. They have a high surface-to-volume ratio compared to bulk sold carbons, but they can be used as mats or bundles without any binders or additives. Basically, carbon fibers are used in those applications where it is not feasible to utilize carbon nanomaterials due to their discrete nature, but a higher surface area is required compared to bulk solid carbons. Carbon fibers are of particular interest for fabrication of flexible devices that demand bendability. Electrical, electrochemical, and other properties of carbon fibers are comparable with other forms of carbon, but they vary depending upon their manufacturing process. Lithographically patterned non-graphitizing carbon for chip-based devices can only be heat-treated to 1200 °C due to limitations related to substrate deformation. Spun carbon fibers, however, can even be heated to graphitization temperatures (i. e., >2500 °C) as they are substrate-free.

In this chapter, mainly the top-down (spinning) techniques of carbon fiber fabrication are described. Chemical structure and properties of the primary carbon fiber precursors are also discussed along with their spinning behavior. Details of CVD can be found in Chapter 4. Microstructure, properties, and classification of carbon fibers, both spun and vapor grown, are detailed afterwards. Since spun fibers are also derived from high molecular weight hydrocarbons (e. g., polymers), their heat-treatment is carried out in a very similar manner as other polymer-derived carbons. The mechanism of microstructural evolution at fiber scale may however differ. Mechanistic aspects and influence of various parameters are also discussed in this chapter. Finally, the secondary techniques employed for patterning or directly using carbon fibers for devices are detailed.

5.2 Spinning techniques

Fiber spinning is a set of techniques based on pulling or drawing fibers out of a liquid or semisolid mass using mechanical and/or electrical forces. Based on the spinning set-up and parameters, several different types have been developed. Common spinning techniques used at industrial scale are electrospinning, melt spinning, wet spinning, dry spinning, and dry–jet wet spinning. Among these, electrospinning is most suitable for producing nanoscale fibers. However, these fibers may exhibit fiber-to-fiber diameter variation. Other techniques, such as melt spinning and wet spinning, yield more uniform fibers, but their diameters are typically >1 μm. Consequently, electrospun fibers are more extensively used for small-scale devices and other research purposes, whereas other (micro)fibers find applications in composite-making and large-scale manufacturing.

Another parameter that influences the selection of spinning process is the physicochemical properties of the precursor material, such as viscoelasticity, drying time, and loss of volatiles during dying. Other than from polymers, carbon fibers can also be produced from petroleum pitches. Pitches are complex aromatic hydrocarbons that are often a mixture of more than one organic compounds. There are also other elements, such as sulphur present in them, which can influence the carbonization process. Certain research applications, such as implantable bioelectronic devices, can be highly sensitive to the purity of carbon fibers. In such cases, it is advisable to prepare the fibers from a precursor with a well-known chemistry and carbonization mechanism. If the carbon fibers are commercially procured, one can either conduct an elemental and surface analysis, or get a confirmation on the fiber purity and quality from the vendor.

It should be noted that melt spinning techniques were used before electrospinning at an industrial scale for fiber manufacturing. However, they were not suitable for materials having a very high molecular weight, or those with strong polar bonds [139]. An additional electrical force acting against the internal material forces proved to be very helpful in such cases. The use of electrical forces in water droplet formation was already studies by G. I. Taylor in early 1960s in the context of thunderstorm generation [222]. Taylor and other authors also reported the formation of ultrathin fibers when the material was passed through conducting tubes [223]. These concepts were combined with spinning, which led to the technique of electrospinning. Extensive theoretical and experimental work has been conducted on electrospun polymer and polymeric carbon fibers, and various new aspects, including nozzle designs, have now become available.

The primary technique described below is electrospinning. In addition to the concepts presented here, several review articles and books on electrospinning are available. One can also refer to precursor-specific books on electrospinning. Note that electrospun polymer fibers are also directly used (without carbonization) in many biomedical, drug delivery, filtration, and device applications [204].

> Notable works on the development of industry-scale spinning techniques, as well as modification of precursor materials, such as polyacrylonitrile (PAN) for carbon fiber production, have been carried out by the research groups of E. Fitzer [68, 65], M. Inagaki [109], and O. P. Bahl [17, 18].

5.2.1 Electrospinning

A basic electrospinning set-up is shown in Figure 5.3. The fibers collected onto electrode 2 are in the form of a mat. When the distance between two electrodes is **basic working principle**: Spinning (drawing or pulling) fibers out of a polymer solution or melt using electrical forces.

Materials used: Viscoelastic polymer solutions or melts.

Instrument set-up: The polymer solution of interest is filled inside a syringe connected to an electrically conductive needle. The needle itself is connected to a power supply, i. e., treated as an electrode. The second electrode is a conductive substrate, where the fibers are collected. A high voltage difference is created between these two electrodes, while the solution in the syringe is gradually pushed out with the help of a pump. The polymer droplets that accumulate onto the needle-tip experience the high voltage, which leads to the formation of a conical shape (known as the Taylor cone), which ultimately escapes the droplet in the form of a fiber. Initially this fiber is thick, but it continues to thin and longitudinally split due to strong electrical forces. By the time the fibers reach the collector electrode, they usually have nanoscale diameters.

Optimization parameters: molecular weight, viscosity, electrical conductivity, and solvent (if any) used with precursor, applied voltage, distance between electrodes, temperature, humidity, size of capillary, type of collector, horizontal or vertical setting, etc.

Far-field Electrospinning, **d = 5 to 20 cm**

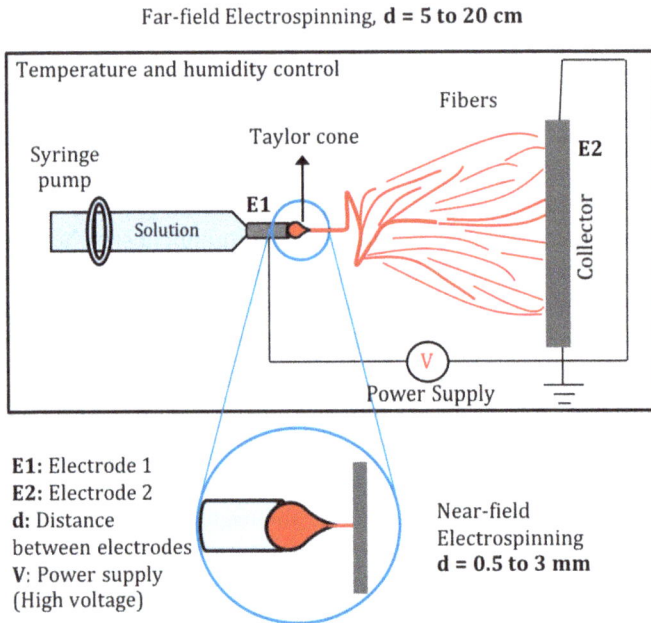

Figure 5.3: Schematic diagram of standard (far-field) and near-field electrospinning.

5.2.1.1 Taylor cone formation

To obtain fibers during electrospinning, one needs to initially push the polymer/melt through the needle (electrode 1) using a syringe pump. The first droplets partially protrudes through the needle outlet. When the electric field is applied, there is a competition between the surface tension, viscosity, and electrical forces. Gravity is yet another force acting on this droplet, which becomes more prominent in the case of a vertical electrospinning set-up. Altogether, the droplet starts to release fibers only when the electrical forces supersede the combined effect of all other forces acting in the opposite direction. Before the fiber jet initiation, the droplet deforms into an ellipsoid followed by a cone-like shape, as shown in Figure 5.4. This phenomenon is known as Taylor cone formation. Based on the applied field, viscosity and electrical conductivity of the solution, and direction of gravitational forces, the shape of this cone may have minor variations. In most cases, the fiber jet is initially single, which splits into multiple jets or simply thins down. The length of the single jet may also vary depending upon material and process parameters.

According to Taylor, the voltage at which the maximum instabilities develop (in other words, the minimum voltage required for fiber jet initiation) can be expressed as follows [223, 139]:

$$V^2 = 4\frac{d^2}{l^2}\left(\ln\frac{2l}{r} - 1.5\right)(0.117\pi r\gamma),$$

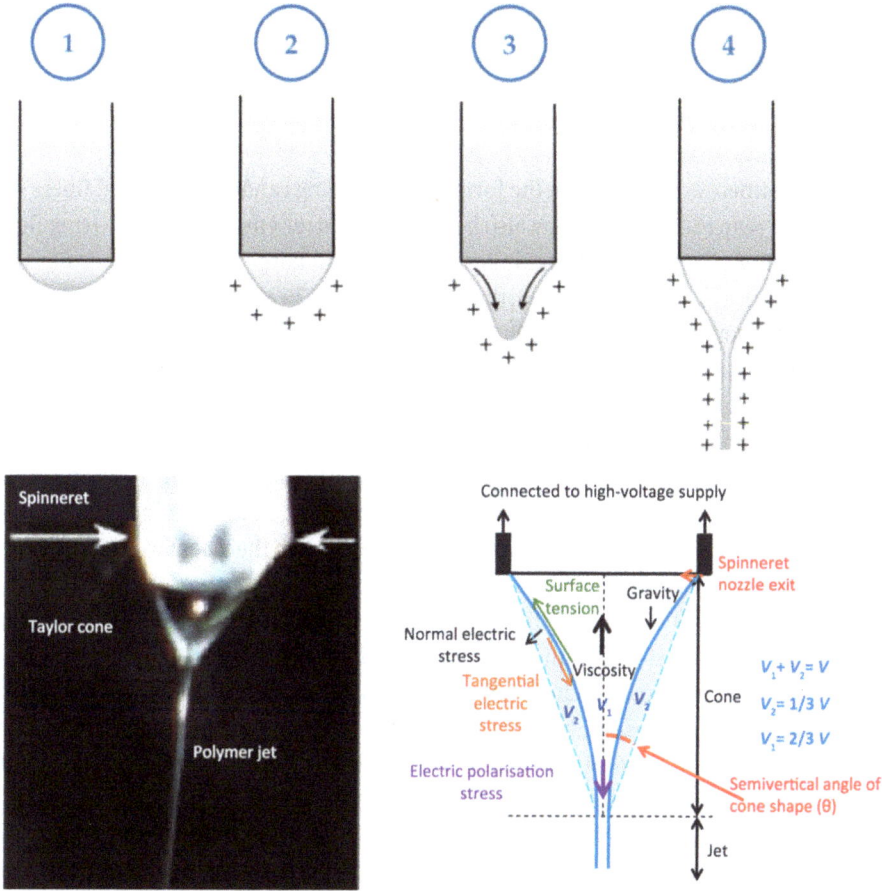

Figure 5.4: Taylor cone formation with respect to increasing voltage (stages 1–4 shown in top panel). In the bottom panel, a digital photograph of a Taylor cone and multiple forces acting on the droplet along with their directions are illustrated. Regions V_1 is the volume of the jet, and V_2 is the empty space in the cone caused by electrical polarization. Top panel is reprinted from [217] and bottom from [252].

where V (in kV) is the applied voltage, d is the distance between the electrodes, l and r (in cm) are length and radius of the capillary (metal needle), and y (in dyn cm^{-1}) is the surface tension of the material. Once the jet is initiated, the fiber diameter decreases with an increase in the applied voltage.

> If the fiber jet simply converts into droplets without bending; this is known as Rayleigh instability. This does not concern electrospinning and is only evaluated in electospraying.

In addition, jet velocity, the pressure on droplet, and electrical conductivity of the solution influence the Taylor cone shape and jet initiation. With the use of a syringe pump, one can maintain a constant pressure and material flow rate. In many studies, therefore, these terms are omitted from the voltage calculations.

5.2.1.2 Bending instabilities

Electrospun fibers are received in the form of mats having randomly oriented fibers on the collector plate. Some fibers may also be bent, twisted, or entangled. These irregularities are attributed to bending instabilities. Due to a high charge density on fiber surface and the presence of high electric field, the motion of fibers becomes arbitrary after a certain distance from the capillary [259]. The surface charge density further increases as the fibers become thinner, which causes them to move even more randomly. Some fibers also bend, owing to the stresses generated by interaction of the surface charge with the external field. The physical instabilities caused by an increase and rapid changes in electric forces acting upon the fibers are known as bending instabilities. It is very difficult to control the motion of the fibers when they are experiencing such an instability. It induces an irregular whipping motion in the fibers, resulting in their bending and twisting before they are collected.

As a result of bending instability, a range of possibilities in terms of jet initiation and fragmentation exist [223], which are manifested in fiber morphologies. The electrospun or sprayed geometries include fibers; droplets of different shapes (e. g., spherical, ellipsoidal); fibers with beads, etc.; fibers ejected at an angle, and length and diameter variations. Several theoretical investigations on electrospinning have been carried out for specific material, equipment, and process parameter needs [259, 146, 95].

There have also been certain efforts to minimize the effect of bending instabilities and obtain aligned fibers. The electrical conductivity of the material influences the charge accumulation patterns. One can tune the conductivity of the polymer solution, where possible, by adding salts. Another option is to simply collect the fiber before the onset of the instability. This means that the electrodes should be placed very close to each other. This type of electrospinning is known as near-field electrospinning or NFES, as described below.

> ⚠ Electrospun fibers are very rarely used in large-scale manufacturing applications, where the uniformity and mechanical properties of fibers are most crucial. They are rather popular in various research applications and in textile industry.

5.2.1.3 Near-field electrospinning (NFES)

In NFES, the collector electrode is placed on a movable precision stage and the distance between two electrodes is 500 μm–3 mm (before the onset of bending instabilities). The pattern or design is controlled by moving the stage, which is generally automated. Since the fiber jet is straight in this region, this technique can be used for direct-writing of patterns onto a substrate. NFES is an additive technique. It is similar to printing (particularly, FDM-type 3D printing) in some respects, except that an additional electrical force is used for drawing the material.

NFES has certain advantages over far-field electrospinning; for example, it requires lower voltages, the designs can be patterned with precision, simultaneous writing can be performed with more than one material, and the fibers can be stacked on top of each other for fabricating 3D geometries. Owing to these, NFES has been used for various biomedical, photonic, and sensor devices. However, the use of NFES for carbon fiber fabrication is limited. This is because the polymer jet in NFES is usually thick. For obtaining thin fibers before bending instabilities, one needs to use polymers with a high viscoelasticity (also see Section 5.4.1). Common polymers used for carbon fiber fabrication (i. e., those with a high carbon content) do not feature a very high viscoelasticity. In fact, they may undergo plastic deformation during spinning itself. There are some options, such as high-molecular weight polyacrylonitrile (PAN), that can be patterned via NFES. This technique has a lot of potential and due to its automation, it is also compatible with digital manufacturing. Other names used for NFES are electrohydrodynamic printing, direct-write electrospinning, and e-jet printing [169].

5.2.1.4 Collector modification

To improve the alignment of fibers, or for selectively collecting them in a pre-defined region, certain modifications can be made in the collector design. Electric field line between the two electrodes can be directed in such a way that fibers go to selected areas. One can also use rotating collectors, which help in aligning the fibers through mainly mechanical forces (pulling, stretching). Fibers can also be collected on a thin metal disc, where they tend to deposit vertically in stacks. The electrospinning set-up can be either vertical or horizontal. The advantage of a horizontal set-up is that it prevents any thick fibers and droplets from reaching the collector. All these variations are shown in Figure 5.5.

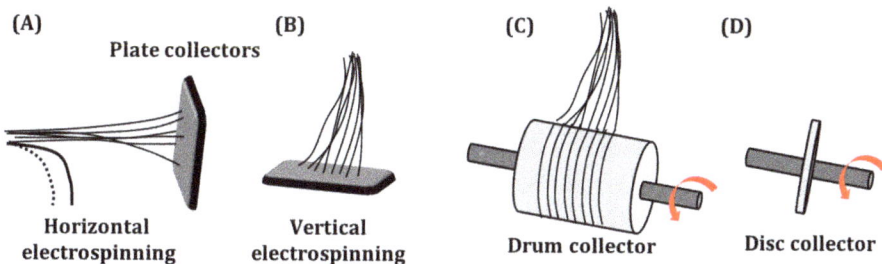

Figure 5.5: Different types of collectors used in electrospinning. (A) and (B) show plate type collectors for horizontal and vertical spinning set-ups. (C) and (D) are illustrations of drum and disc type collectors respectively.

5.2.2 Melt-spinning

Melt spinning is more commonly employed for microscale carbon fiber fabrication, especially for large-scale manufacturing applications. Most of the commercially available fibers are either vapor-grown (nanoscale) or prepared via melt spinning (microscale). Electrospun fibers are typically not sold in bulk powder form due to the lack of diameter uniformity. Melt spinning is a versatile technique that can be used for drawing fibers out of a range of materials, including ductile metals and alloys, thermoplastic polymers, petroleum pitches, glass, etc. The basic principles of melt spinning are described below.

Basic working principle: Drawing fibers out of low melting point hydrocarbons.

Materials used: Petroleum pitches, polymer melts, glass.

Instrument set-up: A typical melt-spinning set-up is shown in Figure 5.6. The top part of the set-up consists of an extruder, which is a temperature controlled chamber for containing the heated precursor. At the bottom of this chamber is a spinneret (top-view shown separately in the figure), which is a metal plate with holes or openings. The melted precursor is pushed through these openings with the help of a slight pressure. The fibers being extruded out of the spinneret are allowed to fall down by the gravitational force and are additionally cooled by air (known as quenching). This quenching

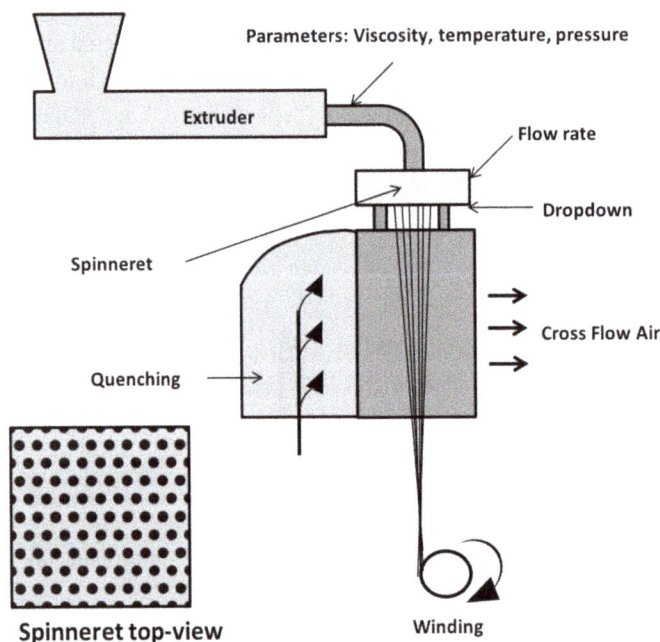

Figure 5.6: Schematic of a typical melt-spinning set-up. The spinneret (top-view shown in inset) contains microscale holes, which allow the melted pitch to pass through. Copyright ©1989, Elsevier Ltd. Adapted with permission from [58].

enables drying of the fibers, which are then collected onto winding wheels. Fiber diameters can be controlled by the design of the spinneret (diameter of the openings), temperature, pressure, and material's viscoelasticity. A similar set-up is used for obtaining fibers from melted glass, which is used for glass-fiber reinforced plastics (commonly known as fiberglass). In the case of glass, the extrusion chamber is typically a box made of platinum alloys, preinstalled with the openings in its bottom surface, known as *bushings*. Bushings and spinneret can have various designs, depending upon the precursor. For example, nozzles or simple circular holes.

5.2.3 Other spinning techniques

PAN can also be spun using wet-spinning and dry-jet wet spinning processes. Briefly, in wet spinning PAN fibers are passed through a container filled with water or other liquids (known as spin bath), which causes solidification of the fiber. For this purpose, the outlet of a needle connected to polymer-filled syringe on a syringe pump, is passed through the spin bath. Solidified fibers are subsequently collected and wound. In dry-jet wet spinning, fibers are initially dried by blowing a jet of air, and then passed through the spin bath. Several modifications in the composition of the liquids in the spin bath, such as addition of salts in water or use of organic solvents, have been carried out to further optimize these processes for carbon precursors. In addition to these techniques being more suitable for making microscale, or even thicker carbon fibers for composite preparation, they have also occasionally been used in device applications. One important advantage of wet spinning is the possibility of fabricating a single fiber, which is useful in certain sensor applications. Moreover, some biomedical implants and larger device components may utilize composites rather than direct fibers, where one may utilize carbon fibers prepared via wet-spinning.

5.3 Carbon fiber precursors

There are three primary precursors used for industrial carbon fiber manufacturing: PAN, petroleum pitch, and Rayon (processed cellulose). For small-scale or research applications, carbon fibers have been produced from spinning of various high carbon-containing polymers and their hybrids. Additionally, certain natural fibers, such as jute, wool, bamboo fibers, etc., can be directly heat-treated and converted into carbon fiber. Similar to other carbon materials, both spinning and carbonization parameters influence the fiber properties. The most crucial role is played by the chemistry of the precursor, which needs to be fine-tuned prior to carbonization. Depending upon the chemical bonding, aromaticity, and cross-linking abilities of the precursor, a suitable precarbonization stabilization step is generally practiced. Physicochemical aspects of selected precursors are discussed as follows:

5.3.1 Polyacrylonitrile (PAN)

PAN is a nitrogen-containing aliphatic polymer with the empirical formula $(C_3H_3N)n$. This is the most commonly used carbon fiber precursor owing to its high carbon content and good spinnability. PAN is compatible with wet spinning, melt spinning, and electrospinning. It comes in a powder form and can be mixed with solvents, such as N,N-dimethyl formamide (DMF) or dimethyl sulfoxide (DMSO), in different weight to volume ratios (typically between 7–13 %). This enables a control over solution viscosity, which in turn influences the fiber diameter. Usually polymers with a high aromatic fraction yield more graphitic carbon materials, but in the case of PAN this is not the case. In fact, during carbonization, aliphatic PAN molecule undergoes a three-dimensional network formation, which ultimately results in turbostratic carbon sheets with a high fraction of six membered rings. Electrospun PAN fibers have a soft cloth-like appearance. If the entire mat is converted into a carbon fiber mat after stabilization in air, it becomes relatively stiff. However, for many flexible device applications such mats can sustain thousands of bending cycles. A PAN fiber mat after stabilization and carbonization is shown in Figure 5.7 [204, 241].

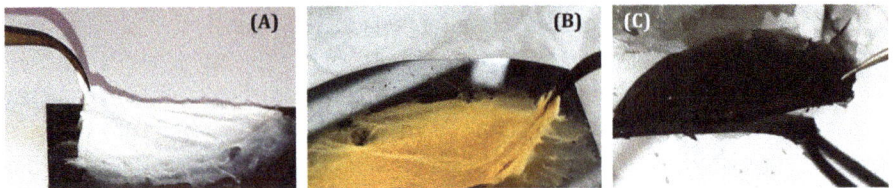

Figure 5.7: (A) Electrospun PAN fiber mat on a silicon. Same mat after stabilization in air is shown in (B) and after carbonization at 900 °C in (C). Copyright ©2019 John Wiley and Sons. Reprinted with permission from [241].

5.3.1.1 Electrospinning of PAN

Prior to electrospinning, a clear solution of PAN is prepared in DMF, which is typically 10 % weight by volume. Notably, certain DMF has lately been recognized as a potential human health hazard on long-term or continuous exposure. As a result, alternatives to DMF are being explored worldwide. Some examples include cyclohexanone, benzaldehyde, dimethyl isosorbide, γ-valerolactone, cyclopentyl methyl ether, and glycofurol, etc. Combinations of DMF with water and other solvents have also been utilized for making PAN solution. PAN in DMF, however, is still widely used at commercial scale. Typical electrospinning parameters for a 10 % solution of PAN are operating voltage: 10–15 kV, when the electrodes are at a distance of 10 cm at room temperature and 50 % humidity. In some cases, especially if the fibers are to be used for their mechanical performance, an additional mechanical stretching step may follow the spinning.

5.3.1.2 Stabilization and carbonization

Post spinning, PAN fibers are heated between 220–290 °C in air for a minimum 1 hour for the purpose of thermal oxidation, commonly known as stabilization. It has been proposed that three primary reactions—cyclization, dehydrogenation, and oxidation—take place during this process. Stabilization can be performed in an isothermal fashion, or by implementing a step-wise temperature increase, or by creating gradient temperature along a tubular furnace, and then passing the fibers through it. For research purposes, isothermal stabilization is most common, although other methods have also been used.

During this process, PAN undergoes a series of changes, leading to the formation of a ladder-like molecule [65], as shown in Figure 5.8. The first step is cyclization. At this stage, aliphatic PAN molecules fuse together to form cyclic intermediate compounds with or without the loss of water (dehydrogenation). There is some oxygen gain by the material during stabilization. Finally, long chains of cyclic intermediates are generated. When the fibers are subsequently heat-treated in an inert environment for carbonization, planar polyaromatics are formed by a sort of cross-linking between these oxidized cyclic fragments along with the loss of HCN and N_2. Above 700 °C, 2D polyaromatic fragments are

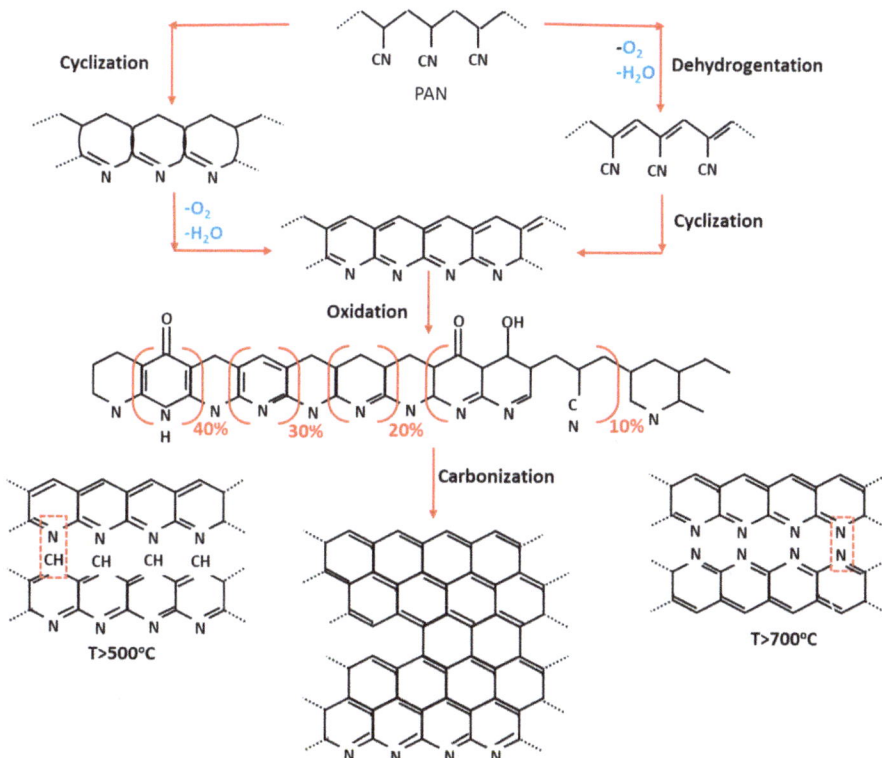

Figure 5.8: Formation of intermediates during stabilization and carbonization of PAN fibers. Copyright ©1986, Elsevier Ltd. Adapted with permission from Fitzer et al. [65].

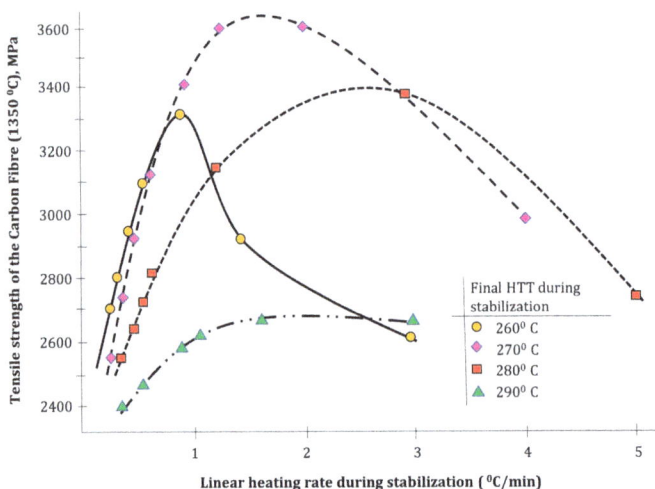

Figure 5.9: Effect of stabilization (thermal oxidation) temperature on tensile strength of carbon fibers prepared at 1350 °C of PAN fibers. Copyright ©1986, Elsevier Ltd. Adapted with permission from Fitzer et al. [65].

developed in the material that serves as the skeleton that ultimately converts into turbostratic carbon. PAN fibers are typically carbonized between 900 and 1400 °C. For most research applications the temperature is 900 °C. PAN-derived carbon fibers are known for their high ultimate tensile strength rather than modulus. A detailed description of mechanical aspects of various carbon fibers is provided in Section 5.5.2.

Stabilization temperature influences the tensile strength of the carbon fiber obtained afterwards, as plotted in Figure 5.9 [65]. Stabilization time can also affect the physicochemical properties of the resulting carbon fibers, such as their density, elemental composition, and electrical conductivity [233]. A longer stabilization enhances the ladder-molecule formation, therefore, the material becomes denser prior to carbonization. Though carbon prepared after extended stabilization times exhibits an improved layer stacking and slightly higher electrical conductivity, it features a reduced total carbon content. One should select the stabilization conditions as per the application requirements.

PAN-derived carbon is turbostratic, but it is hard to clearly identify it as a nongraphitizing carbon. In the case of fibers, the defects (non-six-membered rings, voids, etc.) are easily annealed out owing to their high surface area compared to solid carbons. This promotes a better stacking of carbon layers. Unlike other non-graphitizing carbons, there is a lower probability of closer porosity and fullerene-like structure in fibers. The possibility of containing defects and pores becomes even lower in the case of nanoscale fibers due to their very high available surface area. As a result, even though PAN-derived carbon fibers don't feature ABABA arrangement, their properties (especially mechanical) are reasonably different from other non-graphitizing carbons.

During the initial stabilization process, the chain-like molecules of PAN experience some stretching caused by dehydrogenation. Hence, stabilization parameters, such as temperature and stabilization rate, influence the tensile strength of resulting carbon fibers. One can select a suitable combination of stabilization conditions, as shown in Figure 5.9, for achieving a certain tensile strength in the fibers. It is important to note that PAN-derived carbon fibers are known for their tensile strength rather than Young's modulus. Their turbostrtic nature facilitates microstructural rearrangement when a load is applied. For example, the carbon sheets can stretch along fiber as a result of strain.

The change in tensile strength of PAN-derived carbon fibers with respect to temperature is also interesting. As it can be observed in the plots shown in Figure 5.10, the tensile strength of these fibers increases up to 1350 °C carbonization temperature. But above that, the tensile strength actually decreases. The Young's modulus continues to increase, but this increase is not very significant after 1350 °C. This can be attributed to the formation of discrete graphite crystallites at higher temperatures. Due to this behavior, PAN fibers are almost always carbonized below approximately 1350 °C when they are used in industrial manufacturing applications.

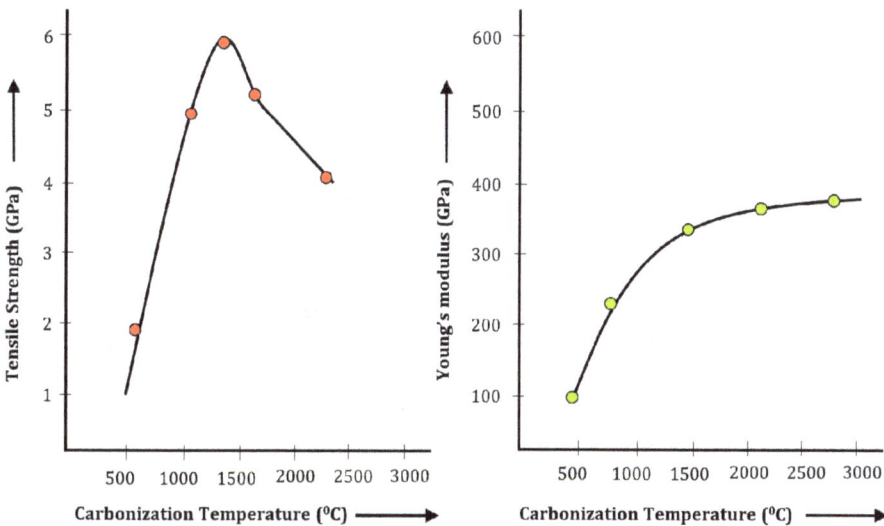

Figure 5.10: Ultimate tensile strength and Young's modulus of PAN derived carbon fibers with respect to their carbonization temperature. Note that fibers with up to 7 GPa tensile strength and 700 GPa modulus have been developed. This plot is drawn based on the availability of temperature-property correlation data. The strength and modulus numbers used here are only for the purpose of projecting the trend.

5.3.2 Pitch

Pitches are a complex mixture of polynuclear aromatic hydrocarbons (PAHs) that are naturally obtained during coal and petroleum processing. Petroleum pitches are obtained either naturally or generated during processing of coal or other petrochemicals in the refineries. At room temperature, their physical state can range from solid to semisolid or high viscosity liquids. Pitches carry a high fraction of carbon (in many cases, >90 %) along with oxygen, nitrogen, and sometimes sulphur.

Natural pitches are of two types: petroleum and coal-tar. In addition, pitches can be synthesized by polymerization of PAHs (e. g., naphthalene, PVC). Though they all have similar physicochemical characteristics, there are certain differences in terms of their origin and chemical composition. Petroleum pitches are produced during polymerization of aromatic decant oil, which is in turn generated as a byproduct of catalytic cracking of the heavy gas oils (components of crude oil) [14]. Coal tar pitches, on the other hand, are a byproduct of coal distillation, performed for preparing metallurgic cokes. Some coal-tar pitches contain a residual solid carbon fraction (generally in the form of particles), which are known as quinoline insolubles (QIs). Details of QI removal and other preprocessing techniques used for pitches are beyond the context of this book, but several articles and books are available for interested readers [14, 41]. In general, petroleum pitches are considered to be safer to work with, as coal-tar pitches may contain certain PAHs that pose health hazards.

Pitches are normally isotropic, i. e., they have no preferred orientation or stacking of PAHs. Consequently, the carbon fibers derived from this isotropic pitch have the same properties in all directions. In terms of microstructure, they are similar to nongraphitizing carbons. Isotropic pitch can be converted into what is known as mesophase pitch by heat-treatment. Mesophase pitch has anisotropic properties due to its liquid crystal-like structure, resulting from partial stacking and chain alignment of constituting PAHs. Mesophase pitch yields graphitizing carbon. The so-called graphite fibers available at industrial scale are mostly derived from mesophase pitch at temperatures >2500 °C. Increased graphitization leads to a higher Young's modulus.

The viscoelastic properties and low melting points of pitches make them compatible with melt-spinning, which remains the most common technique for preparing pitch fibers [49]. Similar to PAN, pitches are also stabilized in the presence of air between 200–400 °C (see Figure 5.11). The exact recipe differs based on pitch composition. Electrospinning of pitch and mixtures of pitch and PAN has also been reported. Such hybrid fibers may display a higher surface area and hybrid properties, which are useful in electrode fabrication. Pitch-derived carbon fibers have microscale diameters with a

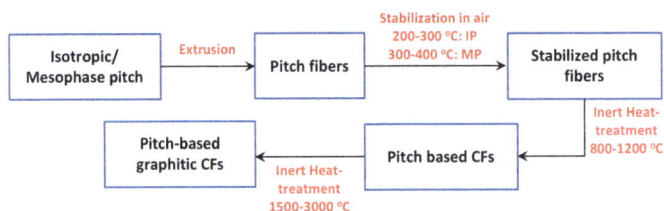

Figure 5.11: The process of stabilization and carbonization of pitch for carbon fiber manufacturing.

smooth surface and high uniformity. Although their primary application lies in composite (CFRP) manufacturing, they have been used as electrode materials for energy storage, adsorbers, and filters [41]. Some commercially available pitch-derived fibers have also been utilized in sensors and biomedical devices. Pitch fibers and related products are also used directly (without carbonization) in electrochemical applications.

> Mesocarbon microbeads to MCMBs is a particulate carbonaceous solid derived from coal-tar and petroleum pitches via heat-treatment. This material is highly graphitic, and is gaining popularity as a relatively inexpensive electrode material for batteries and supercapacitors. It is also used as a precursor for manufacturing high-density carbon in large quantities.

5.3.3 Rayon and natural fibers

Regenerated cellulose is called rayon or viscose rayon. It is prepared by treating natural cellulose-rich materials, such as wood with NaOH, CS_2, and other chemicals, if required. This processing generates a tarry-viscous intermediate material, which has the perfect properties for melt spinning. Spun fibers are then solidified in acidic media. Many variations of Rayon fibers featuring a range of diameters are widely available in the textile industry. Rayon can be prepared from agricultural and forestry waste, hence it can serve as an inexpensive carbon fiber precursor. Rayon-derived carbon is isotropic in nature owing to a random orientation of cellulose crystallites in the precursor itself.

Other than Rayon, various natural fibers, including processed bamboo, jute, wool, and keratin-based fibers can be carbonized. As a rule of thumb, natural materials undergo charring process during their carbonization. However, occasionally this may not hold true, especially if natural precursor is pretreated with chemicals. Moreover, in the case of fibers, charring process may be expressed differently than the solid carbon. Certain characteristics of the fibers after carbonization can provide cues on their carbonization mechanism. Cellulose-derived fibers, for example, have less smoothness in their surface compared to polymeric fibers that undergo coking.

During carbonization of cellulose, an intermediate material called levoglucosan is formed. Levoglucosan then disintegrates to either yield tarry products, or volatile hydrocarbons along with solid carbon. It has been observed that these different products do not convert into one another or mix during the heat-treatment. Nonetheless, solid carbon backbone of cellulose is replicated in the final carbon, thus confirming a charring mechanism. Notably, cellulosic does experience certain softening around 230–255 °C, but it is insufficient to change the morphology of the fiber or to be considered coking.

Another natural precursor that yields interesting carbon fibers is wool. Certain sheep wools are hollow, and the hollow carbon fibers derived from them tend to be extremely lightweight [190]. Waste and recycled wool have been used as a precursor in many such studies. Other than sheep, goat and rabbit hair have been carbonized and characterized. Most of them are either hollow or porous, and thus, applicable in electrochemical device fabrication [272].

5.3.4 Lignin

Lignins are polymeric hydrocarbons that make the rigid cell walls in wood and other plant materials. Chemically, they are primarily phenolic polymers with a variable degree of polymerization. It is well known that phenolic resins are good sources of carbon, but if the oxygen content is very high in the polymer, there is a possibility of carbon loss due to the formation of oxides. Lignins, however, can yield good carbon fibers if they are adequately processed and polymerized. For this purpose, powdered lignin is mixed in water or a suitable organic solvent followed by electrospinning. Lignins are not always able to covert into fibers due to lack of long chains in their chemical structure. As a result, they often yield droplets (as in electrospraying) rather than fibers. To address this problem, lignin mixture can be fractionalized for removal of low molecular weight components [191]. Besides, one can mix it with a carbonizing polymer or perform co-electrospinning. Lignin fibers are stabilized in the presence of air prior to carbonization between 190–250 °C for enhancing cross-linking and preventing fiber deformation during further heat-treatment. Although the exact chemical structure of lignin is dependent on its source (e. g., hardwood or softwood), one proposed structure is shown in Figure 5.12 [13]. Other than electrospinning, lignin fibers can also be prepared via melt spinning and wet spinning.

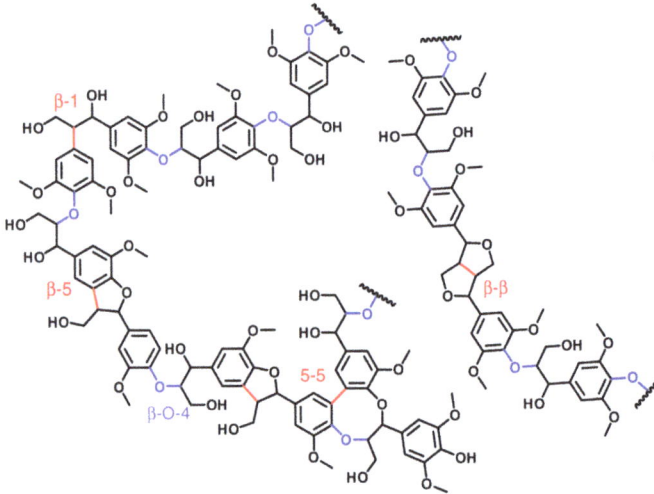

Figure 5.12: One possible chemical structure of lignin. Reproduced from [13].

Morphology of lignin-derived carbon fibers is influenced by the lignin concentration in the solution and the presence of co-polymers, if any. Generally, they feature a smooth surface, and they undergo a coking process. They can be further activated for increasing the surface area. Compared to PAN-derived fibers, lignin-based fibers are much cheaper and show comparable mechanical and electrical properties. However, extensive extraction and pretreatment processes, as well as the presence of some toxic chemicals produce during processing, render lignin-derived carbon fibers less popular. With a growing emphasis on green chemistry, the interest in lignin have been revived, and several advancements in term of processing are being introduced. Blending lignin with other polymer precursors, such as PAN, poly-ethylene oxide, polypropylene, poly-lactic acid, etc., leads to an improved spinnability as well as a higher tensile strength [40].

In addition to aforementioned precursors, synthetic phenol-formaldehyde resins, such as SU8, have also been electrospun and carbonized [209]. Lignin contains furan rings and can also serve as a co-polymer with furan resins. These lignin-based furan resins are being investigated as an inexpensive and green carbon source [273]. Besides, polyimide, carbohydrate, poly(p-xylene tetrahydroth-iophenium chloride) (PXTC), PAN with a range of additives and co-polymers, and various recycled and waste polymers have been utilized as carbon precursors for electrospun fibers [170].

5.4 Precursor requirements

Electrospinning is used for making a range of polymer fibers, but not all of them yield a good quality carbon. When selecting a precursor material, one needs to ensure a good spinnability and a good carbon yield. Accordingly, all process parameters need to be optimized. Some desirable properties of a carbon fiber precursor are discussed here.

5.4.1 Viscoelasticity

If a material features both viscosity (a property of liquids) and elasticity (property of solids), it is known as a viscoelastic material; this property is called viscoelasticity. For a polymer to convert into thin and long fibers, viscoelasticity is an essential property. Most polymers contain either (i) partially stacked sheets or fragments (Figure 5.13(A)) or (ii) long and partially entangled chains (Figure 5.13(B)), along with a semidilute region with floating (unentangled) chains or fragments in a solvent or liquid medium. The stacked fragments and entangles chains behave like an elastic solid, while the remaining part of the polymer (floating fragments, chains or solvent, if any) exhibits the properties of a viscous liquid. The overlapping and entangled chains are essential for obtaining fibers. In their absence the polymer converts into droplets, i. e., gets electrosprayed [6]. For electrospinning to take place, there is a certain threshold value of polymer concentration that is required for the formation of solid-like (elastic) species in the polymer.

There are various ways to mathematically represent viscoelasticity. In mathematical model, the elastic component can be denoted by a spring (using a linear stress-strain relationship) and the viscous component using a dashpot. The two components can then

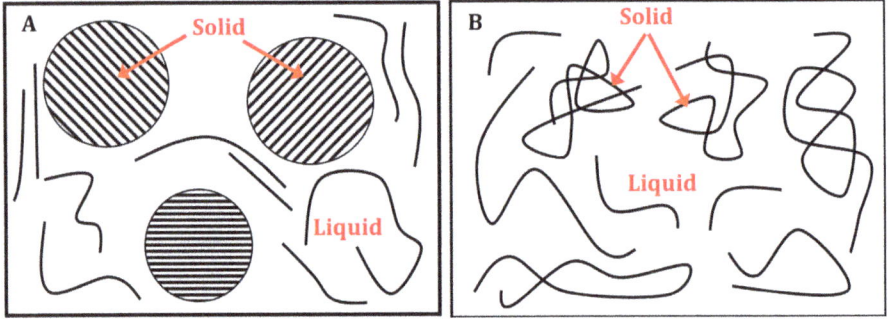

Figure 5.13: Viscous and elastic components in viscoelastic polymers. (A) A semicrystalline polymer with well-defined crystalline and amorphous regions. (B) A polymer composed of entangled chains forming small elastic regions or micro-lumps, while the remaining surrounding material with floating chains or solvent behaves like a true liquid.

be arranged in series, parallel, or a combination of the two to achieve the properties that most closely define the characteristics of a given system. One example of a mathematical representation of viscoelastic polymers, known as the Maxwell model, is described here. Subscripts s and d denote spring and dashpot, respectively.

$$\sigma_s = E\varepsilon_s,$$

and

$$\sigma_d = \eta \frac{d\varepsilon_d}{dt},$$

where E is the Young's modulus, σ and ε are stress and strain, and η is the viscosity constant. If the spring and dashpot are connected in series, the stress will be same in all components, whereas the total strain will be the addition of the strains in individual components. This means,

$$\sigma_{\text{tot}} = \sigma_s = \sigma_d,$$

and

$$\varepsilon_{\text{tot}} = \varepsilon_s + \varepsilon_d.$$

From the two relationships above, one can derive the following equation:

$$\frac{d\varepsilon_{\text{tot}}}{dt} = \frac{1}{E}\frac{d\sigma_s}{dt} + \frac{\sigma_d}{\eta}. \tag{5.1}$$

Equation (5.1) is known as the Maxwell's model of viscoelasticity. Other commonly used models are Kelvin–Voigt and standard linear viscoelastic (SLS) models, where the spring and dashpot are connected in parallel (Kelvin–Voigt) or as a combination of series and parallel components (SLS).

From here one can understand that a high molecular weight polymer will easily form tangled shapes, and hence feature a higher elasticity. Such polymers indeed yield longer fibers, and researchers have come up with high molecular weight PAN. These materials are yet to be commercialized.

5.4.2 Carbonizability

Carbonizability can be defined as the ability of any polymeric hydrocarbon to yield both high quality and quantity of carbon. Carbon quality is heat-treatment temperature-dependent and is measured in terms of crystallinity in most cases. Quantity, on the other hand, is simply a measure of carbon yield (the weight % value of carbon obtained after carbonization). As a rule of thumb, polymers should have a carbon backbone and a possibility of aromatic structure formation during their heat-treatment. A low or easily removable oxygen content is also a general requirement in the case of all carbon precursors, otherwise it leads to the formation of oxides (CO_2 and CO) at higher temperatures.

To improve carbonizability of a given polymer, a high fraction of cross-linking and elimination of any volatile chemical entities prior to carbonization is essential. This is the primary reason for performing stabilization, which is more common for carbon fibers rather than solid carbons. This is due to the fact that release of volatiles and polymer chain stretching is more convenient in fibers owing to their high surface area.

It is clear that a rigid carbon backbone with plenty of aromatic chains or sheets in the polymer make it more carbonizable. But such chemical structures, particularly partially stacked ones, can also make the polymer less ductile. A perfect spinnable polymer is the one that features a high ductility and optimum viscoelasticity during its spinning, but can acquire aromaticity through stabilization, cross-linking or other pretreatment pathways. One reason for the popularity of PAN in carbon fiber fabrication is indeed the fact that it is aliphatic during spinning, but semiaromatic already during initial stages of carbonization.

5.5 Properties of carbon fibers

5.5.1 Electrical conductivity

Electrical resistance, R (Ω), of a single carbon fiber is calculated as follows:

$$R = \rho \frac{L}{\pi r^2},$$

(5.2)

where ρ (Ω m) is the specific resistivity of the material, L (m) is the length of the fiber, and r (m) is the radius of the fibre. The total differential of R as a function of ρ, L, and r

can then be written as

$$dR = \frac{\partial R}{\partial \rho} d\rho + \frac{\partial R}{\partial L} dL + \frac{\partial R}{\partial r} dr. \tag{5.3}$$

From equations (5.2) and (5.3), we get

$$dR = \frac{L}{\pi r^2} d\rho + \frac{\rho}{\pi r^2} dL - \frac{2\rho L}{\pi r^3} dr. \tag{5.4}$$

Accordingly, any change in resistance of a carbon fiber due to the influence of an external parameter, for example in a sensor device, can then be calculated as follows:

$$\frac{dR}{R_0} = \frac{d\rho}{\rho} + \varepsilon(1 + 2v), \tag{5.5}$$

where ε is the strain, and v is the Poisson ratio. In (5.5), the first term $\frac{d\rho}{\rho}$ is due to material's inherent properties, whereas the term $\varepsilon(1 + 2v)$ is dependent on fiber geometry [103]. This equation is used in various sensor applications as well as for correlating mechanical and electrical properties of carbon fibers. In electrochemical measurements (AC circuits), impedance (Z) is measured instead of resistance (typically DC circuits). When the phase angle is 0, $R = Z$.

Similar to other polymeric carbons, the heat-treatment temperature influences all material properties of the fiber. Graphitic carbon fibers, of course, feature a lower electrical resistance compared to turbostratic ones. Presence of defects also plays an important role in determining the properties of the fibers. Typical electrical conductivity values for carbon fibers produced under different conditions are provided in Table 5.1.

5.5.2 Mechanical properties

When the tensile strength is being calculated for a bundle of fibers, one needs to consider the total surface area of the bundle and count the number of fibers. The cross-sectional area and diameter of the bundle can be measured using a planimeter, whereas the fiber counting can be done through optical microscopy or electron microscopy images. Various image processing software are available for assistance in counting.

The standard method of evaluating mechanical properties, particularly the tensile strength and Young's modulus of the fibers is described in the ASTM standard D3544-76. It is difficult to directly load fibers in a universal testing machine (UTM), due to their small size. Therefore, a special slot, made of cardboard or wood with clips for holding the fibers (as shown in Figure 5.14) is employed for placing the sample. According to this method, the Young's modulus of a fiber, E, is calculated as follows:

$$E = \frac{L}{C \times A},$$

Table 5.1: Electrical conductivity of carbon fibers prepared under different conditions. PAN: polyacrylonitrile.

S. No	Precursor	Fiber diameter (μm)	Carbonization temperature (°C)	Electrical conductivity (S/m)	Ref.
1	Mesophase pitch	13.942	1000	7.04×10^5	[269]
2	PAN	0.15 ± 0.05	2200	7.59×10^3	[86]
3	PAN	0.24 ± 0.1	1800	5.1×10^3	[86]
4	PAN	0.45 ± 0.1	1000	5.32×10^2	[86]
5	PAN	5.3 ± 0.1	2400	1.14×10^5	[193]
6	Pyrolyzed polydopamine	120–150	1200	2.1×10^5	[268]
7	PAN + coal	0.262 ± 0.1	800	0.541	[119]
8	SU-8	0.198 ± 0.01	900	1.84×10^3	[201]
9	Coal extracted tar	0.5 ± 0.1	1000	1.57×10^3	[151]
10	PAN + tar	0.2 ± 0.1	1000	1.726×10^3	[151]
11	Rayon	500 ± 1	2600	1.0×10^5	[129]
12	PAN	300 ± 1	2400	1.4×10^5	[129]
13	Pitch	17.2 ± 0.1	1000	6.0×10^4	[182]
14	Cellulose	–	800	2.53×10^4	[78]
15	Lignocellulose	–	900	6.6×10^3	[248]

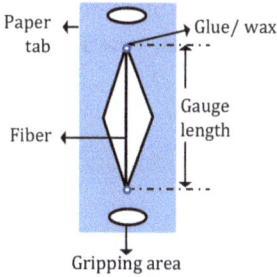

Figure 5.14: Sample holder attachment used for measuring mechanical properties of carbon fibers on a universal testing machine. Copyright ©2012 The authors. Reproduced from [102].

where,

$$C = C_s - C_a,$$

and the indicated compliance,

$$C_a = \frac{I \times H}{F \times S}.$$

C_a is calculated from the load-displacement curve, whereas C_s is the system compliance, which is taken as the y-intercept of the C_a vs specimen length graph. A typical load-

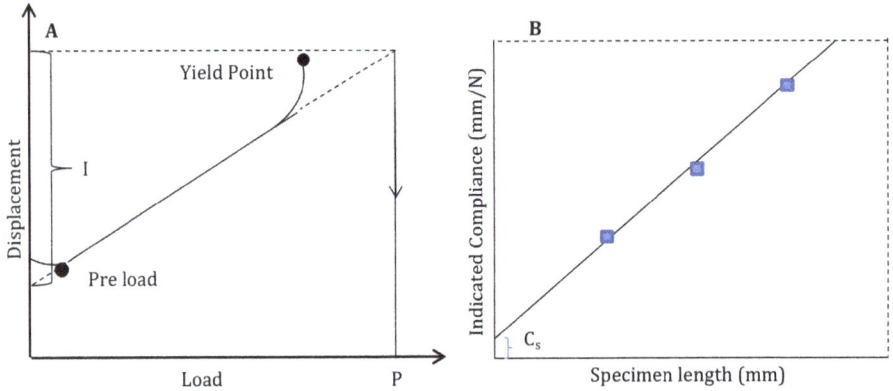

Figure 5.15: (A) Typical load-displacement curve, and (B) plot showing indicated compliance with respect to fiber length for carbon fiber samples analyzed on a UTM in a sample holder similar to that in Figure 5.14. Data based on ASTM D3379-75 standard. Copyright ©2012 The authors. Reproduced from [102].

displacement curve and the plot of C_a with respect to the specimen length are shown in Figure 5.15. Other symbols are described as follows: L = gauge length, mm; C = true compliance, mm/N; A = average cross-sectional area of the specimen, mm^2; C_s = system compliance; C_a = indicated compliance; I = full-scale elongation, mm; F = full-scale load, N; H = cross-head speed, mm/min; S = chart speed, mm/min. Ultimate tensile strength is simply calculated as a ratio of failure load and cross-sectional area. The area is calculated using the following relationship:

$$A = \frac{\Sigma a_f \times 10^{-6}}{N \times M_f^2},$$

where a_fs are the diameters of individual fibers in millimeter, N is the number of fibers in the bundle, and M_f is the magnification factor of the image used for fiber counting.

If there is a temperature variation during the measurements, stretching of fibers at high temperatures may align the carbon sheets and chains along the fiber axis and lead to an increase in the stiffness (Young's modulus) and tensile strength of the fibers. As a general rule, turbostratic carbon fibers have high value of tensile strength, and graphitic carbon fibers have higher Young's modulus (higher stiffness).

Problem: Calculate the Young's modulus of a carbon fiber bundle if it contains 10 fibers of diameters 0.020, 0.020, 0.015, 0.030, 0.045, 0.025, 0.050, 0.020, 0.030, 0.040 mm. Take the true compliance of the system as 0.25, magnification as 40, and gauge length as 30 mm.
Solution: We know that,

$$A = \frac{\Sigma a_f \times 10^{-6}}{N \times M_f^2}.$$

Accordingly,

$$A = \frac{(0.020 + 0.020 + 0.015 + 0.030 + 0.045 + 0.025 + 0.050 + 0.020 + 0.030 + 0.040) \times 10^{-6}}{10 \times 40^2}$$
$$= 1.84 \times 10^{-11} \, m^2.$$

It is given that $L = 30 \, mm$, and $C = 0.25 \, mm/N$.
Young's modulus $E = L/CA$.
Therefore,

$$E = 30/(0.25 \times 1.84 \times 10^{-11}) = 6.5217 \times 10^{12} \, Pa = 6521.7 \, GPa$$

5.5.3 Thermal conductivity

There are a number of experimental methods used for measuring the thermal conductivity of a material that include steady-state, time-domain, and frequency-domain techniques. In the case of carbon fibers, experimental measurements of thermal conductivity of individual fibers or bundles can be cumbersome. Hence, the following general formula can be used for this purpose:

$$\lambda = \frac{1261}{\rho}, \tag{5.6}$$

where λ $(Wm^{-1}K^{-1})$ is the thermal conductivity, and ρ is the resistivity. Equation (5.6) is based on the assumption that the product of ρ and λ remains constant at the given temperature.

5.6 Classification of carbon fiber

The most common way to classify carbon fibers is based on their manufacturing process: bottom-up or top-down. One can also classify these fibers according to their microstructure, tensile strength, or electrical conductivity. In some cases, the fibers are postprocessed, for example, chemically activated, which can be used for differentiating them from the untreated fibers. Some common classification pathways are discussed below.

5.6.1 Classification based on manufacturing process

There are two primary processes for making carbon fibers: (i) CVD (bottom-up) and (ii) spinning of polymeric hydrocarbons (top-down), followed by heat-treatment. Spinning is further divided into electrospinning and melt-spinning. In the CVD process, a suitable light hydrocarbon (e. g. methane, acetylene, alcohol) is pyrolyzed at temperatures generally above 900 °C, which leads to its cracking or disintegration. This results in the formation of some carbon species, which are present in the resulting vapors or smoke. Similar to other nanomaterials, these carbon species are subsequently collected onto catalyst substrate. Fibers prepared this way are known as VGCFs (see Chapter 4 for details).

In the top-down method, heavy hydrocarbons (polymers, melts, petroleum pitches) or their mixtures are spun using either electrospinning or melt spinning. These spun fibers are then carbonized via heat-treatment, which typically involves pyrolysis and carbonization steps. In some cases, the fibers are converted into polycrystalline graphite. Such a heat-treatment entails temperatures >2000 °C. In all cases, an inert or reducing environment is maintained, which protects the fibers from oxidizing. Figure 5.16 is a schematic representation of different processes used for carbon fiber manufacturing, which are also used for comparing and classifying the fiber type.

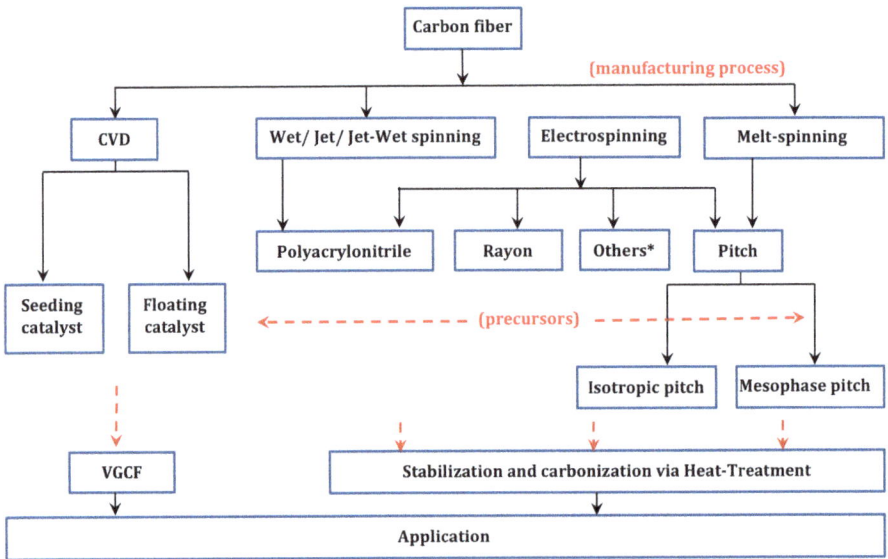

Figure 5.16: Manufacturing pathways for carbon fibers using common precursors. Precursors for VGCFs are gaseous hydrocarbons.
Acronyms: CVD: Chemical vapor deposition; VGCF: vapor-grown carbon fiber.
*Other precursors include, but are not limited to, lignin, SU8, wool, jute, and various natural and synthetic fibers. Only the most common precursors are shown in the illustration.

5.6.2 Classification based on microstructure

Similar to other carbon materials, carbon fibers display a varying graphitic content, depending upon their fabrication conditions, which results in different microstructures. Based on their microstructure, carbon fibers are divided into two main categories: (i) turbostratic fibers (generally carbonized <1500 °C), and (ii) graphite fibers (carbonized >2000 °C). In some research articles, the term amorphous carbon fiber has been used to denote highly disordered fibers. Turbostratic fibers have crystallite sizes of <5 nm, and contain several point defects that may be annealed out at high heat-treatment temperatures. Similar to other carbon materials, their interlayer spacing is >0.335 nm, which can be characterized via XRD and TEM.

> Similar to other solid carbon materials, microstructural models for carbon fibers have also been sug- **!**
> gested. Most models indicate the development of a fibril-like morphology. In the case of melt-spun fibers,
> the microstructure is influenced by the size and shape of the holes in spinneret, in addition to carboniza-
> tion temperature and fiber diameter.

5.6.3 Classification based on physicochemical properties

Carbon fibers are electrical conductors in most cases. The conductivity, however, may greatly vary depending upon the graphitic content and fiber diameter. In some cases, the fibers are highly disordered, and hence are very poor conductors. Such fibers are generally used for their mechanical or surface properties, rather than electrical or electrochemical. The electrical conductivity of carbon fibers is typically within the range of 10^3–10^5 S/m [209]. Fibers are also called *wires* in some electrical devices when a current passes through them.

Electrical and electrochemical properties are often used together in electrochemical devices. Carbon's conductivity is much lower compared to metals that are traditionally used for making device components. However, the electrochemical performance of carbon and the possibility of obtaining high surface area structures, such as fibers, gives carbon an advantage over metals. Like other carbon materials, carbon fibers feature a mostly inert surface. They can be activated via physical or chemical activation methods if needed. The electrochemical properties are associated with any material's response within an environment that can induce the transfer of electrons within the material. This is performed in the presence of current flow and often beyond a certain value of current, the material starts to change its chemical properties, or in other words, it becomes unstable. The window in which the material can be safely used is known as the electrochemical stability window. For carbon fibers this window is wider than that of common metals. They also have a high electrochemical surface area, which makes them highly suitable for electrode applications in, for example, Li-ion batteries [156].

Industrial carbon fibers designed for manufacturing applications are classified or "graded" based on their mechanical properties. Two important parameters, the elastic (Young's) modulus and the UTS, are used for representing the mechanical aspects of individual and bulk carbon fibers. The elastic modulus indicates the slope of the stress-strain plot in the elastic region of the materials. Ultimate tensile strength, on the other hand, is the measure of load the fibers can sustain prior to fracture. These properties can be measured for a single strand of fiber, or for a bundle of fibers, depending upon the process requirements.

Commercially available carbon fibers are available mainly in two grades: high performance (HP) and general purpose (GP). HP fibers have high UTS and high Young's modulus. GP grade fibers have low UTS and low modulus. Typically, HP grade fibers feature a UTS >4000 MPa and a modulus >400 GPa. GP fibers have their UTS and modulus in the range of 200–1000 MPa and 50–100 GPa, respectively. In addition to these grades, many other combinations of UTS and modulus are plausible. Based on them, fibers can be designated high strength type, high modulus type, intermediate strength/modulus type, and in some cases, high strain withstanding type, as shown in Figure 5.17. Generally, VGCFs have a much higher strength and modulus compared to spun fibers. They predominantly exhibit a nanoscale diameter. Thinner fibers are mechanically stronger, because they cannot contain large defects in them. VGCFs are classified as ultrahigh-strength and ultrahigh-modulus type (not shown in Figure 5.17).

It has been reported that the tensile strength of individual VGCFs can reach up to 30 GPa. This is due to a high graphitic content. Unlike spun fibers, vapor-grown fibers are prepared bottom-up. They carry much less defects and impurities. They are grown atom-by-atom, which allows a sufficient time for their ordering and stacking during fab-

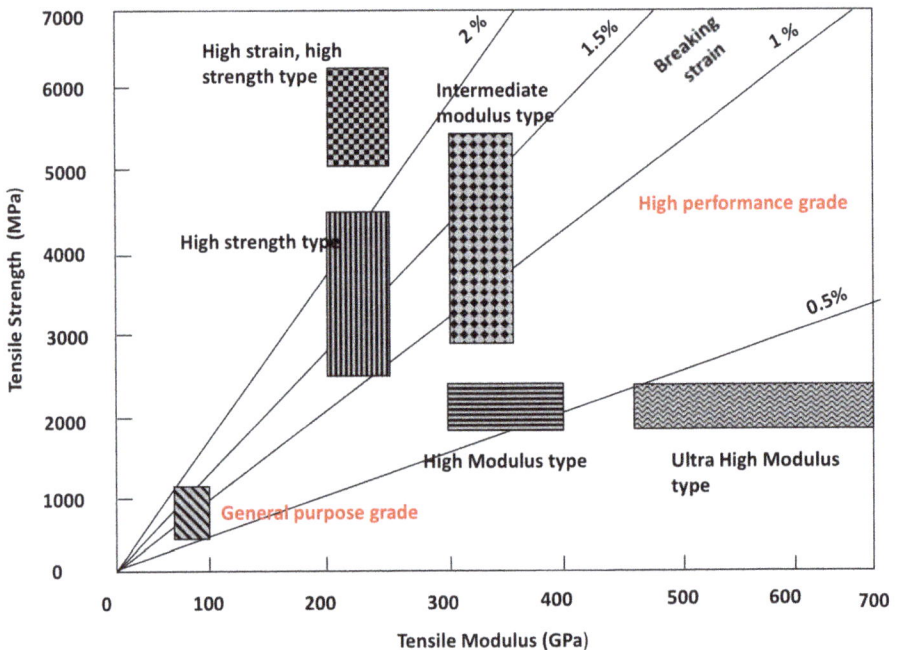

Figure 5.17: Classification of carbon fiber based on mechanical properties. Copyright ©2000 Elsevier B.V. Adapted with permission from [109].

rication. But since they are relatively short, their use in manufacturing applications, particularly in woven form, is limited.

> Carbon fibers can be activated using physical or chemical activation methods similar to porous carbons. Activated carbon fibers are extremely high surface area materials used in adsorbent and filters. **!**

5.7 Carbon fibers in devices

Carbon fibers have been used in a wide range of electrochemical energy storage and sensor devices, wearable flexible electronics, smart textiles, and bioelectronic medicine [36, 37, 216, 241]. Though their electrical conductivity is lower than most metals, graphene and carbon nanotubes, they compensate for it with their tensile strength and binder-free patternability, which enables their use in a pristine form. Their electrochemical stability (typical value for fiber mats prepared around 900 °C: $-0.9V - +1.1V$ [241]) is much better than metals and comparable with other polymeric carbons (for a detailed comparison, see Chapter 2). Like other carbon materials, surface of carbon fibers is inert, unless intentionally activated. This renders them biocompatible, especially for *in vitro* applications. Some *in vivo* carbon implants, such as tissue growth templates, have also been developed and tested. They are excellent supports that promote the tissue growth, however, their disintegration inside the body is in the form of fragments, which may not naturally excrete. In smart textiles, carbon fibers are obviously popular, as they utilize the fabrication principles of textiles, but feature a very important property: electrical conductivity. Carbon fiber fabrics or patches can serve as active sensors in clothing.

Biomedical devices may also entail cell culture or other forms of interaction with biological tissue. In such cases, the high surface area fiber mats can also serve as a cell culture scaffold. In the case of membranes for pollution control, equally cut carbon fiber mats can be arranged in a layered fashion. There have also been reports of using the mat in a crumpled form, connected with epoxy resins to electrical circuitry. Here the main techniques of carbon fiber patterning, especially those demanding complex microfabrication, are detailed.

Unlike printable carbon nanomaterials or lithographically patterned carbon, carbon fiber can be embodied in devices in many different ways. One can perform sensing utilizing only the tips of the fibers (single or bundles), or pattern a non-woven electrospun mat using lithographic techniques. Mats or stacks of mats can be used directly as electrodes, without any patterning. Yes another option is to use woven mats, for example, in functional tissue implants. VGCFs or chopped spun fibers can be printed onto various substrates similar to carbon nanomaterials. Some of these pathways are shown in Figure.

After preparation of fibers, surface treatment and activation can be performed, if needed, prior to further processing. Fibers already exhibit a high surface area compared to bulk forms of carbon. If microporosity is introduced, these materials become an excellent choice for electrochemical devices. Surface functional groups also enhance their wettability.

5.7.1 Tip-sensitive carbon fiber microelectrodes

Tip-sensitive microelectrodes are the ones that utilize trimmed and/or sharpened tips of carbon fibers for sensing purposes. These electrodes are common in chemical and biosensing. Both single and multimodal sensing can be performed with such electrodes. Among these, glass capillary-type electrodes are most common, and even commercially available. Combining the fundamental concepts of glass capillary electrodes with advanced microfabrication techniques enables chip-type carbon fiber electrodes that can be easily connected to printed circuit boards (PCBs). Both of these pathways are explained below.

5.7.1.1 Carbon fiber in glass capillary

Glass capillary microelectrodes are the most common type of carbon electrodes used in biosensing. To fabricate them, individual filaments (or sometimes filament bundles) are inserted inside borosilicate glass capillaries having a typical outer diameter of 1.5–2.0 mm. During fiber insertion, a fast evaporating liquid, such acetone, is filled inside the capillary. This helps the fiber filling using capillary action and also serves as a lubricant. One end of the capillary is then pulled via heat-assisted micropipette pulling methods such that some part of fiber protrudes out. Due to their high thermal resistance and good tensile strength, fibers generally don't break and can be connected to the electronic chip for recording through the other end of the capillary with the help of low vacuum. Depending upon the device requirements, single or multiple capillaries can be used in one electrode, as shown in Figure 5.18.

The next step is to make electrical contact with the fibers at the other end of the capillary for spark etching purposes. This must be done with utmost care, as a poor contact can lead to high impedance and electrode failure. Once a good electrical contact is established, fiber tips are sharpened by electric spark etching. Chemical and electrochemical etching can also be used. After achieving desirable fibers, a permanent electrical contact is finally made with the help of silver paste and/or placing metal wires along the length of the fiber. A more detailed stepwise description of glass capillary microelectrode fabrication is provided by Miller and Pelling [158].

These microelectrodes can be used directly (unmodified) or with a thin coat of polymer onto the fiber surface. Application of specific surface coatings or even fiber decoration with other nanomaterials have also been reported. This electrode design is highly suitable in sensors that are designed for detecting biochemicals. Multibarrel type sensors can also perform stimulation of the cells, followed by signal recording (e. g. in neural senors) at the same location.

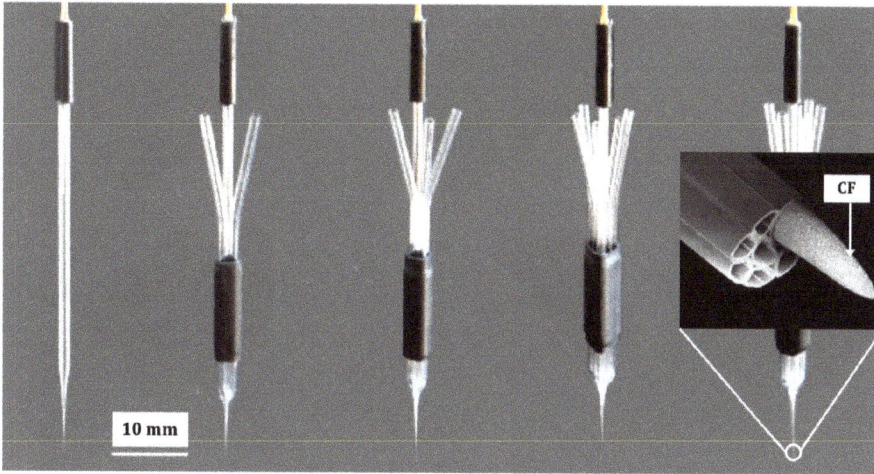

Figure 5.18: Carbon fiber microelectrodes prepared by inserting fibers in glass capillaries. First image shows a single-barrel microelectrode, and subsequent images show an increasing number of barrels. In the last image, a magnified view of the capillary tip containing one carbon fiber is presented in the inset. Image courtesy Dénes Budai, Kation Scientific.

5.7.1.2 Microfabricated tip-sensitive electrodes

Mounting carbon fibers along metal traces of customized PCBs is yet another way of using carbon fiber tips as electrodes [130]. The general idea is to manually place carbon (micro)fibers along the metal, and then cover this region with an epoxy for insulation. Prior to placing, fibers are often coated with a thin layer of a biocompatible polymer(s), depending upon the requirements of the device. Tips of the fibers are exposed via chemical or electrochemical etching of polymers at the end. In addition to insulation, epoxy (or any cross-linkable insulating polymer) helps in preventing the fiber movement and ensuring their contact with the metal. Gold and silver are commonly used metals in such designs, as they are compatible with PCBs and also commonly available. Using this manufacturing scheme, Patel et al. [184] fabricated multimodal carbon fiber electrodes with 16 individual carbon fibers on to a PCB (fabrication steps shown in Figure 5.19). These microelectrodes have been employed in implantable neural interfaces, but can potentially be applied to various sensing platforms.

In biological applications, a reliable performance and stability of the electrode material within the harsh biological environment becomes an essential factor. Several sensors utilize noble metals, such as gold and platinum, in biosensor devices. The advantages of noble metals include a high electrical conductivity and possibility of patterning thin films (few nanometer thickness). These metals are generally corrosion-resistant, but the pH of biological fluids and constant interaction with biological tissue leads to their early corrosion and limits their performance. This is where a good electrochemical stability in both acidic and basic media becomes the most crucial advantage of carbon

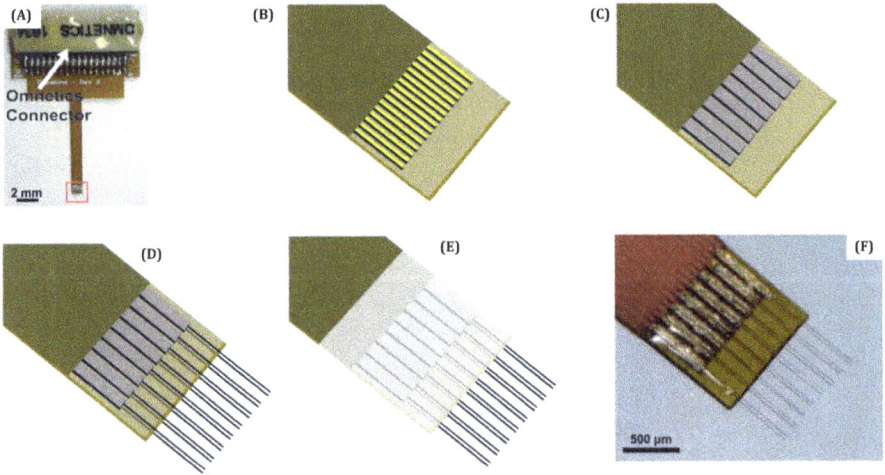

Figure 5.19: Fabrication steps involved in flexible carbon fiber microelectrode array. (A) Layout of a fully populated flex array with Omnetics connector at the top and carbon fibers at the bottom (red box). (B) Illustration of the bare gold traces at the bottom of an unpopulated flex array. (C) Deposition of silver epoxy between every other trace creates a conductive well for the carbon fibers to rest within. (D) Hand placement of the carbon fibers on both sides of the flex array after which the silver epoxy is oven cured. (E) The exposed trace-silver epoxy-carbon fiber bonds are covered with a UV-cured insulating epoxy. (F) Image of a fully populated flex array with 16 carbon fibers, 8 on each side, at a pitch of 132 μm. ©2020 IOP Publishing Ltd. Reproduced with permission from Patel et al. [184].

materials. To minimize the influence of the biological environment, sometimes protective coatings are applied on to fibers. For example, in the case of an implantable neural sensor, a polymer coating for protein protection is necessary.

5.7.2 Surface-sensitive carbon fiber electrodes

If the entire surface of the fibers are used as the sensing element, the sensing electrode is denoted as a surface-sensitive electrode. In solid carbons, such as glass-like or activated carbon, it is always the entire exposed surface that is functional as an electrode, but using all available surfaces in fibers or tubes is not straightforward. This is because of their random orientation, entanglement and the presence of binders, if any. However, if a mat of pristine fibers (typically nonwoven) can be patterned into an electrode shapes, most of the available surface area of the fibers can be utilized for sensing.

5.7.2.1 Surface sensitive mats

Vomero et al. reported a combination of microfabrication processes to fabricate such electrodes. First a PAN fiber mat was prepared by electrospinning and carbonization. This mat was placed on to a silicon wafer and embedded in polyimide to result in a car-

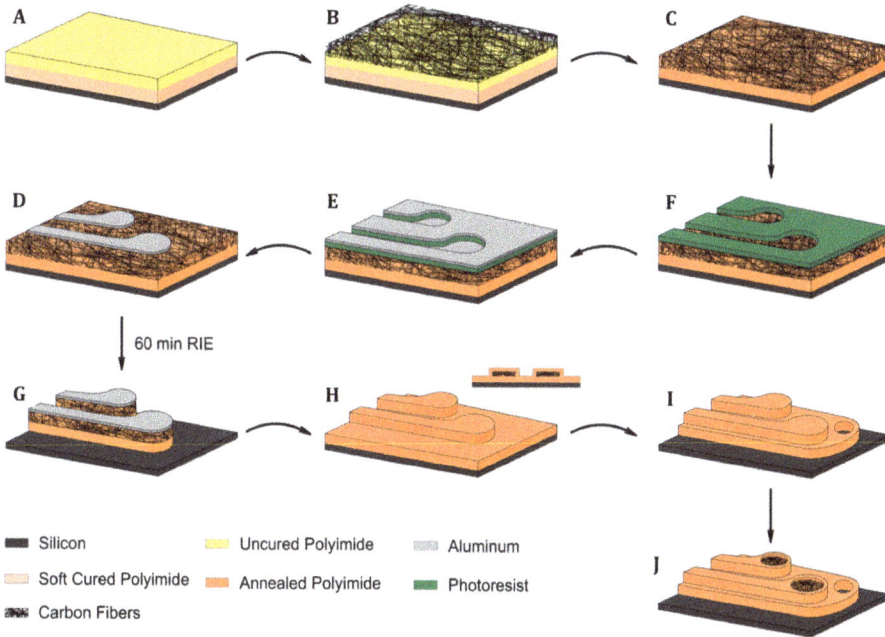

Figure 5.20: Fabrication protocol for carbon fiber based neural probes reported by Vomero et al. (A–C) Embedding electrospun carbon fiber mat into soft-cured/ uncured polyimide followed by curing, (D–F) patterning aluminum mask (positive design) onto carbon fiber/ polyimide surface via resist lift-off, (G) patterns after reactive ion etching, (H) aluminum mask removal and insulation with polyimide (see inset for cross-section). Creation of (I) device outline and fenestrations, and (J) recording site openings by lithographic masking followed by reactive ion etch of polyimide and resist removal. Copyright ©2019 John Wiley and Sons. Reproduced with permission from [241].

bon fiber / polyimide film-like geometry. This film was then patterned using lithography and etching, as shown in Figure 5.20. Briefly, a metal mask was first fabricated using lift-off process (photolithography of opposite pattern, sputtering of metal, removal of resist). The entire film was subsequently etched through deep reactive ion etching (DRIE) such that the electrode patterns under the metal remain unetched. Now the metal was removed, and the patterns were insulated with polyimide on top, leaving only the recording sites exposed. These carbon fiber / polyimide electrodes were connected to the PCB and used as implantable neural electrodes.

5.7.2.2 Surface sensitive single fibers
There are some microfabrication techniques where individual fibers have been used in devices or device components. In one such process, a single polymer fiber was electrospun on top of photolithographically patterned shapes, and this entire assembly for carbonization in one step [206]. This single fiber can be selectively heated via Joule heating and used as a substrate for CVD. Similar structures were used by Thiha et al. [225] for

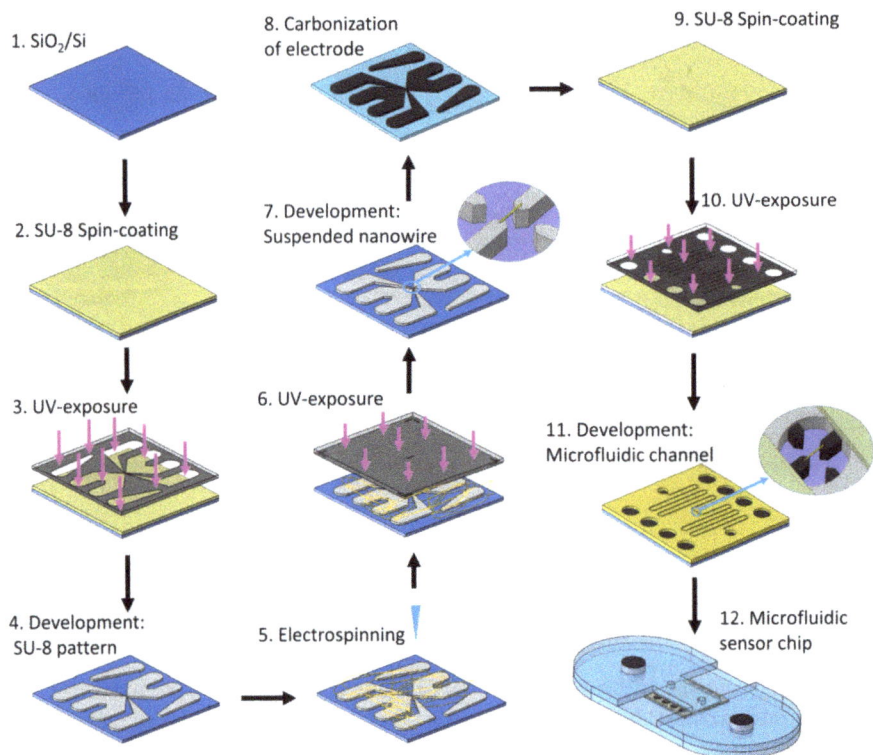

Figure 5.21: Schematic illustration for fabrication steps of carbon nanofiber biosensor. Insets show the magnification of suspended nanowire area. Copyright ©2018 Elsevier B.V. Reproduced with permission from [225].

biosensor fabrication. A schematic of their fabrication scheme is provided in Figure 5.21. This device was based on the principle that the attachment of biomolecules (bacterial cells in this case) to the fiber causes changes to its electrical resistance, which can be measured.

The primary challenge in manufacturing such a device is it low manufacturing yield. Electrospinning is not generally suitable for single fiber fabrication, as the fiber jet inevitably splits into multiple fibers due to bending instabilities. Moreover, there is little control over fiber diameter in this process. Collection of one fiber may not always be possible onto the electrodes, though one can break the extra fibers using FIB milling. This challenge can potentially be addressed by the use of NFES, which is capable of single fiber writing.

The aforementioned techniques are suitable for high-precision detection, for example, of biomolecules and trace quantity of chemicals. In such chemo- and biosensor devices, the microelectrodes need to collect data from specific locations with a high accuracy. Hence, a patterned electrode with well-defined sizes of recording sites is essential.

But there are several applications where patterning of fibers may not be required. It is simply the surface are of the fibers in their bulk form that is availed and utilized. This can be the case in some energy storage devices, where a bundle of carbon fibers or a crumpled mat can be directly connected to an external contact pad, and then partially insulated such that only the carbon fiber is exposed in the electrolyte.

5.7.3 Direct use of carbon fiber mats

Carbon fiber mats (electrospun) can be used directly, i. e., without any micro-patterning in membrane technology, cell and tissue culture substrates, and energy storage devices [143, 161]. There are a wide range of membranes, and their applications range from hydrogen storage and molecular sieves to water filtration. In all membranes, a high surface area is of paramount importance. Although carbon fibers already have a high surface area compared to solids, it can be further enhanced via activation processes. There are several physical and chemical activation methods that are routinely used in porous and activated carbon industry. Physical activation (porosity creation) typically involves heat-treatment of fibers at 800–1000 °C under an oxidizing agent, such as air, steam, CO_2, etc. In chemical activation, dehydrating agents, such as phosphoric or sulfuric acid, zinc chloride, metal hydroxides, or carbonates, are added to the fiber before their carbonization. This is less common for fibers, as it may lead to a compromise in their tensile strength. Another option of fabricating porous fibers is to add a polymer in PAN (or any other precursor), which has a very poor carbon content, and thus, it almost completely disappears after the heat-treatment. The same can be done for manufacturing hollow fibers, if the core is made up of a low-carbon polymer, such as PDMS. Activation with specific surface functional groups can be carried out similar to other carbon materials, depending upon the application. These fibers are then used as woven or nonwoven mats as membranes.

Other than porous and hollow carbon fibers, solid carbon fiber mats can serve as an electrode in their *as fabricated* form in supercapacitors [226], batteries [155], and various sensors. One must ensure a good electrical contact and subsequent insulation of exposed metal in such designs.

5.8 Advances in carbon fibers

Carbon fiber for electrodes have been modified in various ways in the recent past [162]. One can add nanoparticles of a range of metals in the PAN solution prior to electrospinning for achieving the desirable properties. Another option is to decorate the fibers with nanoparticles or other nanomaterials (attach only to the top surface post-carbonization) for specific functionalities.

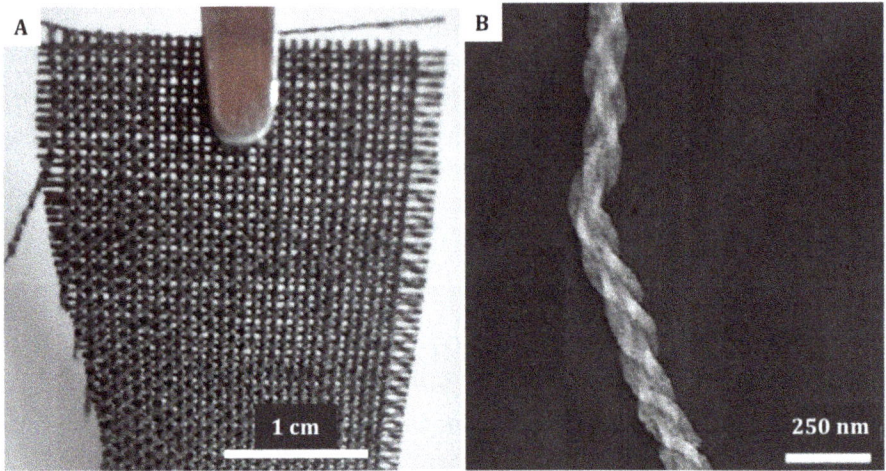

Figure 5.22: (A) Digital image of carbon cloth. (B) SEM micrograph of a CNT rope. Copyright for part (B) ©2014 Elsevier Ltd. Reproduced with permission from [270].

Preparation of carbon cloth (a woven carbon fiber mat or a carbon fabric; a photograph is shown in Figure 5.22(A)) is yet another step forward. Although woven carbon fibers are widely available for composite manufacturing (laminates and pre-pregs), carbon cloth does not contain any resin coating, which is often the case with laminates used in composite manufacturing. Even if the laminate does not contain additional resin, small quantities of it are used as lubricant during the weaving and braiding processes. Carbon cloth for device applications is prepared under highly controlled parameters and carbonized at 1500–2000 °C [144] to feature uniform properties, particularly electrical conductivity. Consequently, it can serve as a good substrate on which film, dendrite, and flower-like structures of various metals, metal oxides, etc., can be grown via electrodeposition, hydrothermal routes, and other deposition techniques. Forming electrical contacts with carbon cloth is also relatively easy compared to as-spun mats. This material nowadays is commercially available and is gaining a rapid popularity in carbon device fabrication.

In terms of advances, rope-like shapes formed by CNTs are emerging as a new type of carbon fiber (see Figure 5.22(B)). In the early 90s, there was a surge in fullerene- and CNT-related research, with numerous efforts being made worldwide to optimize the arc evaporation process for synthesizing various curved carbon structures. It was already reported by Thess et al. [224] in 1996 that CNTs tend to self-assemble in helical ropes. In the last two decades, the formation of CNT ropes has been refined to yield helical ropes containing multiple strands tied together to yield superstrong crystalline carbon ropes [263]. Obviously, these ropes have a higher electrical conductivity as well. In fact, it has been proposed that ropes of certain SWCNT ropes exhibit superconductivity [63], and can thus be employed in highly sophisticated electronic and magnetic devices. Although a large number of applications, including fabrication of electric cables, construction material, and spacecraft, have been envisioned with CNT ropes, due to a complex and expensive preparation method with low manufacturing yields, at this point their use is limited to devices, smart textile, and other applications, which entail small material quantities. CNT ropes have already been employed in neural sensors [105], strain sensor, [257], gas sensor [140], and preparation of composite materials [220].

1. How are hollow carbon fibers different from carbon nanotubes, if they are?
2. Mention at least two strategies for fabricating neat-parallel PAN fibers via electrospinning.
3. What is the advantage of horizontal electrospinning over vertical? Suggest possible applications of disc-type fiber collector used in an electrospinning set-up.
4. What is the reason behind electrical property variation in carbon fibers obtained using different manu-facturing processes and precursors?
5. What is CFRP? How is this material different from carbon fiber?
6. What is the need for considering two characteristics—ultimate tensile strength and elastic modulus—for defining the mechanical properties of carbon fibers? Why are high strength but low modulus fibers chosen for certain applications?
7. Which type of carbon fibers, in your opinion, are most suitable as for electrode fabrication in electro-chemical applications?
8. List at least three possible micropatterning methods for making a carbon sensors using electrospun carbon fiber mats.
9. What is Rayon? Should melt-spinning be preferred over electrospinning for Rayon fiber fabrication? Support your answer with some research articles.
10. PAN-derived carbon fibers continue to dominate the carbon fiber manufacturing industry, despite the availability of various other inexpensive precursors. Why, in your opinion?
11. What is the difference between isotropic and mesophase pitch?
12. Which precursor material is most suitable for the commercially available, so-called graphite fibers for manufacturing applications.
13. Calculate the ultimate tensile strength of a carbon fiber bundle if it contains 10 fibers of diameters 0.1, 0.3, 0.3, 0.1, 0.6, 0.9, 1.0, 0.2, 0.3, 0.4 mm. Take magnification factor of the system as 400. The bundle fractures at 1 N load.

6 Flexible and futuristic carbon devices

6.1 Flexible electronics

Devices whose properties and functions remain unchanged during and after bending multiple times are known as flexible (or bendable) devices. If a flexible device is integrated with wearable electronic gazettes, such as smart watches, it may be denoted as a wearable devices. Individual components of such devices and the entire assembly are tested to withstand thousands of bending cycles during their fabrication. It is ensured that the material properties and response of the overall device don't change as a consequence of bending, stretching, twisting, or any motion that induces a curvature. Flexible device technology is expected to have a large share in the global market in the near future, because these devices are conformable, lightweight, portable, and relatively more energy-efficient compared to a rigid device in the same category. In certain applications, such as photovoltaics, flexibility can bring additional advantage, such as absorption of sunlight incident at all angles. A flexible biological implant (e. g., a neural interface) can display a much higher number of contact points with the cells and tissues in the naturally curved and uneven environment. A sweat sensor in a smart watch can access even very small quantities of sweat microdroplets from the entire surface of the wrist and provide a quick analytical response. Altogether, flexible electronics and sensors are of tremendous benefit over their rigid counterparts. An image showing some examples of wearable sensors, their sensing mechanisms, target objects, and contact forms is shown in Figure 6.1 [262]. Other popular applications of flexible electronics include organic light-emitting diodes, thin-film transistors, radio frequency identification tags, and a variety of energy storage devices. Notably, flexible electronics also require flexible batteries, capacitors, catalytic beds, etc. As a result, a multidimensional research is on flexible materials and associate manufacturing processes are being conducted worldwide.

There are three important components of a flexible device: substrate, functional material(s), and response circuitry. In addition, they may contain certain passive materials, adhesive joints, software platforms, and of course, packaging for a complete operation. For a robust device design, it is important that all of these components can withstand the bending cycles with robustness, and that they are electrically and mechanically compatible with each other. In general, they should a exhibit good adhesion with each other without any significant interface effect, and form good ohmic contacts without much work function difference. The functional materials in the context of this book are carbon and its derivatives. Some manufacturing processes pertaining to carbon-based flexible shapes (e. g., printed carbon nanomaterials) have already been described in other chapters, but some additional ones will be detailed here. Before we get to the carbon part, it is essential to understand some general concepts and pertaining to flexible substrates.

https://doi.org/10.1515/9783110620634-006

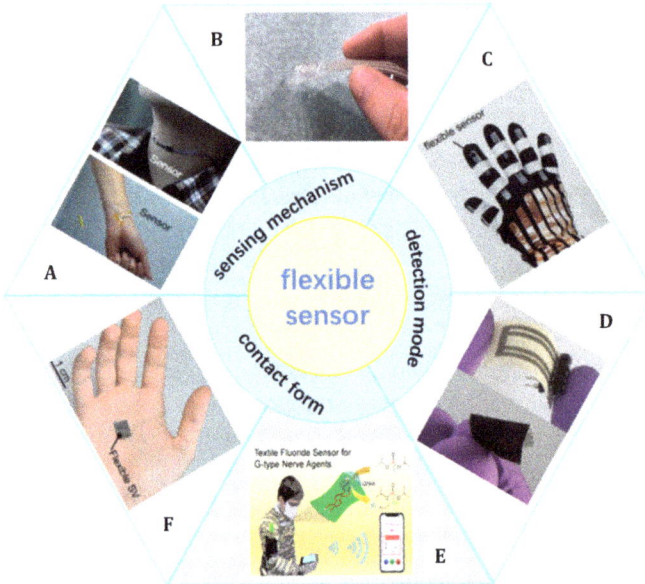

Figure 6.1: Flexible sensors based on different sensing mechanisms, detected objects, and contact forms: (A) a resistive strain sensor attached to the skin. Copyright ©2014, Springer Nature Ltd. Reproduced with permission from [126]. (B) A wearable triboelectric sensor for a dynamic prosthetic fit. Reproduced from [42]. (C) A highly stretchable, tough, and sensitive strain sensor; Copyright ©2020 American Chemical Society. Reproduced with permission from [266]. (D) Organic chemical sensors for the detection of chemically corrosive vapors. Copyright ©2012 American Chemical Society. Reproduced with permission from [12]. (E) Contact sensors for biodetection of flexographically printed fabrics made of fluorine-containing G-type nerve agents. Copyright ©2021 Elsevier B.V. Reproduced with permission from [81]. (F) A noncontact sensor for a physiological monitoring system based on the wearer's palm. Reproduced from [152]. Overall image published in [262].

6.1.1 Flexible substrates

Flexible substrates are sheets of organic or inorganic bendable materials that are used as base for patterning miniaturized structures. Common organic flexible substrates are polymers such as polyether ether ketone (PEEK), transparent conductive polyester, polyethylene terephthalate (PET or PETE), polyethylene naphthalate (PEN), biaxially oriented polypropylene (BOPP), polyetherimide (PEI), polyimides, poly-lactic acid (PLA), and various fluropolymers. A silicon-based polymer, polydimethylsiloxane (silicone or PDMS) is also a common biocompatible substrate, which is commercially used for manufacturing biological implants, contact lenses and other accompaniments to human body. Inorganic materials include ultrathin glass (primarily used in flexible LEDs) and thin metal foils. These materials can be used alone (single layer) or multiple coated layers. Occasionally, thin films of amorphous or crystalline functional oxides are grown on the substrate that provides it further stability and adds certain functionality. Once the substrate is prepared, structures are patterned onto it, employing various micro-fabrication techniques, and contact pads to the external electrical circuitry are laid out. These contact pads are typically sputtered thin metal films. Thermosetting polymers are used as glue where required. Flexible printed circuit boards are also becoming rapidly available, and have already been integrated in small equipment, such as a camera. Other than polymers, ultrathin glass, paper, various fabrics, and carbon cloth can also be used as the base material for flexible and printed electronics. New composite and hybrid materials that can be woven into fabric-like morphologies are being rapidly developed. In many cases, the additive in these composites is a carbon material, given the electrical conductivity and biocompatibility of carbon.

The substrate material may be chemically cleaned and functionalized prior to patterning a structure onto it [43]. Common surface treatment methods include exposure to plasma, electron/ion beam, laser, UV/O_3 and gamma-rays of different wavelengths. Among the chemical pretreatment processes, the most common ones are carboxyl functionalization,

Denneulin et al. [43] reported a primer formulation composed of monomers (tetrahydrofurfuryl acrylate and ethoxylated trimethylolpropane triacrylate) and photoinitiators (2-hydroxy-2-methyl-1-phenyl-propan-1-one, and bis(2,4,6-trimethylbenzoyl)-phenylphosphineoxide), which could be cured with the UV light after spreading onto a paper substrate, prior to carbon nanomaterial printing.

The selection of the pretreatment process depends upon the type of film or ink being deposited onto it. For example, for an aqueous ink, introducing hydrophilic groups via plasma or chemical treatments may facilitate better anchoring, whereas polymer-based hydrophobic inks may benefit from suitable primers. Surface roughening generally helps in all cases.

> One must differentiate between the terms flexible, organic, and biodegradable while describing a device having a soft substrate. Most of the synthetic polymers are organic, but not biodegradable. Also, flexible devices may be fabricated onto inorganic substrates, such as ultrathin glass.

6.2 Manufacturing of carbon-based flexible devices

As shown in Figure 6.2, there are four main pathways of fabricating patterns onto flexible substrates using carbon materials. These are (i) pattern transfer, (ii) micropatterning of carbon fiber, (iii) laser-induced carbonization, and (iv) printing using carbon-based inks. Whereas pattern transfer and printing can be carried out on any flexible substrate (polymer, glass, others), laser-assisted carbonization is limited to carbonizable polymeric substrates. In the case of carbon fiber patterning, a polymer is most commonly used for providing support to the patterns. However, one could also use any other material for this purpose. Among these schemes, patterning of carbon fibers and printing of carbon inks have already been described in the previous chapters of this book.

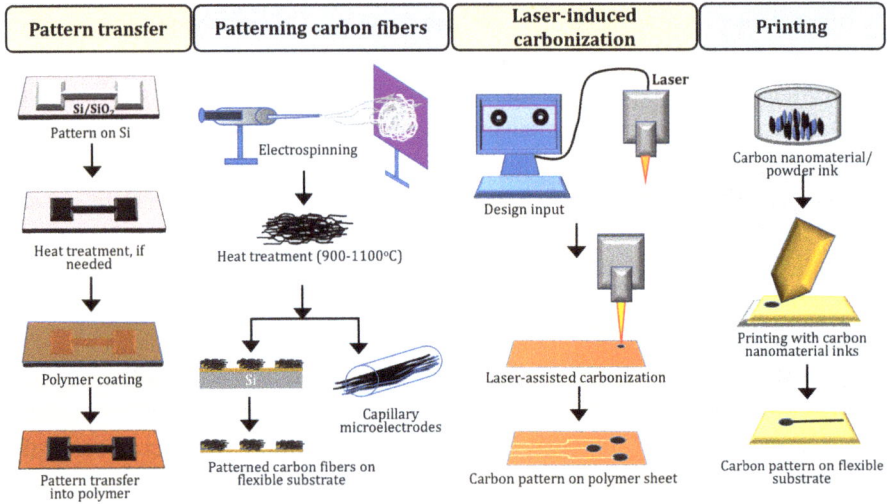

Figure 6.2: Four primary pathways of patterning carbon on to flexible substrates. Adapted from [50].

The remaining two techniques, namely pattern transfer and laser-induced carbonization, are discussed here.

6.2.1 Pattern transfer

This process is primarily based on the idea that a pattern can be first prepared onto a rigid substrate, and then transferred to a flexible one. Most microfabrication techniques are established for rigid substrates, such as silicon. Elevated temperatures are often involved while working with carbon materials, which is essential for the substrate material to be thermally stable. Most flexible substrates do not feature thermal stability at temperatures above 300 °C. Hence, it is more convenient to pattern rigid substrates. Nonetheless, depending upon the application requirements, one can also pattern a shape onto a flexible substrate, and then transfer it to another flexible one.

In the context of carbon materials, pattern transfer is commonly used in the following two ways: (i) patterning polymeric carbon on a silicon wafer via lithography and transferring this into polymer, and (ii) growth of carbon nanomaterials and transferring this material (with or without patterning) into polymer.

Such patterns, embedded in durimide (a biocompatible fully imidized polyimide often used for passivation coatings), were employed as neural probes for electrocorticography (ECoG) by Nimbalkar et al. [171]. In their fabrication process, first SU8 photoresist was lithographically patterned onto SiO_2/Si substrate. This was carbonized at 900 °C to yield a material similar to glass-like carbon. This type of fabrication process is described in detail in Chapter 3. To make flexible devices, these carbon patterns need to be trans-

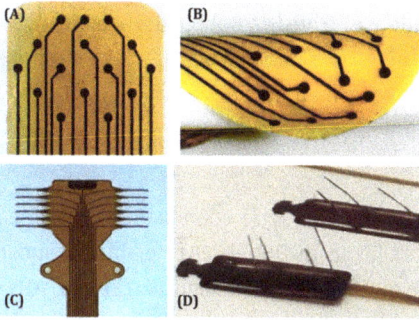

Figure 6.3: (A, B) A cortical microelectrode array and (B, C) spinal microelectrode array fabricated by transferring a polymeric carbon pattern into flexible polyimide sheet. Images adapted from [50] and [171].

ferred into a polymer, which can be cross-linked or cured via heat-treatment or UV light. It is easier to transfer such patterns when there is a thick enough SiO_2 layer right underneath the carbon patterns. SiO_2 can be selectively etched in a HF acid bath. Prior to etching SiO_2, a layer of suitable polymer must be coated on top of carbon pattern. In the reported work [171], first a durimide layer was coated onto a carbon substrate, with an additional coat of a transfer substrate. After removing the carbon pattern from Si substrate via SiO_2 etching, design-specific contact windows were created through simple microfabrication steps.

Pattern transfer in this fashion enables the use of thin films and other patterns consisting of non-graphitizing bulk solid carbon without any binder as microelectrodes. Various design modifications are possible, given that the initial patterning is done using photolithography. The main challenge of this fabrication scheme is a low manufacturing yield. Some patterns may develop cracks during their transfer from one to another substrate, which is detrimental to their performance in a flexible device. In the case of high aspect ratio (tall) features, nonuniform shrinkage during carbonization (i. e., conversion from photoresist to carbon via heat-treatment) can pose additional design issues, but in general this is not significant in the electrodes fabricated for biosensing purposes. The heat-treatment parameters may also be slightly modified, for example, the rate of temperature increase can be reduced. This leads to a slow annealing of the volatiles during pyrolysis and yields carbon with improved mechanical properties.

Other than polymeric carbon, pattern transfer is a widely used technique for thin carbon films, such as single- or few-layer graphene, which have relatively weaker bonds with the substrate. Graphene may or may not be micropatterned prior to transfer, based on the application requirements. CNT and CNFs prepared via CVD can also be transferred in a similar fashion.

Graphene is most commonly grown on Cu or Ni substrates, which serve as catalysts. An inexpensive way of obtaining single- to few-layer graphene is through CVD on a Cu or Ni foil. Graphene can be peeled or removed from these foils in various ways, followed by its collection on another substrate. In this process, some sacrificial or support substrates may also be used. Some such techniques include wet transfer, bubble-assisted transfer (electrochemical or non-electrochemical), dry transfer, roll-to-roll transfer, and support-free transfer. Each of these techniques is briefly described below.

6.2.1.1 Direct wet or dry transfer

In most of the transfer processes, first a layer of PMMA is coated onto deposited graphene, which serves as a sacrificial support layer. In the absence of this support, graphene may experience mechanical damage and wrinkles during the transfer. After PMMA coating, the next steps is to detach the graphene from the other side, i. e., from the metal substrate, where it was deposited during CVD (typically Cu or Ni). In wet transfer process, graphene layers are removed from Cu or Ni using ionic solvents, such as ammonium persulfate (for Cu) or ferric chloride (both Cu and Ni). These solvents partially dissolve the metal, thus enabling graphene removal and transfer into PMMA layer [234]. The ionic etchants are typically aqueous, hence, water is the medium during this kind of transfer. Other than PMMA, examples of support substrates include paraffin wax, camphor, poly(bisphenol A carbonate), polydimethylsiloxane (PDMS), thermal release tapes, and polymers blends, such as that of ethylene propylene diene monomer (EPDM) and polyaniline (PANi). Each has its own advantages in terms of chemical purity and mechanical characteristic of the removed graphene. Thermal expansions or shrinkage in the support substrates can cause wrinkles in the film. Some support substrates also induce doping, for example of N, B, or F, in graphene layers. Depending upon the material requirements, one can utilize this fact and achieve the desirable doping during the material transfer itself.

PMMA is the most common support substrate because of the following reasons: (i) it can be easily dissolved in acetone, which enables a convenient removal of graphene layers and subsequent transfer to other substrates, and (ii) it is used as a resist for electron-beam lithography (EBL), which can be used for microscale patterning of graphene for devices. The details of EBL are provided in Chapter 1. The standard procedure of this PMMA-assisted transfer is as follows: 100–500 nm thick layer of PMMA is spin-coated onto metal/graphene and baked at 60 °C. If there is any graphene or carbon deposition on the backside of the metal, it can be removed by oxygen plasma etching. This assembly is then immersed in a suitable etchant for relatively long duration (e. g., overnight). Afterwards, this sample is cleaned using rinsing with DI water and, if needed, with RCA. Finally, the graphene / PMMA is transferred onto the desired flexible (or rigid) substrate.

Graphene can also be transferred onto unconventional materials, such as amorphous fluoropolymers (commercially available with the name Cytop) and inorganic substrates, such as indium-tin oxide and glass (typically for photovoltaics applications) via wet transfer. The primary challenge in this process is a residue-free transfer of structurally intact graphene. If the growth substrate is nonmetal (e. g., Germanium), the graphene layer may exhibit weaker bonds and can be removed from it without a liquid medium using a thermal release taps and some sacrificial support layers [164].

6.2.1.2 Bubble-assisted transfer

In the bubble-assisted transfer process, the graphene is detached from the metal with the help of controlled release of bubbles (generally of O_2 or H_2) in a liquid medium that contains the metal / graphene/ PMMA assembly. Bubbles are typically generated through an electrochemical reaction, which works well with metals. The metal / graphene/ PMMA serves as both anode and cathode during the electrochemical bubbling process. However, if the substrate material is electrically insulating or this process is used for secondary transfer (i. e., PMMA to another polymer), electrochemical bubble generation does not work. Chemical reactions containing strong oxidizing agents, such as H_2O_2 can be used in such cases for producing bubbles. In principle, these bubbles help in mechanical delamination of graphene from the substrate. Bubble-assisted transfer is inexpensive, scalable, and relatively residue-free, but the main challenges are defect generation and unwanted contamination from the liquid medium.

6.2.1.3 Roll-to-roll transfer

Roll-to-roll fabrication processes are used for large-area patterning of flexible materials. Graphene films spanning over several inches on a copper foil can be mounted on to cylindrical rollers, followed by their transfer. The basic principles of pattern transfer remain the same, except it is now performed using rollers, etch bath, and mechanical pressing, as shown to Figure 6.4. It has been shown that in this process, best results are obtained when the transfer rates are low, heating is conducted gently, and the roller pressure is high. In most processes, the primary transfer is carried out on a thermal release tape followed by PET sheets for inexpensive device fabrication.

In roll-to-roll manufacturing, an assembly of cylinders is utilized for patterning features in a continuous fashion onto a flexible substrate. Roll-to-roll set-ups of different designs are used for printing, pattern transfer, hot-embossing, nanolithography, and various other applications.

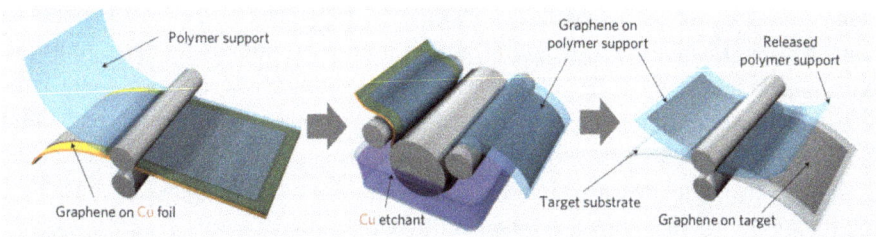

Figure 6.4: An illustration showing roll to roll transfer of graphene film grown onto a copper substrate. The process steps include adhesion of a polymer support on graphene film, etching of copper from below, and dry transfer-printing on a another flexible substrate. Adapted from [16]. Copyright ©2010, Springer Nature Ltd.

6.2.2 Laser-patterned carbon

Carbon obtained via laser-assisted pyrolysis of polymer sheets is a new addition to carbon materials family. This type of carbon is specifically developed for flexible devices. The fundamental principle is that certain polymers can be converted into usable carbon via laser-induced pyrolysis, instead of the traditional in-furnace heat-treatment. Depending upon the laser wavelength, the features patterned on the polymer can be as small as 5 μm. Since most of the automated laser set-ups accept CAD designs, a wide range of shapes can be directly *written* via this technique. Similar to traditional carbonization, the polymer should have a high carbon fraction and preferably an aromatic backbone for yielding a good quality carbon on laser-induced pyrolysis.

Laser-patterned carbon (LP-C) is always porous, because the annealing of pyrolysis byproducts (several light volatile hydrocarbons) is extremely rapid. This is not the case in traditional pyrolysis, where the heating rates are typically below 10 °C per minute. For obtaining high quality carbon and graphite, the heating rate may be kept as low as 1 °C per minute. Similar to all synthetic carbons, the exact microstructure and chemical functional groups on the surface of LP-C depend strongly upon the nature of precursor. Certain heteroatoms, such as N, often get integrated in the structure of the carbon itself, leading to the formation of, for example, pyridinic and pyrrolic N rings within the carbon network. S or other heteroatom containing polymers may exhibit a significantly different chemical composition after their laser-assisted carbonization. Though there is still a lot of research work required to understand the mechanism of laser-assisted pyrolysis and the microstructural evolution of laser-carbonized materials, several applications have already been demonstrated. The primary application areas for LP-C are supercapacitors, antennas, sensors, and some bioelectronic devices.

6.2.2.1 History and nomenclature

Laser processing is commonly used for plastic sheet or metal cutting, metal welding, drilling, piercing, etc., in mechanical workshops. Certain laser operations in manufacturing may also involve microstructural optimization, surface engineering, or finishing. Accordingly, a minimum (threshold) laser energy per unit area is required. Some of the early laser carbonization reports are indeed related to incidences where instead of (or along with) the planned operation, such as cutting or engraving, some residues of carbon were observed on the surface of a polymer. Although this type of carbon was considered impure and undesirable, this know-how ultimately led to the controlled laser-assisted patterning of carbon on flexible polymer sheets. Subsequent investigations were aimed at obtaining carbon with a superior electrical conductivity using UV radiation with a nanosecond pulse or continuous wave [153]. A combination of photochemical and photothermal mechanisms was proposed for such a conversion, which was supported by a model of single photon and multiphoton absorbance of chromophores. Eventually, carbonization of polyimide sheets (in the form of Kapton^{TM}) was also carried out using

IR and femtosecond lasers [153]. With time, more control as well as understanding of polymer-laser interaction evolved to yield patterned carbon materials.

The potential of this general method to obtain carbonized patterns was recognized around the year 2010 with the rising interests in carbon-based printed and flexible electronics. Prior to laser-carbonization, flexible carbon electronics was either based on nanomaterials or involved multiple fabrication steps. One crucial trait of carbon, its porosity, which is often very useful for energy storage purposes could not be fully utilized. In addition to polyimides and other polymers, laser-reduction of graphene oxide (LR-GO) for flexible electronics added yet another dimension to this technology, and inspired the research on direct laser-writing of carbon films. The increase in popularity of flexible electronics around 2010 provided impetus not only to LP-C but to all flexible device patterning processes that utilize carbon materials. Today, a large variety of polymers and other carbon precursors are used for carbon-writing using laser beams in different morphologies and areal spread.

Porous carbon material produced via laser-assisted carbonization has been referred to as laser-induced graphene (LIG), porous graphene, and simply porous carbon in the contemporary literature. However, the terms laser-carbon or laser-patterned carbon (LP-C) are more appropriate, as the material does not exhibit the properties of graphene. Graphene, graphene oxide, and other nanomaterials are characterized by their "discrete" nanoscale structural units and the properties obtained as a result of their layered structure, edge dynamics, etc. Laser-patterned carbon, on the other hand, exhibits properties similar to bulk disordered porous carbon, but with a gradient across the film (along its depth). Additionally, LP-C may contain heteroatoms (e. g., N) depending upon the precursor that may even be bonded, and thus change the layer separation and incorporate non-six-membered rings. LP-C is therefore a broader term that (i) entails all carbon materials derived from a carbonaceous precursor using the laser energy, and (ii) conveys the idea that the material is always *patterned* directly by carbonizing the substrate itself. An even more general term is laser-carbonized materials (LCM), which also encompasses composite or hybrid materials produced in the presence of additives or reactants. For more specific cases, such as in the context of those LCMs that have an intrinsic functionalization, e. g., laser-pattered nitrogen-doped carbon, more suitable terminology may be used. Similarly, for laser-processing technology, the term laser-induced carbonization or simply laser-carbonization provides a good description and covers all materials studied. In this book, the term LP-C is used for addressing the material of interest.

6.2.2.2 Mechanism of laser-assisted carbonization

Unlike most polymeric carbons produced in ovens or furnaces, laser-assisted carbonization is not a result of thermal energy alone. It is a relatively more complex process that entails photochemical and thermal/photothermal effects. There are additional secondary chemical reactions that may or may not depend upon the laser energy directly.

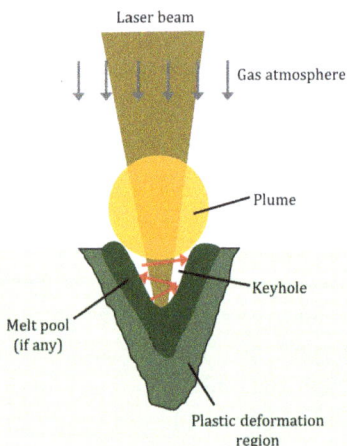

Figure 6.5: An illustration showing a proposed mechanism of laser interaction with polymer during laser-assisted carbonization [51]. Red arrows in the keyhole indicate reflections of the laser beam. Gas atmosphere is not essential but generally some oxygen is present when the process is performed under ambient conditions.

A brief description of this process based on the state-of-the-art is provided as follows: When a high-energy laser beam of a wavelength within the absorbance range of the polymer (preferably at maximum absorbance) is incident upon a hydrocarbon, it provides sufficient energy to evaporate a small quantity of the material (typically small and volatile hydrocarbons) and forms a keyhole on the surface of the sheet (see Figure 6.5). Around this keyhole, the material may be molten, along with plastic deformation surrounding it owing to the energy impact from the laser and/or the so-called heat-affected zone. Within the keyhole, the laser beam may undergo multiple reflections and is eventually (re)absorbed. Above the keyhole, a plasma (plume) consisting of ionized organic/carbon clusters, volatile gases and/or other manifestations of colloidal carbon is generated. Above the threshold (minimum laser-energy required for sufficient material evaporation), this plume is visible through the naked eye. The exact role of plume involving the carbonization process is yet to be fully understood. However, based on common knowledge of plasma-assisted processes (e. g., plasma enhanced CVD), one can envision a plausible lowering of activation energy for various pyrolysis associated chemical reactions.

After the absorption of the high energy of the laser beam and keyhole formation, competing processes, such as melting, bond-dissociation, plasma absorption, and high gas pressure occur simultaneously. Some of these are dominant on the top surface only, and others are equally effective around the keyhole. Modeling the mechanism of laser-carbonization process is extremely challenging owing to the complexity of the process, consideration of all reaction parameters, and constantly changing materials properties. Due to this, any polymer or metal processing with laser beams requires a comprehensive knowledge about the underlying beam-matter interactions. This is also valid for carbonization.

It has been known from industrial laser processing that every material shows a different response to the exposure of a laser beam due to inherent materials characteristics. To quantify them, the direct laser beam–material interactions need to be considered. This mainly includes the primary light-matter interactions, such as absorptivity and reflectivity. These two parameters depend on the material's chemistry and surface properties, and define the so-called laser coupling as per equation (6.1),

$$P = \alpha P_A + \beta P_R, \tag{6.1}$$

where α and β are the absorptivity and reflectivity, respectively, and $\alpha + \beta = 1$. P is the incident laser power, P_A is the portion of power absorbed, and P_R is the portion reflected.

Post absorption, the thermal properties of the materials, such as heat capacity, heat conductivity, melting point, and evaporation point, play decisive roles in energy conversion and transport within the material. For example, different melting points would lead to a highly variable heat dissipation patterns during the reaction due to enthalpic effects during melting. Finally, gases or volatile species released during the laser beam-induced (pyrolysis, combustion, and other) reactions form a colloidal plasma / plume, which may get further ionized by the incident laser energy. This plume interacts with the laser beam and reacts with the product material. Now, the evaporation rate(s), the density of evaporated products, their reactivity, and their interaction with the laser beam need to be understood. Another critical parameter especially in post-pyrolysis carbonization reactions is the reaction atmosphere, i. e., whether the laser-irradiation was carried out in air or in a chamber filled with an inert gas. Despite the fast reaction rates induced by the laser, the reaction does get influenced by the presence of gases such as oxygen, ammonia, nitrogen, or hydrogen surrounding the sample. As the interaction of these gases are mostly with the material surface, one can observe different surface chemistries under inert environments as opposed to the reaction in air [153].

In summary, the primary material characteristics essential to understand and optimize laser-induced carbonization are [51] the following:
– absorptivity / reflectivity of the laser beam (reflectivity very low in organic materials)
– thermal properties: heat capacity and heat conductivity
– melting point and evaporation point
– interaction with the products (absorption, reflection, e. g., plasma, gases, dust, or melt)
– reaction atmosphere

Laser parameters
During laser irradiation, the high photonic energy is typically absorbed by the material, and then ultimately converted into heat. In fact, lasers are used as a tool for localized

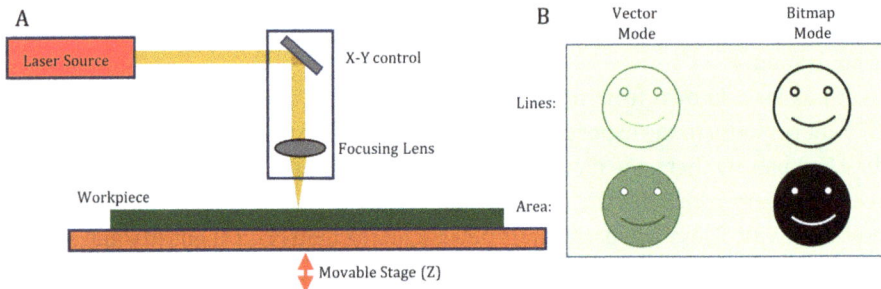

Figure 6.6: An illustration showing the working principle of a processing laser; B) Sketch illustrating the difference between vector and bitmap mode. Figure reproduced from [51].

heat generation in various applications. A laser source generates a beam having a tunable intensity, which is directed by a mirror system and focusing lens(es) onto the target material or part thereof (see Figure 6.6(A)). The position and speed of the laser beam is controlled by a movable stage along the x-y plane. The focal length and distance depend upon the specifications of the lens and stage. Most laser set-ups can be operated in two possible modes: vector and bitmap. In vector mode, the x-y-control follows the lines (or paths) in vector graphic, whereas in bitmap (or image) mode, a bitmap having a certain resolution (in dots per inch or dpi), is scanned over the sample. This process is also known as rastering, and hence, this mode is commonly known as the raster mode (Figure 6.6(B)). Vector mode is used when the design entails a high precision and power control, whereas the bitmap mode is more suitable for covering larger patterned areas (e. g., in the case of engraving or in some instances, laser-carbonization). If the vector mode is applied to carbonize larger areas, the line separation must be defined (Figure 6.6(B)). Importantly, the interaction cross-section is dependent on the absorptivity and thermal diffusivity of the material for a given incident laser wavelength and energy. As a result, the line separation varies with the material.

The temperature achieved during laser-carbonization is adjusted by the energy input, which in turn depends upon the power density and laser exposure time (generally known as reaction time). In comparison to conventional (furnace) heating, laser-assisted heating is extremely fast and energy-intensive. The reaction typically takes place on the time scale of milliseconds, as opposed to several hours allowed in a furnace. To understand this rapid heat-generation, some fundamental aspects that differentiate laser-heating from conventional Joule heating need to be understood. For simplification purposes, let us consider a laser just as a source of bundled photonic energy. For now, any additional effects, such as those caused by the plasma, have been ignored. Lasers are among the highest power density sources of energy. This is a result of the condensation of photons of a certain wavelength in a laser beam. Consequently, the power density and associated wavelength are the two most relevant parameters during laser-carbonization. The power density depends on the beam diameter (spot size) in the focus

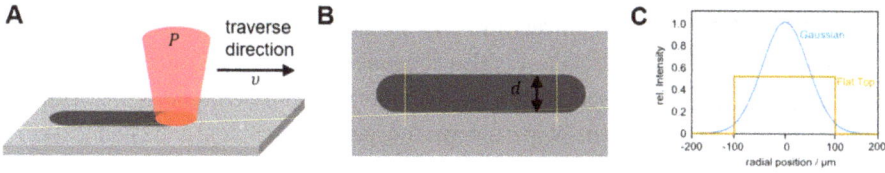

Figure 6.7: A) Laser-beam scanned over the surface of a workpiece; B) Laser-modified area determined by the scanning speed and the beam diameter; C) Graph illustrating the Gaussian or flat-top beam profiles of a laser beam. Figure reproduced from [51].

of the laser. The beam diameter is determined by the focusing lens and the wavelength of the laser. For example, a 1 W laser that is focused to a spot size of 0.2 mm results in a power density of 3200 $W.cm^{-1}$ according to equation (6.2),

$$\frac{1W}{\pi(0.01cm)^2}. \tag{6.2}$$

Another important parameter in laser-processing is the fluence (or energy density). Fluence is the energy delivered by the laser per unit area, and is a direct function of the laser power and spot size. On a laser set-up, the fluence F can be adjusted by setting the incident power P and the scanning speed v (see Figure 6.7A). The fluence is mentioned as lineal (FL, in $J.cm^{-1}$), or areal (FA in $J.cm^{-2}$) formats. The values are calculated directly from the preset output power, scanning speed, and spot size according to equations (6.3) and (6.4) [51]:

$$F_L = \frac{P}{v}, \tag{6.3}$$

and

$$F_A = \frac{P}{v.d}. \tag{6.4}$$

From this relationship, one can deduce the appropriate fluence values for a laser-carbonized line (see Figure 6.7B). The effective energy distribution additionally depends upon the beam profile, which can be either Gaussian or non-Gaussian (flat-top) in terms of its intensity distributions (Figure 6.7C). The beam-profile strongly influences the distribution of energy density delivered to the polymer film. For Gaussian beams, often the peak energy density is provided as it is significantly higher than the average energy density. Line-shaped beams with cylindrical lenses may also be utilized to create a larger focal area onto the film.

The incident power (effective power received by the polymer film) can be measured directly with an intensity meter, similar to the UV lamp intensity in photolithography. This is recommended, as machine manufacturers provide a specific set of specifications for their set-up. Often, in commercial laser equipment, the effective output power is a fraction (%) of the maximum power. However, the actual output may scale nonlinearly for a given set of parameters, since it is also dependent on the scanning speed. As previously discussed, the laser exposure time is also important, and should be set according to material properties, such as its heat capacity, thermal conductivity, and chemical composition.

Yet another important beam property is its operation type, i. e., whether it is continuous wave (CW) or pulsed. In pulsed mode, the beam energy per pulse has a significant effect on the properties of resulting carbon. Repetition rates in pulsed lasers may range between Hz and GHz, depending upon the type of laser. In many commercial laser-engraving machines, the repetition rate is given either in pulses per distance (e. g., per inch: PPI) or pulses per time (e. g., per second or Hz).

> **!** A beam propagation speed of 500 ms mm^{-1} (2 mm s^{-1}) and a pulse repetition rate of 1000 Hz, result in a repetition rate of 500 pulse/mm. If the propagation speed is changed to 50 ms mm^{-1}, the repetition rate becomes 50 pulse/mm. Evidently, a change in laser speed can potentially modify the microstructure, morphology, and properties of the resulting carbon.

In summary, the most crucial laser parameters responsible for laser-induced carbonization are [51]
- photon-energy/wavelength of the laser,
- incident power,
- processing speed,
- pulse sequence, repetition rate,
- spot size and focal position.

There is still a lot of scope when it comes to systematic mechanistic studies based on a statistical range of laser-carbonized materials and process parameters. A comprehensive model for laser-carbonization based on different precursor materials and respective laser parameters will be of high utility for a better understanding the mechanisms of laser-carbonization, ultimately leading to an improved control over the carbon properties.

6.2.2.3 Precursors for laser-carbonization

Polyimides in a compressed, sheet form are the most common polymer films used as starting material or precursors for laser-assisted carbonization. Polyimides are common substrates for flexible electronics owing to their biocompatibility, thermal stability and availability at a reasonable cost. Polyimides have also been known as carbon precursor in curable liquid and pressed sheet form. When carbonized in a furnace, these sheets yield a graphitizing-type carbon, whereas the cured liquid typically converts into non-graphitizing carbon. This has been attributed, by Inagaki and Meyer [227], to stress-induced graphitization. The preparation method of polyimide sheets induces a layered arrangement of polyimide sheets, thus making it more suitable for graphitization [227]. It is not clear if LP-C can be graphitized at elevated temperatures (>3000 °C), because the current laser-carbonization process gives little to no freedom of temperature control during the operation. Moreover, as discussed earlier, there are additional factors, such as plasma effect, that may be contributing to carbonization of the material.

Besides polyimide, polyaramid, polybenzimidazole, polyaniline, poly(vinyl alcohol), cellulose, lignin and its derivatives, melamin, waste natural organic, sugars, and various combinations of organic and inorganic substances with polyimide have been used as precursor for laser-carbonization [51]. Other than polymers, laser-reduced graphene oxide (LR-GO) has shown compatibility with the laser-patterning process. LR-GO is a special case, as the laser-induced reaction of GO is considered a deoxygenation or a reduction, rather than a classic carbonization. This could also be a reason for the use of terminology LIG to denote the thus obtained material.

6.2.2.4 Laser-patterned carbon: applications

To date, the four primary application areas of LP-C include energy storage, sensors, electrocatalysis, and antennas. The high surface area of LP-C combined with electrical conductivity renders them immediately attractive for energy storage applications, such as electrodes for electrochemical double layer capacitors (EDLCs). In the case of antenna applications, metals such as Cu or Ag come with the problem of oxidation, corrosion, difficult disposal, and limited availability. Most carbon materials in their pristine form are electrically and electrochemically stable and do not easily corrode. A comparison of properties of various device-friendly carbon materials is provided in Chapter 2. LP-C, due to its patternability combined with good electrical conductivity and corrosion resistance, is a highly suitable material for antenna fabrication. In the last 5 years, there has been a surge of publications on LP-C based antenna and transducers. Further reduction of sheet resistance and uniform pore distribution are the key optimization parameters in this context.

LP-C microelectrodes are used in a variety of sensors, including both chemical and biosensors. As discussed in other chapters, most carbon materials are suitable for sensor fabrication because of their tunable surface properties combined with stability. Here the hierarchical porosity of LP-C [153] is an added advantage, as it can be utilized for size-based detection of large molecules. LP-C, in general, can be used as a catalytic or support bed for nanoparticles and quantum dots. There are numerous proven applications of small-scale particles and dots, but many of them are solution-based. For example, detection of trace ions in water is often carried out using carbon quantum dots, but via a suspension of quantum dots colorimetric fashion. The need for inexpensive solid-state devices, particularly for point-of-care analysis demand a high-surface-area support for the particles. This support must not affect the sensing mechanism in any negative fashion, should not react with the analytes, and have a corrosion resistance. All of these characteristics are present in LP-C. In a recent study by Devi et al., LP-C was used as a support material for carbon nitride quantum dots for fluoride ion sensing [48]. Interestingly, LP-C participated in the sensing and further improved the sensitivity of the device.

Common pathways for microfabrication of carbon-based flexible sensors are discussed in a review article by Devi et al. [50]. As a viable alternative to other device-friendly forms of carbon, LP-C offers several benefits, such as large surface area patterning, tunable surface chemistry during fabrication itself, scalability, low cost, and most importantly, a rapid one-step fabrication of monolithic carbon patterns. LP-C also provides the additional advantage of bendability and rollability, which comes in handy when it comes to conformable biosensors. Sensors derived from other carbon materials (graphene CNTs, CNFs) require additional steps of pattern transfer or printing onto flexible substrates, after their synthesis, to facilitate a complete microdevice fabrication. This increases their cost and probability of manufacturing defects. LP-C not only offers a single-step fabrication, it even reduces the number of post-processing and packaging steps owing to its already functional large surface area. Its porous surface can be further utilized to support other active nanoparticles to increase the sensitivity of the device towards the analyte, which increases its utility in sensor fabrication manifolds.

> **!** The line-width and smallest patternable feature size in laser-assisted carbonization depend upon the diameter (spot-size) of the laser beam used, which is influenced by the laser-wavelength. This in turn depends upon the absorbance characteristics of the precursor polymer. While attempting laser-assisted carbonization of a new polymer, its absorbance spectrum is one of the first tests, which provides the information essential for laser-selection.

6.3 Heteroatom-containing carbon

Heteroatoms such as H, N, S, O are present in addition to C in many organic polymers. When these materials are heat-treated and used as carbon precursors, often some heteroatoms are not entirely removed. In fact, small quantities (<1%) of these atoms can be present in carbon materials, even after carbonization at 2000 °C. Industrial carbon materials, such as porous activated carbons, are often only heated up to 1000 °C. Microfabricated carbon patterns, such as resin-derived glass-like carbon, for chip-based devices are also carbonized at temperatures as low as 900 °C. These unintentional heteroatom impurities, especially when they are <1%, are rather undesirable. However, if a specific heteroatom can be introduced in carbon's structure or graphitic network itself, it can play a functional role. Such materials are called heteroatom-containing carbons. Activated carbon derived from biomass or other natural precursors is a good example of N-, and in some cases S-containing carbon. N- and S-containing carbon nanoparticles, quantum dots, and soot are also popular functional carbons. Other common dopants are more electronegative atoms compared to carbon such as B and P. Notably, heteroatoms with a higher electronegativity, but comparable atomic size, cause a positive charge density on adjacent carbon atoms, which improves the overall electrocatalytic activity (in particular, oxygen adsorption and charge transfer) [7].

It is essential that the heteroatom is present in a controlled quantity and can be prepared reproducibly. It is also important that one knows the mechanism of heteroatom's contribution in the desired property and the maximum fraction of the heteroatom that keeps carbon's structure intact. If the heteroatom is added externally, e. g., by doping or mixing, it should be clearly understood whether the activity is due to heteroatom alone, or if it is a result of the combination of carbon and the other element.

In many electrochemical applications, traditionally some carbon black is added during electrode fabrication to increase the electrical conductivity of the mixture. Such materials or mixtures are not considered heteroatom-containing carbons. In the last few years, the popularity of heteroatom-containing carbons has significantly increased, and their effects have been investigated both experimentally and theoretically. Not only bulk carbon materials, but even nanomaterial, are often functionalized or doped with heteroatoms such as N. Nitrogen functionalization of fullerenes and CNTs are good examples of the same. Here we will briefly discuss N- and S-containing carbons and their common manufacturing processes.

6.3.1 N-containing carbon

N-containing carbon can generally be divided into three types: (a) carbon materials with doped or functionalized with N, (b) carbon materials with bonded (inherent) N, and (c) carbon nitrides. A schematic shown in Figure 6.8 summarizes sp^2 N-containing carbon materials.

Heteroatom-doped carbon materials belonging to type (a) contain N in extremely small quantities (ppm or ppb levels). N doping can be carried out employing N-plasma treatment, ion implantation, and CVD, and even some physical methods, such as ball-milling [7]. N co-doping, along with other elements such as S, B, P, is also practiced particularly with carbon nanomaterials for enhancing electrocatalytic activity. Futuristic rechargeable batteries, including metal-air, Li-S and Na-ion, have already started replacing elemental carbon with N-containing carbon due to its improved electrochemical activity and specific surface area.

Within type (a) itself, the chemical functionalization pathway relies on the attachment of active chemical groups, and their exact fraction may depend upon the specific requirements. Common functional groups include amine, amide, nitro, cyano, nitrile, nitrogen oxide, etc. In these N-functionalized carbon materials, the carbon matrix typically facilitates electron exchange or mobility, whereas N contributes more towards chemical reactions with the target molecules. Lately, various carbon nanomaterials have been doped with N to incorporate special functionalities. Some crystalline carbon materials, such as diamonds that contain N as a defect or crystal vacancy, may also be placed in this category. The N-vacancy diamond features amazing magnetic and optical properties, and is extensively used in optical sensing applications. It is not discussed in detail here, but interested readers may refer to the available literature [192, 199].

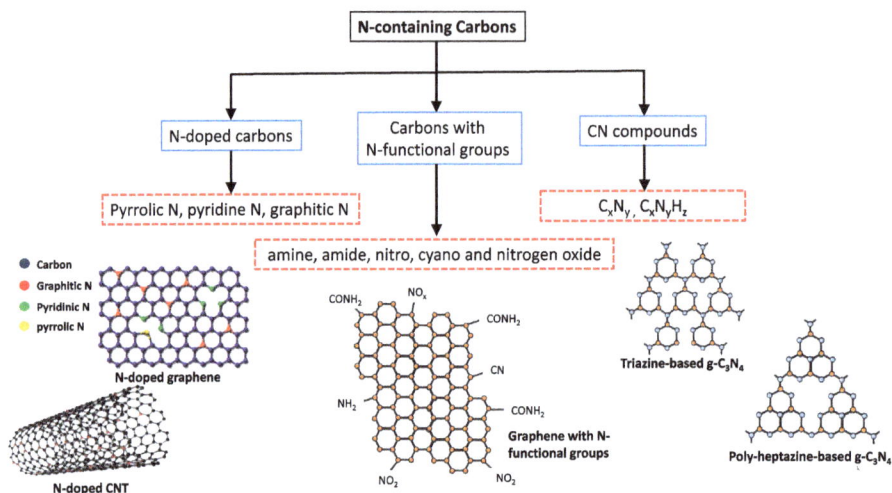

Figure 6.8: Schematic representation of different types of N-containing carbon materials. Overall diagram reproduced from [51]. N-doped CNT: Copyright ©2007 EPL (EDP Sciences database). Reproduced with permission from [133]. N-doped graphene: Copyright ©2009, American Chemical Society. Reproduced with permission from [251].

Type (b) or bonded N-containing carbon covers a wide range of materials, especially those derived from N-containing solid precursors. In these materials, the N is chemically bonded within the carbon matrix such that it changes the properties of the matrix material itself. However, the change is not enough to compromise the identity or basic properties of carbon. A good example of this type of material is LP-C, described earlier in this chapter. LP-C derived from polyimide contains N in its (crystal) structure. With the same laser parameters, the same fraction and chemical structure of these N-containing carbon rings (that may be five-membered) can be reproduced. It has been reported that this N, for example, in the form of pyridine rings, can cause a slight polarity shift within the carbon network. This induces inherent electrocatalytic and chemical activities in the carbon material. Other similar examples include certain biomass-derived non-graphitizing carbons and PAN-derived fibers carbonized at low temperatures.

Besides aforementioned materials, graphitic carbon nitride (gCN), especially in the form of nanoparticles and quantum dots have lately gained a lot of popularity, in particular owing to their photocatalytic activity. Carbon nitrides (described here as type c) can potentially be placed in category (b); they are studied separately because of their unique properties. The first attractive feature of carbon nitrides is the frequently used prefix, *graphitic*. As per the IUPAC, the term graphitic may be used for materials that exhibit a detectable peak for graphite 002 planes in their X-ray diffractograms. If we take a pure graphite crystal and start replacing one carbon atom with one nitrogen atom, the separation between the 002 planes will change, but not enough to completely destroy the layered structure of graphite. In fact, if the number of N atoms is small enough, the separation may indeed be smaller compared to many non-graphitizing carbons. One would still be able to observe the characteristic 002 peak around 2θ value of 26–28. This is indeed the origin of the prefix graphitic for carbon nitrides.

The interesting fact here is that gCNs come in a range of stoichiometry (ratio of C and N), rather than one unique material. If one starts replacing C with N in a graphite framework, initially one can just call it N-containing carbon. However, as the fraction of N increases, at some point a carbon nitride (CN) will form. But this CN (or gCN) may not have the C_3N_4 stoichiometry. If more and more C atoms are replaced with N, finally the stoichiometrically balanced C_3N_4 will be formed. All of these materials together are called gCNs. One can also designate them as C_xN_y if the exact structure is not known. Some gCNs may also contain H atoms, as can be seen in Figure 6.8.

Carbon nitrides are most commonly synthesized using organic precursors having nitrogen in them at relatively lower temperatures (200–400 °C) compared to carbonization. One popular pathway of CN synthesis is using autoclave (high pressure) synthesis, which is based on the use of laboratory-scale hydrothermal or solvothermal reactors at temperatures around 200 °C. Different combinations of organic precursors can be mixed inside such closed reactors. The volatile products generated during the reaction and/or dissociation of precursors causes an internal pressure in the sealed pots. The chemical reactions may take place in the presence of water (hydrothermal), other solvents (solvothermal) or no solvent (solvent-free), depending upon the chemistry and the desired output. Note that unlike elemental carbon, gCNs are only stable up to 700 °C in their crystalline form. The autoclave-assisted reaction typically yields a mixture of CN quantum dots along with slightly larger particles, flakes, and residual reactants. Quantum dots are then purified via standard chromatographic or other purification methods and utilized for their inherent catalytic characteristics.

As stated above, carbon nitrides are more commonly used in the form of quantum dots, whereas other bulk solid N-containing carbons are directly used as powders, with or without binders, in device applications. An example of gCN quantum dots-based fluoride ion sensor is shown in Figure 6.9. In this device, gCN quantum dots were first coated onto a LP-C, a form of porous carbon described above. This assembly was then dipped in water samples having different concentrations of F^- (ions). One can observe the change in the intensity of color with an increase of F^- (ion) concentration. One can plot a calibration curve based on known concentrations and respective colors (measured by absorbance at a certain wavelength), and accordingly back-calculate the concentration of analyte in an unknown sample.

Primary application areas of N-containing carbons include electrode material in rechargeable batteries, direct use in electrocatalysis, photocatalysis and chemical catalysis, and sensing [48, 271]. Printing methods (described in Chapter 4) are commonly used for patterning various N-containing carbons.

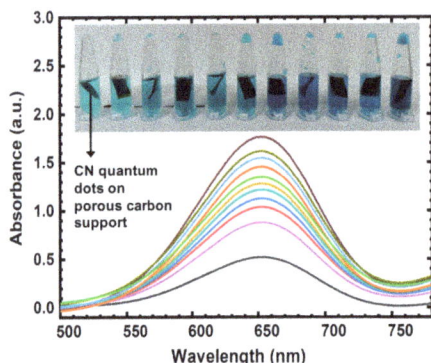

Figure 6.9: Digital images showing change in color of water containing increasing concentrations of F⁻ (ions), along with the absorbance values. Reproduced with permission from [48]. Copyright ©2023 Wiley-VCH GmbH.

6.3.2 S-containing carbon

Similar to N, S-containing carbons can be prepared via heat-treatment of S-containing organic precursors. Some such precursor materials include thiophene- and thienyl-based monomers, as well as biomass (waste peels, etc). Common acids and salts, such as such as $Na_2O_3S_2$, Na_2SO_4, H_2SO_4, Na_2SO_3, etc., may also be mixed with the precursors prior to their carbonization [44]. Salts of S lead to internal etching during carbonization, which increases the porosity in the material, besides the integration of S. S can also be mixed or chemically attached as a post-treatment to already prepared carbon materials for their sulphur-activation. For this purpose, elemental S and widely available S compounds are used as activating agents. Hydrothermal and solvothermal methods have also been utilized for S-containing carbon production. Additionally, CVD, flame pyrolysis, microwave synthesis, and sol-gel synthesis have been reported to yield S-containing carbon in different various [200, 44].

These materials tend to have a very high specific surface area owing to microporosity. The presence of S, similar to other heteroatoms, increases the layer separation in the carbon material, thus making it suitable for intercalation of ions. The pore structure can often be tailored by changing precursors or taking a combination of precursors. Recent advances in Li-S batteries and fuel cells have revived the interests of carbon community in S-containing carbons. Other major application areas include wastewater treatment, supercapacitors, solar hydrogen production, sensing, mercury absorption, etc [200, 122, 128]. F- and S-containing thin-films are widely used as industrial coating materials. In this context, a lot of research has been carried out which is beyond the scope of this book at present.

Co-doping of different heteroatoms may lead to further development of hybrid carbon materials with manifold functionalities. In principle, by selecting the right combination of precursors and heat-treatment or synthesis conditions, researchers can design

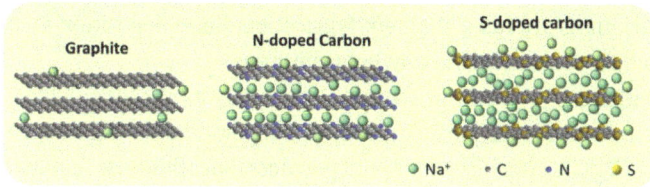

Figure 6.10: Schematic diagrams for Na$^+$ (ion) storage in graphite, N-doped carbon, and S-doped carbon. Figure reproduced from [194].

their own carbon materials. Co-doping of N and S is rapidly gaining popularity, particularly aimed at Na ion battery applications. Notably, addition of S in carbon can increase the layer separation up to 0.4 nm, thus facilitating an easy intercalation of Na ions during charging and discharging. A comparison of layer separation for N and S-containing is provided in Figure 6.10.

6.4 3D printing of carbon

The fundamental principle of 3D-printing carbon shapes is similar to lithographic microfabrication, i. e., carbonization of polymer patterns. It is irrelevant whether a polymer was lithographically patterned or 3D-printed, as long as it satisfies the following conditions: (i) polymer converts into carbon without burning or deformation, and (ii) the substrate (on which the shape is printed) can withstand at least 900 °C temperature used for carbonization. Though this may sound straightforward, often condition (i) is very difficult to achieve in the case of 3D shapes. Most synthetic polymers undergo coking mechanism during pyrolysis, i. e., the pass through a liquid or semisolid phase. They also experience shrinkage due to mass-loss. Though a 2D pattern (e. g., lithographically patterned SU8) may only deform near its base, a complex 3D structure may significantly deform at points that experience more thermal and mechanical stresses. These stresses are influenced by shrinkage-induced pulling, change in material's physical state (liquid or semisolid), and of course load distribution in a given design. Moreover, most polymers that are compatible with 3D printing do not carbonize well. They either yield extremely low fractions of carbon, or contain a lot of unsaturated bonds and/or oxygen that leads to a complete collapse of structure and/or burning of the material. Though there are many examples of 3D printed carbon shapes, the main challenges and their potential solutions are explained below. First, one needs to know a few basic aspects of 3D printing.

3D-printing is an additive manufacturing technique, where multiple layers of materials are printed on top of each other. Some forms of 3D printing may involve material removal, but this is only excess or sacrificial material; the fabrication process still remains bottom-up, that is, based on layer-by-layer or voxel-by-voxel manufacturing. 3D printing is commonly divided into five categories, stereolithography, fused deposition modeling (FDM), selective, laser sintering (SLS), e-beam melting and laminated object manufacturing. Other variations are fused filament fabrication, fused deposition modeling, continuous fibre reinforcement, and digital light synthesis.

Among these, stereolithography and FDM are more commonly used with polymers. SLS is based on sintering (fusion of small particles at lower temperatures), where the energy is provided via a laser. This is often used for metal powders that otherwise feature a high melting point. Some ceramics can also be processed using SLS. E-beam melting is a relatively expensive technique, where an electron beam is utilized to melt the material for a layer-by-later processing. This is more commonly employed for nanoscale metal-based shapes. Laminated object manufacturing works for larger (centimeter scale) composite shapes, where layers of materials (e. g., paper) are glued on top of each other, and then a shape is cut out.

In the case of stereolithography, a liquid polymer (typically a thermosetting resin) is filled in a pot and cross-linked a UV laser. The exact wavelength of the laser can be in the UV or DUV region. One may also use other (non-UV) lasers if that is suitable for cross-linking. The light then cross-links it layer by layer as per the supplied CAD design. This process employs thermosetting photopolymers or photoresists, such as SU8 and some acrylate-based resists (sold as inks). Micro-/nanoscale 3D printing can be performed using UV lasers that utilize a two-photon polymerization process for cross-linking. Details of two-photon lithography (also known as direct laser writing or DLW) are provided in Chapter 3. After printing, these polymers can be carbonized with relatively less deformation. A nanoscale carbon lattice fabricated using DLW, followed by carbonization at 900 °C, is shown in Figure 6.11. Note the sacrificial spiral anchors at the bottom of the pattern, which were fabricated to ensure that the main shape is not adhered to the substrate. During pyrolytic shrinkage only these anchors are stretched, thus preventing the remaining pattern from deformation.

The challenge is that the currently available stereolithography-based printers are typically compatible with specific resins or inks (often nontransparent), which may not give the best results on carbonization. They are generally capable of printing only on transparent substrates, such as glasses, which may not withstand the high carbonization temperatures. There are, however, examples of printing carbonizable resists on opaque Si. One can also use sapphire or quartz substrates.

The FDM type of 3D printing is based on using filaments of thermoplastic materials that can be melt-deposited in a layered fashion. Certain sacrificial polymers are also deposited along with the target material, which are washed away in a suitable solvent after printing. Most commonly used printing materials in FDM are poly-lactic acid (PLA) and acrylonitrile butadiene styrene (ABS). Thermoplastic polymers, such as nylon, polypropylene, polyethylene, and various other polymers are also compatible with FDM [187]. However, owing to their thermoplastic nature, they start to melt and deform during the initial stages of heat-treatment (already <300 °C). Plus, their carbon yields are low. Therefore, FDM printed materials are not directly carbonizable. One possibility is to incorporate carbon fibers or carbon nanomaterials (e. g. CNTs, graphene) in the

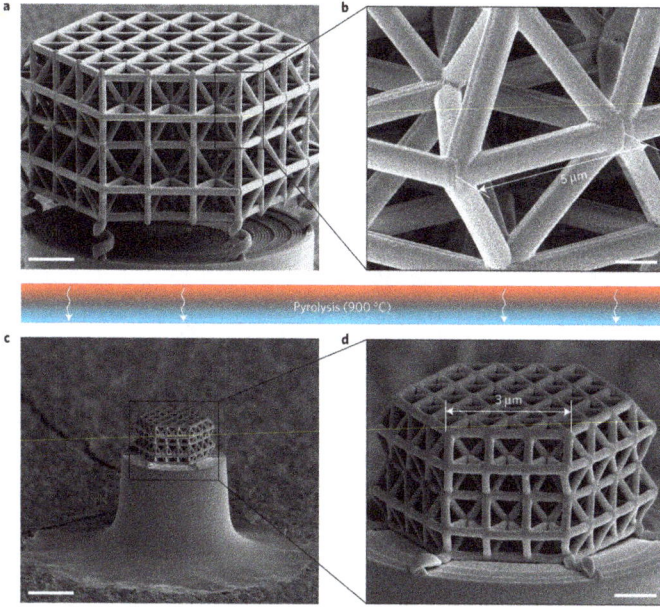

Figure 6.11: Scanning Electron Micrographs of a nano-scale carbon lattice printed via two-photon lithography. Scale bars: (a, c): 5 μm, and (b, d): 1 μm. Copyright ©2016, Springer Nature Ltd. Reproduced with permission from [21].

precursor, which is then printed via FDM. This is indeed known as continuous fiber reinforcement and is widely used for printing composite materials [187]. Notably, the final printed shape is not entirely carbon. It does contain the carrier polymer as a binder.

Other efforts to achieve a carbonized 3D-printed polymer shape include inserting cotton inside the printed shape during heat-treatment, and addition of metal nanoparticles and other reinforcements in the precursors, coating of printed shapes prior to heat-treatment, etc [111]. Development of advanced carbonizable resists and composites is another aspect with an extensive ongoing research.

Deformation-free carbonization of 3D-printed polymer shapes still remains a challenge. Any polymer that undergoes coking (i. e., passes through a soft or liquid phase during pyrolysis) experiences some degree of deformation. In 2D or 2.5D patterns, it is easier to minimize the deformation by optimizing the designs. But the same requires extensive stress-analysis (usually software-assisted) for 3D designs, and it needs to be carried out for each individual shape.

6.5 Waste-derived carbon

As briefly discussed in Chapter 3, pyrolysis of organic materials is extensively used for waste treatment worldwide. Most of the commercial pyrolysis reactors, however, utilize both combustion and pyrolysis, rather than only pyrolysis. They typically utilize a limited quantity of oxygen in the reactors, which is not sufficient for a complete combustion of the biomass. Nonetheless, both pyrolysis and combustion processes occur together. In general, it is extremely complicated and impractical to separate these two processes in a waste treatment reactor. Oxygen can lead to ash formation, which limits the applications of the resulting carbon materials and drastically reduces its economic value. This is one of the reasons that entirely pyrolysis-based waste management plants suffer from lack of sustainability.

State-of-the-art pyrolysis plants are designed to maximize the yield of syngas or pyrolysis oil, and reduce the quantity of solid carbon / ash. Due to partial pyrolysis, some solid carbon (biochar) does form in many processes, but since the process is performed at ≤650 °C, the resulting carbon is considered low-grade. This implies that it contains a lot of impurities and features only very short range order with almost insulator-like properties. Its applications are typically limited to soil porosity improvement and absorbers for environmental pollutants. There is a significant number of research articles on utilizing waste-derived carbon in high-end applications such as electrode manufacture. However, the fact that the preparation of such carbon demands higher pyrolysis temperatures (generally ≥900 °C) and strictly inert environment, hinders the commercial viability of such processes.

To address such issues, integration of pyrolysis with other waste-treatment technologies, such as microbial bioprocessing, has been attempted. In this process, the waste biomass is first treated with fungi, bacteria, or both (co-cultured) for producing any plausible valuable chemicals (e. g. lactic acid, ethanol). The remaining biomass at the end is then converted into carbon via pyrolysis. An illustration of this scheme is presented in Figure 6.12. In this process, microbes with annotated genomes were selected to improve the breakdown of large cellulose molecules and in turn, the fraction of the valuable products. Since pyrolysis entirely destroys the structure of the initial biomass, one may also use genetically engineered streams of bacteria for increasing the product yield. Importantly, the biggest challenge in using genetically engineered organisms is their disposal after the bioprocess. High-temperature degradation through pyrolysis is one of the safest ways to handle them.

> **!** There are significant differences in the electrical, electrochemical, and surface properties of carbon prepared from waste polymers with or without inert atmosphere. Carbon obtained in partial presence of oxygen contains ash and is relatively lower-grade with limited applications. Use of inert atmosphere, however, adds to the manufacturing cost. In commercial waste-treatment plants, one needs to find the right trade-off.

Whereas carbon obtained from biodegradable waste typically displays a high surface area due to charring mechanism (also see Chapter 3), nonbiodegradable synthetic polymers are more difficult to handle when present in mixed waste. Often the urban solid waste is a mixed mass of both degradable and nondegradable polymers. Each polymer may have its own chemical structure, pyrolysis mechanism, and phyisicochemical prop-

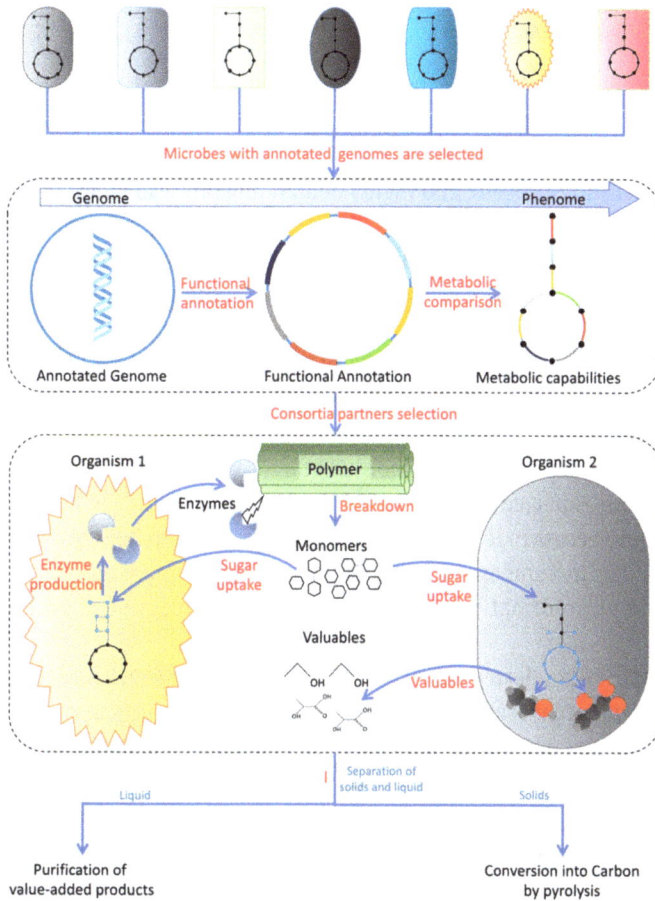

Figure 6.12: A schematic diagram showing the microbial bioprocess steps for the breakdown of cellulose molecules present in biomass, followed by a complete degradation of the residues via pyrolysis. The pyrolysis performed under inert environment enhances solid carbon production, which features a high surface area and can be further activated. Copyright ©2022 Elsevier Ltd. Reproduced with permission from [118].

erties. It becomes difficult to utilize the products in such cases, because the exact composition of the product is hard to control. This holds true not only for biochar, but also for pyrolysis oils and syngas. Filtration, fractionalization, and condensation may purify the product to a great extent, but after all, they all add to the overall cost of the process. Advanced pyrolysis methods integrated with bioprocesses, plasma treatment, and electrochemical processes may pave the way for futuristic waste management. A good understanding of pyrolysis and carbonization could be of great help in the development of not just large quantities of solid carbon but also waste-carbon-based devices.

6.6 Future of carbon technology

The future of any technology is influenced by the demands of the society, new material development, and support technologies. The very concept of manufacturing has experienced several modifications in the last 100 years. From heavy machinery, metal-based materials, very high temperature process, and highly demanding physical labor; today manufacturing has advanced to become highly automated and associated with synthetic materials. Digital manufacturing, the integration of artificial intelligence and machine learning, and the use of high-performance composite and hybrid materials has entirely revived the idea of making things. Functional devices that can take decisions and perform general tasks, such as providing medicine to a patient, have already come into existence. Human-free manufacturing units that design and synthesize their own materials are not so far-fetched anymore.

The availability of electronic prints and online study materials, such as in the form of free video lectures, is also expected to contributed significantly to new methods of flow of information. An easy access to learning material prevents researchers from repeating errors, and thus helps in a smooth technological progress. This may also lead to occasional confusions and misinformation, but ultimately the technology shall evolve to handle these hurdles.

Coming to the influence of modern technological developments on carbon device fabrication: it is the only element that can exist in the outer space and on Earth in its solid or colloidal form. It can form stable compounds and withstand extreme temperature and pressure conditions. The extensive database available on carbon materials makes carbon an attractive material for machine learning. Digital manufacturing and Internet of Things (IoT) are already connected to carbon devices, such as sensors in some way or the other.

In the next hundred years or so, it is expected that manufacturing techniques will become material-specific. As a result, they will heavily rely on material development. These materials will entail functionalities and easy manufacturability via advanced digital manufacturing techniques. The size of the devices will be largely in the micro- or nanodomain. Further size reduction is also anticipated. Manufacturing processes at nanoscale are already moving from top-down to bottom-up approaches. Many of them are inspired by natural and biological phenomena. If the size of the functional element of the devices reduces to molecular level, such self-assembly-based manufacturing techniques may become more popular in the years to come. As stated earlier, carbon material development and manufacturing processes go hand in hand, which makes it a promising futuristic material for a variety of applications. By tweaking with its heat-treatment temperature and/or other synthesis conditions, one can tune its microstructure and properties. This is one unique advantage of carbon, rarely observed in any other element. There are, though, certain challenges in carbon manufacturing that are described below.

6.6.1 Interdisciplinary approach

Another trend in the current science and technology that will strongly influence the future of device-making is the interdisciplinary approach of problem solving. The term interdisciplinary is used for any research work that requires experts from different areas or branches of science and technology for its completion. Material science and engineering is interdisciplinary by nature, as materials are developed for applications that are usually beyond the expertise of the developer. However, it is essential to explore as many applications of a new material as possible to exploit its full potential.

Carbon materials, such as CNTs and graphene, have been utilized directly or as additives in a large number of applications, ranging from single molecule detectors to very large-scale manufacturing. Desired material properties and purity, scale of production, and device or shape manufacturing may significantly differ in each case. Such aptness and versatility has become a necessity for any new-age material. It is the demand of the material that enables its commercial production and subsequently a smooth supply chain.

6.6.2 Biodegradable electronics

The next logical step of flexible electronics is biodegradable electronics. In general, biodegradable polymers are extremely important for general purpose applications and various such polymers are being developed by scientists around the world. Such polymers can also be used for printing of electronic devices, which will be the ultimate environment-friendly technology. Biodegradable electronics have already found applications in drug delivery [147]. Needless to say, carbon materials will play an important role in such devices, since carbon is already biocompatible, even if not easily degradable.

6.6.3 Low-cost and sustainable manufacturing

At many instances in this book, the reader must have felt that while high purity carbons are excellent device materials, acquiring the desired purity levels entails expensive processes. Laser-assisted carbonization is one way to achieve carbon at a low temperature, but its chemical composition, morphology, and physicochemical properties are strongly process- and precursor-dependent.

The cost of devices can be decreased by (i) reducing precursor cost, and (ii) reducing the energy requirements of the carbon manufacture. A potential third option is to simply utilize carbon materials of relatively low purity (90–95 %) in advanced applications. Among low-cost precursors, in this book we have discussed inexpensive furan resins, some biomass-derived materials, commonly available small hydrocarbons, and petroleum byproducts, such as pitch and coke, which can be used for carbon production. More research is needed for designing organic compounds with a high carbon yield without compromising their manufacturability. Among existing low-cost manufacturing methods, autoclave-assisted (hydrothermal, solvothermal, solvent-free) syntheses and laser-assisted carbonization have an immense potential in micro-/nanofabrication. These processes are based on inexpensive precursors, which is an added advantage. LP-C can be attempted on a wide variety of precursors and the laser wavelength can be optimized to decrease the critical dimensions (line width in this case).

The cost of carbon is strongly dependent on two factors: purity and crystallinity. Even in the case of non-graphitizing carbons and carbon fibers, reducing the 002 layer separation to approach that of graphite has always been the goal of every carbon manufacturer. Despite the development of numerous new carbon forms, graphite still remains the gold standard and is one of the most expensive carbon materials when synthetically produced. Its electrical conductivity, beautifully organized anisotropic structure, electron transport pathways, and intercalation ability still remain unmatched. But in miniaturized devices having micro- and nanoscale dimensions, it is very challenging to use graphite without destroying its structure. Its high thermal and chemical stability renders it difficult to pattern. As a result, single layers separated from graphite (i. e., graphene) are preferred in micro-/nanofabrication. In the last two decades or so, more attention has been paid to increasing the L_a in graphite-related carbon materials, rather than reducing the L_c. In the 1960s, a material was not even considered a usable form of carbon if it had a layer separation >0.344 nm. But nowadays, carbons with up to >0.4 nm are designed and developed for advanced energy storage applications. This has definitely changed the perspective of many carbon scientists. Importantly, owing to its versatility, newer application areas and revised material characterization parameters have only increased the popularity of carbon.

It is often difficult for researchers working in the field of carbon to accept lower purity carbon materials in the carbon family, which I must admit, is also the case with me. Carbon materials have always experienced problems with their nomenclature, and advanced hybrid materials further complicate the issue. Nonetheless, the increasing demands of different aspects of device technology and their large-scale production demands a much wider variety of functional carbon materials. Of course, there is no replacement for elemental forms of carbon when it comes to understanding their fascinating properties and scientific phenomena responsible for the formation of this wonderful element.

1. What are flexible devices? Are all flexible devices wearable? Explain with examples?
2. Do flexible substrates always need to be organic? Give an example of an inorganic flexible substrate.
3. Which top-down and bottom up techniques are most suitable for fabricating carbon devices on to flexible substrates?
4. What is the fundamental principle of laser-assisted carbonization?
5. What changes in the crystal structure of graphite can be expected if <1 % nitrogen atoms are present?
6. What is the difference between a hydrothermal and a solvothermal synthesis? If only solid reactants are used for a reaction inside an autoclave, would it be called hydrothermal or solvothermal?
7. Is S-containing or N-containing carbon better for making a sodium ion battery anode? Justify your answer.
8. Compare laser-patterned carbon and carbon fiber-based flexible devices (refer to Chapter 5 for carbon fiber-based devices).
9. Can quantum dots be used for fabricating wearable devices? If yes, which fabrication techniques should be used for this purpose?
10. Name at least two carbonizable polymers that can be 3D-printed using the current stereolithographic techniques.
11. What is the difference between N-doped carbon and N-containing carbon?
12. Pattern transfer is used for making flexible carbon devices. Explain the similarities and differences in pattern transfer process used for carbon nanomaterials and bulk disordered carbon, such as glass-like carbon.

A Additional questions (all chapters)

1. Classify the following carbon allotropes in terms of their hybridization: C_{70}, graphite, porous carbon, carbon quantum dots, electrospun PAN-derived carbon fibers, laser-patterned carbon, Lonsdalite.

2. Some applications areas are mentioned below (a–g). For each one of them, list at least three possible carbon materials that can be utilized. Also mention which component(s) of the device can be made using carbon, wherever appropriate.
   (a) EDLC type supercapacitor,
   (b) composite material having a polymer matrix,
   (c) electrocatalyst,
   (d) field effect transistor,
   (e) neural sensor,
   (f) 3D-cell culture platform,
   (g) wearable battery.

3. Solar cell technology relies on semiconductor material, which can be excited using a photon. Name at least two carbon materials that can be used as active material (the one that responds to photons) and can be used in solar cells.

4. Mention the (i) raw material(s), (ii) fabrication technique(s), and (iii) carbonization conditions for manufacturing the following:
   (a) A carbon nanofiber suspended on carbon microelectrodes.
   (b) Activated carbon cylinder of 2 cm length and 1 cm diameter.
   (c) A conical shape in micro- to nanometer scale composed of a non-graphitizing carbon.
   (d) A piece of HOPG having the following dimensions: 1 cm × 1 cm × 2 mm.

5. Mention the difference between the following:
   (a) Coking and charring mechanism of polymer carbonization.
   (b) Porous and activated carbon.
   (c) Amorphous and disordered carbon.
   (d) Graphitizing and non-graphitizing carbon.
   (e) Organic and flexible electronics.
   (f) *In-vivo* and *in-vitro* biocompatibility of carbon.
   (g) Electrospinning and melt-spinning.
   (h) Carbonization and graphitization.

6. Explain all application area of the pyrolysis process. How is pyrolysis for carbon manufacturing different from that used in waste degradation?

7. A microfluidic device shown in Figure A.1(A) (top and side view with dimensions) has rectangular channels made of carbon walls and a silicon bottom surface. Its top surface can be made of any polymer. All surfaces are attached without any external glue / binder such that there is no leakage. Mention the raw material(s), fabrication technique(s), heat-treatment conditions, and any other steps involved in manufacturing this device. You are provided with a standard silicon wafer substrate, furnace for carbonization, and any other material / equipment / set-up/ glassware/ chemicals you need. For the top layer you can select any polymer of up to 1 mm thickness as long as you clearly mention its name. Minor error in the shape and dimensions is allowed (for example, if the walls are not perfectly straight, it is OK).

8. In Figure A.1(B) you see carbon nanotube (CNTs) growing on all sides of a carbon fiber. Mention all the steps for making such a structure. You can use any materials (including catalysts) and manufacturing processes.

9. Mention all the steps, materials, methods and parameters to manufacture the following: A hollow polycrystalline graphite cylinder that contains cast iron filling as shown in Figure A.1(C). The top plate has a

https://doi.org/10.1515/9783110620634-007

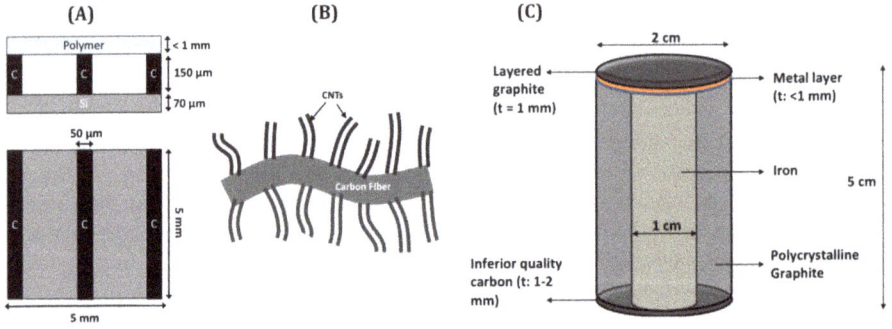

(A)

Polymer ↕ < 1 mm
                150 µm
C   C   C
        Si   ↕ 70 µm

50 µm
↔

C   C   C

5 mm

5 mm

(B)

CNTs

Carbon Fiber

(C)

2 cm

Layered
graphite
(t = 1 mm)

Metal layer
(t: <1 mm)

Iron   5 cm

1 cm

Inferior quality
carbon (t: 1-2
mm)

Polycrystalline
Graphite

Figure A.1: (A) A microfluidic device made of carbon walls and other polymers, (B) CNTs growing onto a carbon (micro)fiber surface in all directions, (C) Hollow graphite cylinder with cast iron filling (t = thickness).

layered graphite structure with a mosaic spread of 0.8°. Between the top graphite plate and the cylindrical structure, there is a thin metal layer (any metal can be chosen). The carbon in the bottom layer should be of lower purity and contain porosity. Note that after filling, iron should not be heated > 900 °C. There should be no glue/ adhesive or weak joints.

Bibliography

[1]   GradeJP932, howpublished = https://www.graphite-eng.com/uploads/downloads/grade_jp932.pdf, note = Accessed: Sept, 2015. 2015.

[2]   E. G. Acheson. Production of artificial crystalline carbonaceous materials, February 1893.

[3]   T. Adinaveen, J. J. Vijaya, and L. J. Kennedy. Comparative study of electrical conductivity on activated carbons prepared from various cellulose materials. *Arabian Journal for Science and Engineering*, 41(1):55–65, January 2016.

[4]   P. M. Ajayan and O. Z. Zhou. *Applications of Carbon Nanotubes*, pages 391–425. Springer Berlin Heidelberg, Berlin, Heidelberg, 2001.

[5]   D. Akinyele, E. Olabode, and A. Amole. Review of fuel cell technologies and applications for sustainable microgrid systems. *Inventions*, 5(3) 2020.

[6]   S. Akkoyun and N. Öktem. Effect of viscoelasticity in polymer nanofiber electrospinning: Simulation using fene-cr model. *Engineering Science and Technology, an International Journal*, 24(3):620–630, 2021.

[7]   W. Al-Hajri, Y. De Luna, and N. Bensalah. Review on recent applications of nitrogen-doped carbon materials in co2 capture and energy conversion and storage. *Energy Technology*, 10(12):2200498, 2022.

[8]   M. H. Al-Saleh and U. Sundararaj. Review of the mechanical properties of carbon nanofiber/polymer composites. *Composites. Part A, Applied Science and Manufacturing*, 42(12):2126–2142, 2011.

[9]   R. Alcántara, J. M. Jiménez-Mateos, P. Lavela, and J. L. Tirado. Carbon black: a promising electrode material for sodium-ion batteries. *Electrochemistry Communications*, 3(11):639–642, 2001.

[10]   M. J. Allen, V. C. Tung, and R. B. Kaner. Honeycomb carbon: A review of graphene. *Chemical Reviews*, 110(1):132–145, January 2010.

[11]   L. Amato, L. Schulte, A. Heiskanen, S. S. Keller, S. Ndoni, and J. Emnéus. Novel nanostructured electrodes obtained by pyrolysis of composite polymeric materials. *Electroanalysis*, 27(7):1544–1549, July 2015.

[12]   S. Ammu, V. Dua, S. R. Agnihotra, S. P. Surwade, A. Phulgirkar, S. Patel, and S. K. Manohar. Flexible, all-organic chemiresistor for detecting chemically aggressive vapors. *Journal of the American Chemical Society*, 134(10):4553–4556, 03 2012.

[13]   E. M. Anderson, M. L. Stone, R. Katahira, M. Reed, W. Muchero, K. J. Ramirez, G. T. Beckham, and Y. Román-Leshkov. Differences in s/g ratio in natural poplar variants do not predict catalytic depolymerization monomer yields. *Nature Communications*, 10(1):2033, 2019.

[14]   B. Apicella, A. Tregrossi, F. Stazione, A. Ciajolo, and C. Russo. Analysis of petroleum and coal tar pitches as large pah. *Chemical Engineering Transactions*, 57:775–780, 2017.

[15]   N. Arora and N. N. Sharma. Arc discharge synthesis of carbon nanotubes: Comprehensive review. *Diamond and Related Materials*, 50:135–150, November 2014.

[16]   S. Bae, H. Kim, Y. Lee, X. Xu, J.-S. Park, Y. Zheng, J. Balakrishnan, T. Lei, H. Ri Kim, Y. I. Song, Y.-J. Kim, K. S. Kim, B. Özyilmaz, J.-H. Ahn, B. H. Hong, and S. Iijima. Roll-to-roll production of 30-inch graphene films for transparent electrodes. *Nature Nanotechnology*, 5(8):574–578, August 2010.

[17]   O. P. Bahl and T. L. Dhami, editors. *Advances in Carbon Materials: Proceedings of the National Conference, Delhi, November, 1999*. Shipra Publications, Delhi, 2000.

[18]   O. Bahl, G. Bhatia, R. B. Mathur, R. K. Aggarwal, T. L. Dhami, and C. Lal. Development of advanced carbon products at national physical laboratory, New Delhi, India. *Journal of Scientific and Industrial Research*, 59:877–892, January 2000.

[19]   D. F. Baker and R. H. Bragg. The electrical conductivity and hall effect of glassy carbon. *Journal of Non-Crystalline Solids*, 58(1):57–69, 1983.

[20]   F. Banhart, J. Kotakoski, and A. V. Krasheninnikov. Structural defects in graphene. *ACS Nano*, 5(1):26–41, January 2011.

[21]   J. Bauer, A. Schroer, R. Schwaiger, and O. Kraft. Approaching theoretical strength in glassy carbon nanolattices. *Nature Materials*, 15(4):438–443, February 2016.

https://doi.org/10.1515/9783110620634-008

[22] A. Bianco, H.-M. Cheng, T. Enoki, Y. Gogotsi, R. H. Hurt, N. Koratkar, T. Kyotani, M. Monthioux, C. R. Park, J. M. D. Tascon, and J. Zhang. All in the graphene family – A recommended nomenclature for two-dimensional carbon materials. *Carbon*, 65:1–6, December 2013.

[23] V. Bochenkov and G. Sergeev. *Sensitivity, Selectivity, and Stability of Gas-Sensitive Metal-Oxide Nanostructures*, volume 3, pages 31–52. American Scientific Publishers, 01 2010.

[24] H. Bockholt, W. Haselrieder, and A. Kwade. Intensive powder mixing for dry dispersing of carbon black and its relevance for lithium-ion battery cathodes. *Powder Technology*, 297:266–274, 2016.

[25] H. P. Boehm, A. Clauss, G. O. Fischer, and U. Hofmann. Das Adsorptionsverhalten sehr dünner Kohlenstoff-Folien. *Zeitschrift für Anorganische und Allgemeine Chemie*, 316(3–4):119–127, July 1962.

[26] H. P. Boehm, R. Setton, and E. Stumpp. Nomenclature and terminology of graphite intercalation compounds (IUPAC Recommendations 1994). *Pure and Applied Chemistry*, 66(9):1893–1901, January 1994.

[27] L. Boumia, M. Zidour, A. Benzair, and A. Tounsi. Physica E: Low-dimensional Systems and Nanostructures, 59:186, 2014.

[28] A. M. Brasil, T. L. Farias, and M. G. Carvalho. A recipe for image characterization of fractal-like aggregates. *Journal of Aerosol Science*, 30(10):1379–1389, 1999.

[29] J.-L. Bredas and J. R. Durrant. Organic Photovoltaics. *Accounts of Chemical Research*, 42(11):1689–1690, November 2009.

[30] J.-P. Brog, C.-L. Chanez, A. Crochet, and K. M. Fromm. Polymorphism, what it is and how to identify it: a systematic review. *RSC Advances*, 3:16905–16931, 2013.

[31] F. Bruner, G. Crescentini, and F. Mangani. Graphitized carbon black: A unique adsorbent for gas chromatography and related techniques. *Chromatographia*, 30(9):565–572, 1990.

[32] F. Bruner, P. Ciccioli, and F. Di Nardo. Use of graphitized carbon black in environmental analysis. *Journal of Chromatography A*, 99:661–672, 1974.

[33] A.-I. Bunea, N. del Castillo Iniesta, A. Droumpali, A. E. Wetzel, E. Engay, and R. Taboryski. Micro 3d printing by two-photon polymerization: Configurations and parameters for the nanoscribe system. *Micro*, 1(2):164–180, 2021.

[34] J. R. Camargo, T. A. Silva, G. A. Rivas, and B. C. Janegitz. Novel eco-friendly water-based conductive ink for the preparation of disposable screen-printed electrodes for sensing and biosensing applications. *Electrochimica Acta*, 409:139968, 2022.

[35] H. Chen, S. Chen, Y. Zhang, H. Ren, X. Hu, and Y. Bai. Sand-milling fabrication of screen-printable graphene composite inks for high-performance planar micro-supercapacitors. *ACS Applied Materials & Interfaces*, 12(50):56319–56329, 2020. PMID: 33280375.

[36] S. Chen, L. Qiu, and H.-M. Cheng. Carbon-based fibers for advanced electrochemical energy storage devices. *Chemical Reviews*, 120(5):2811–2878, 03 2020.

[37] M.-Y. Cho, J. H. Lee, S.-H. Kim, J. S. Kim, and S. Timilsina. An extremely inexpensive, simple, and flexible carbon fiber electrode for tunable elastomeric piezo-resistive sensors and devices realized by LSTM RNN. *ACS Applied Materials & Interfaces*, 11(12):11910–11919, March 2019.

[38] W. Choi, I. Lahiri, R. Seelaboyina, and Y. S. Kang. Synthesis of graphene and its applications: A review. *Critical Reviews in Solid State and Materials Sciences*, 35(1):52–71, 2010.

[39] A. Chuvilin, U. Kaiser, E. Besley, N. Besley, and A. Khlobystov. Direct transformation of graphene to fullerene. *Nature Chemistry*, 2:450–453, 06 2010.

[40] M. N. Collins, M. Culebras, and G. Ren. Chapter 8 – the use of lignin as a precursor for carbon fiber–reinforced composites. In D. Puglia, C. Santulli and F. Sarasini, editors, *Micro and Nanolignin in Aqueous Dispersions and Polymers*, pages 237–250. Elsevier, 2022.

[41] C. Daulbayev, B. Kaidar, F. Sultanov, B. Bakbolat, G. Smagulova, and Z. Mansurov. The recent progress in pitch derived carbon fibers applications. a review. *South African Journal of Chemical Engineering*, 38:9–20, 2021.

[42] V. Dejke, M. P. Eng, K. Brinkfeldt, J. Charnley, D. Lussey, and C. Lussey. Development of prototype low-cost QTSS™ wearable flexible more enviro-friendly pressure, shear, and friction sensors for dynamic prosthetic fit monitoring. *Sensors*, 21(11):3764, May 2021.

[43] A. Denneulin, J. Bras, A. Blayo, and C. Neuman. Substrate pre-treatment of flexible material for printed electronics with carbon nanotube based ink. *Applied Surface Science*, 257(8):3645–3651, February 2011.

[44] S. S. Desa, T. Ishii, and K. Nueangnoraj. Sulfur-doped carbons from durian peels, their surface characteristics, and electrochemical behaviors. *ACS Omega*, 6(38):24902–24909, September 2021.

[45] T. G. Desai, J. W. Lawson, and P. Keblinski. Modeling initial stage of phenolic pyrolysis: Graphitic precursor formation and interfacial effects. *Polymer*, 52(2):577–585, 2011.

[46] M. M. Deshmukh and V. Singh. Graphene — An exciting two-dimensional material for science and technology. *Resonance*, 16(3):238–253, March 2011.

[47] M. Devi, C. Madan, A. Halder, and S. Sharma. Laser-derived porous carbon as a metal-free electrocatalyst for oxygen evolution reaction. *Carbon Trends*, 9:100221, October 2022.

[48] M. Devi, B. Raut, and S. Sharma. Laser-patterned carbon-supported graphitic carbon nitride quantum dots for flexible nanozyme-based fluoride sensor. *Particle & Particle Systems Characterization*, n/a(n/a):2300018, 2023.

[49] M. Devi, S. Rawat, and S. Sharma. A comprehensive review of the pyrolysis process: from carbon nanomaterial synthesis to waste treatment. *Oxford Open Materials Science*, 1(1):itab014, 11 2021.

[50] M. Devi, M. Vomero, E. Fuhrer, E. Castagnola, C. Gueli, S. Nimbalkar, M. Hirabayashi, S. Kassegne, T. Stieglitz, and S. Sharma. Carbon-based neural electrodes: promises and challenges. *Journal of Neural Engineering*, 18(4):041007, August 2021.

[51] M. Devi, H. Wang, S. Moon, S. Sharma, and V. Strauss. Laser-Carbonization – A powerful tool for micro-fabrication of patterned electronic carbons. *Advanced Materials*, page 2211054, February 2023.

[52] A. Di Corcia, P. Ciccioli, and F. Bruner. Gas chromatography of some reactive gases on graphitized carbon black. *Journal of Chromatography A*, 62(1):128–131, 1971.

[53] M. S. Dresselhaus, G. Dresselhaus, and M. Hofmann. Raman spectroscopy as a probe of graphene and carbon nanotubes. *Philosophical Transactions of the Royal Society A: Mathematical, Physical and Engineering Sciences*, 366(1863):231–236, 2008.

[54] M. S. Dresselhaus, G. Dresselhaus, and R. Saito. Physics of carbon nanotubes. *Carbon*, 33(7):883–891, 1995. Nanotubes.

[55] D. R. Dreyer, R. S. Ruoff, and C. W. Bielawski. From conception to realization: An historial account of graphene and some perspectives for its future. *Angewandte Chemie International Edition*, 49(49):9336–9344, 2010.

[56] T. W. Ebbesen and P. M. Ajayan. Large-scale synthesis of carbon nanotubes. *Nature*, 358(6383):220–222, 1992.

[57] T. W. Ebbesen, H. J. Lezec, H. Hiura, J. W. Bennett, H. F. Ghaemi, and T. Thio. *Nature*, 382:54–56, 1996.

[58] D. D. Edie and M. G. Dunham. Melt spinning pitch-based carbon fibers. *Carbon*, 27(5):647–655, 1989. Carbon Fibers and Composites.

[59] M. Endo, Y. A. Kim, T. Hayashi, K. Nishimura, T. Matusita, K. Miyashita, and M. S. Dresselhaus. Vapor-grown carbon fibers (vgcfs): Basic properties and their battery applications. *Carbon*, 39(9):1287–1297, 2001.

[60] M. Endo, T. Hayashi, Y. A. Kim, and H. Muramatsu. Development and application of carbon nanotubes. *Japanese Journal of Applied Physics*, 45(6R):4883, jun 2006.

[61] B. Fadeel, C. Bussy, S. Merino, E. Vázquez, E. Flahaut, F. Mouchet, L. Evariste, L. Gauthier, A. J. Koivisto, U. Vogel, C. Martín, L. G. Delogu, T. Buerki-Thurnherr, P. Wick, D. Beloin-Saint-Pierre, R. Hischier, M. Pelin, F. Candotto Carniel, M. Tretiach, F. Cesca, F. Benfenati, D. Scaini, L. Ballerini, K. Kostarelos, M. Prato, and A. Bianco. Safety assessment of graphene-based materials: Focus on human health and the environment. *ACS Nano*, 12(11):10582–10620, 2018. PMID: 30387986.

[62] G. Fau, N. Gascoin, and J. Steelant. Hydrocarbon pyrolysis with a methane focus: A review on the catalytic effect and the coke production. *Journal of Analytical and Applied Pyrolysis*, 108, 07 2014.

[63] M. Ferrier, A. De Martino, A. Kasumov, S. Guéron, M. Kociak, R. Egger, and H. Bouchiat. Superconductivity in ropes of carbon nanotubes. *Solid State Communications*, 131(9):615–623, 2004. New advances on collective phenomena in one-dimensional systems.

[64]   E. Fitzer. From silicon to carbon. *Carbon*, 16(1):3–16, 1978.

[65]   E. Fitzer, W. Frohs, and M. Heine. Optimization of stabilization and carbonization treatment of pan fibres and structural characterization of the resulting carbon fibres. *Carbon*, 24(4):387–395, 1986.

[66]   E. Fitzer, K. Mueller, and W. Schaefer. Conversion of organic compounds to carbon. In *Chemistry and Physics of Carbon*, volume 7. Marcel Dekker, Inc., New York, 1971.

[67]   E. Fitzer and W. Schäfer. The effect of crosslinking on the formation of glasslike carbons from thermosetting resins. *Carbon*, 8(3):353–364, June 1970.

[68]   E. Fitzer, A. Gkogkidis, and H. Michael. Carbon fibers and their composites (A review). *High Temperatures. High Pressures*, 363–392, 1984.

[69]   W. Francis. *Coal, Its Formation and Composition*. E. Arnold, London, 1961.

[70]   R. E. Franklin. Crystallite growth in graphitizing and non-graphitizing carbons. *Proceedings of the Royal Society of London A: Mathematical, Physical and Engineering Sciences*, 209(1097):196–218, 1951.

[71]   J. Friedl, M. A. Lebedeva, K. Porfyrakis, U. Stimming, and T. W. Chamberlain. All-fullerene-based cells for nonaqueous redox flow batteries. *Journal of the American Chemical Society*, 140(1):401–405, 01 2018.

[72]   J. E. Fromm. Numerical calculation of the fluid dynamics of drop-on-demand jets. *IBM Journal of Research and Development*, 28(3):322–333, May 1984.

[73]   E. Fuhrer, A. Bäcker, S. Kraft, F. J. Gruhl, M. Kirsch, N. MacKinnon, J. G. Korvink, and S. Sharma. 3D carbon scaffolds for neural stem cell culture and magnetic resonance imaging. *Advanced Healthcare Materials*, 7(4):1700915, 2018.

[74]   S. B. Fuller, E. J. Wilhelm, and J. M. Jacobson. Ink-jet printed nanoparticle microelectromechanical systems. *Journal of Microelectromechanical Systems*, 11(1):54–60, February 2002.

[75]   O. Garate, L. Veiga, A. V. Medrano, G. Longinotti, G. Ybarra, and L. N. Monsalve. Waterborne carbon nanotube ink for the preparation of electrodes with applications in electrocatalysis and enzymatic biosensing. *Materials Research Bulletin*, 106:137–143, 2018.

[76]   C. Garion. Mechanical properties for reliability analysis of structures in glassy carbon. *World Journal of Mechanics*, 04(03):79–89, 2014.

[77]   C. Gassner. Galvanic battery, 1886 (Germany), 1887 (USA). Patent No. 37758 (Germany), 373064 (USA).

[78]   J. Gelfond, T. Meng, S. Li, T. Li, and L. Hu. Highly electrically conductive biomass-derived carbon fibers for permanent carbon sequestration. *Sustainable Materials and Technologies*, 35:e00573, 2023.

[79]   M. A. Ghanem, J.-M. Chrétien, A. Pinczewska, J. D. Kilburn, and P. N. Bartlett. Covalent modification of glassy carbon surface with organic redox probes through diamine linkers using electrochemical and solid-phase synthesis methodologies. *Journal of Materials Chemistry*, 18:4917–4927, 2008.

[80]   V. Gopal, A. Venkataraman, L. Babu, and R. Rajan. Preparation of black lyophilic ink using the carbon soot emitted by vehicles. *Environmental Science and Pollution Research*, 28(45):63440–63447, Dec 2021.

[81]   K. Y. Goud, S. S. Sandhu, H. Teymourian, L. Yin, N. Tostado, F. M. Raushel, S. P. Harvey, L. C. Moores, and J. Wang. Textile-based wearable solid-contact flexible fluoride sensor: Toward biodetection of g-type nerve agents. *Biosensors & Bioelectronics*, 182:113172, 2021.

[82]   G. Grau, J. Cen, H. Kang, R. Kitsomboonloha, W. J. Scheideler, and V. Subramanian. Gravure-printed electronics: recent progress in tooling development, understanding of printing physics, and realization of printed devices. *Flexible and Printed Electronics*, 1(2):023002, June 2016.

[83]   F. Grosshans. How to determine the (n,m) dimensions of a carbon nanotube?, https://physics. stackexchange.com/users/373/fr%c3%a9d%c3%a9ric-grosshans, URL (version: 2010-12-03): https://physics.stackexchange.com/q/1589.

[84]   C. Gueli, M. Vomero, S. Sharma, and T. Stieglitz. Integration of micro-patterned carbon fiber mats into polyimide for the development of flexible implantable neural devices. In *2019 41st Annual International Conference of the IEEE Engineering in Medicine and Biology Society (EMBC)*, pages 3931–3934. Berlin, Germany, IEEE, July 2019.

[85]   U. Gulzar, S. Goriparti, E. Miele, T. Li, G. Maidecchi, A. Toma, F. De angelis, C. Capiglia, and R. Zaccaria. Next-generation textiles: From embedded supercapacitors to lithium ion batteries. *Journal of Materials Chemistry A*, 4:16771–16800, 09 2016.

[86] A. Gupta, S. R. Dhakate, P. Pal, A. Dey, P. K. Iyer, and D. K. Singh. Effect of graphitization temperature on structure and electrical conductivity of poly-acrylonitrile based carbon fibers. *Diamond and Related Materials*, 78:31–38, 2017.

[87] R. C. Haddon. Hybridization and the orientation and alignment of pi.-orbitals in nonplanar conjugated organic molecules: pi.-orbital axis vector analysis (poav2). *Journal of the American Chemical Society*, 108(11):2837–2842, 1986.

[88] R. C. Haddon, R. E. Palmer, H. W. Kroto, P. A. Sermon, H. W. Kroto, M. A. Lindsay, T. Grenville, and D. R. M. Walton. The fullerenes: powerful carbon-based electron acceptors. *Philosophical Transactions of the Royal Society of London. Series A: Physical and Engineering Sciences*, 343(1667):53–62, April 1993.

[89] E. Hammel, X. Tang, M. Trampert, T. Schmitt, K. Mauthner, A. Eder, and P. Pötschke. Carbon nanofibers for composite applications. *Carbon*, 42(5):1153–1158, 2004. European Materials Research Society 2003, Symposium B: Advanced Multifunctional Nanocarbon Materials and Nanosystems.

[90] P. J. F. Harris. Fullerene-related structure of commercial glassy carbons. *Philosophical Magazine*, 84(29):3159–3167, 2004.

[91] P. J. F. Harris. Transmission electron microscopy of carbon: A brief history. *C*, 4(1), 2018.

[92] Y. M. Hassan, C. Caviglia, S. Hemanth, D. M. A. Mackenzie, T. S. Alstrøm, D. H. Petersen, and S. S. Keller. High temperature su-8 pyrolysis for fabrication of carbon electrodes. *Journal of Analytical and Applied Pyrolysis*, 125:91–99, 2017.

[93] M. Hatala, P. Gemeiner, M. Hvojnik, and M. Mikula. The effect of the ink composition on the performance of carbon-based conductive screen printing inks. *Journal of Materials Science. Materials in Electronics*, 30(2):1034–1044, Jan 2019.

[94] R. M. Hazen, A. P. Jones, and J. A. Baross. *Carbon in Earth*. De Gruyter, Inc., Boston, 2018. OCLC: 1076808640.

[95] J.-H. He. On the height of Taylor cone in electrospinning. *Results in Physics*, 17:103096, June 2020.

[96] P. He, J. Cao, H. Ding, C. Liu, J. Neilson, Z. Li, I. A. Kinloch, and B. Derby. Screen-printing of a highly conductive graphene ink for flexible printed electronics. *ACS Applied Materials & Interfaces*, 11(35):32225–32234, 2019. PMID: 31390171.

[97] R. D. Heidenreich, W. M. Hess, and L. L. Ban. A test object and criteria for high resolution electron microscopy. *Journal of Applied Crystallography*, 1(1):1–19, 1968.

[98] R. B. Heimann, S. Evsyukov, and L. Kavan. *Carbyne and Carbynoid Structures*. Kluwer Academic Publisher, 01 1999.

[99] G. Hennig, K.-H. Selbmann, and A. Brockelt. Laser engraving in gravure industry. In W. Gries and T. P. Pearsall, editors, *Proceedings Volume 6157, Workshop on Laser Applications in Europe*, page 61570C, Dresden, Germany, December 2005.

[100] J. Heremans and C. P. Beetz. Thermal conductivity and thermopower of vapor-grown graphite fibers. *Physical Review B*, 32:1981–1986, Aug 1985.

[101] W. P. Hoffman, F. J. Vastola, and P. L. Walker. Pyrolysis of propylene over carbon active sites—i: Kinetics. *Carbon*, 23(2):151–161, 1985.

[102] https://www.scirp.org/html/6-1860055_25135.htm.

[103] A. Horoschenkoff and C. Christner. Carbon fibre sensor: theory and application. In N. Hu, editor, *Composites and Their Applications*. InTech, August 2012.

[104] Z. D. Hu, Y. F. Hu, Q. Chen, X. F. Duan, and L.-M. Peng. Synthesis and characterizations of amorphous carbon nanotubes by pyrolysis of ferrocene confined within AAM templates. *The Journal of Physical Chemistry B*, 110(16):8263–8267, April 2006.

[105] Y.-J. Huang, H.-C. Wu, N.-H. Tai, and T.-W. Wang. Carbon nanotube rope with electrical stimulation promotes the differentiation and maturity of neural stem cells. *Small*, 8(18):2869–2877, September 2012.

[106] T. Huggins, H. Wang, J. Kearns, P. Jenkins, and Z. J. Ren. Biochar as a sustainable electrode material for electricity production in microbial fuel cells. *Bioresource Technology*, 157:114–119, 2014.

[107] Z. Husain, A. R. Shakeelur Raheman, K. B. Ansari, A. B. Pandit, M. S. Khan, M. A. Qyyum, and S. S. Lam. Nano-sized mesoporous biochar derived from biomass pyrolysis as electrochemical energy storage supercapacitor. *Materials Science for Energy Technologies*, 5:99–109, 2022.

[108] S. Iijima. Helical microtubules of graphitic carbon. *Nature*, 354(6348):56–58, 1991.

[109] M. Inagaki. Chapter 4 – carbon fibers. In M. Inagaki, editor, *New Carbons - Control of Structure and Functions*, pages 82–123. Elsevier Science, Oxford, 2000.

[110] M. Inagaki. Chapter 2.1 – advanced carbon materials. In S. Somiya, editor, *Handbook of Advanced Ceramics (Second Edition)*, pages 25–60. Academic Press, Oxford, second edition, 2013.

[111] M. Islam, A. D. Lantada, D. Mager, and J. G. Korvink. Carbon-based materials for articular tissue engineering: From innovative scaffolding materials toward engineered living carbon. *Advanced Healthcare Materials*, 11(1):2101834, 2022.

[112] I. M. K. Ismail. Structure and active surface area of carbon fibers. *Carbon*, 25(5):653–662, 1987.

[113] E. Jabari, F. Ahmed, F. Liravi, E. B. Secor, L. Lin, and E. Toyserkani. 2D printing of graphene: a review. *2D Materials*, 6(4):042004, August 2019.

[114] G. M. Jenkins and K. Kawamura. Structure of glassy carbon. *Nature*, 231(5299):175–176, 1971.

[115] G. M. Jenkins. *Polymeric Carbons–Carbon Fibre, Glass and Char*. Cambridge University Press, Cambridge, New York, 1976.

[116] W. B. Jensen. Ask the historian: The origin of the polymer concept. *Journal of Chemical Education*, 88:624–625, 2008.

[117] Z. Jiang, Y. Zhao, X. Lu, and J. Xie. Fullerenes for rechargeable battery applications: Recent developments and future perspectives. *Journal of Energy Chemistry*, 55:70–79, 2021.

[118] C. Joshi, M. Kumar, M. Bennett, J. Thakur, D. J. Leak, S. Sharma, N. MacKinnon, and S. K. Masakapalli. Synthetic microbial consortia bioprocessing integrated with pyrolysis for efficient conversion of cellulose to valuables. *Bioresource Technology Reports*, 21:101316, 2023.

[119] M. Kakunuri, S. Kaushik, A. Saini, and C. S. Sharma. Su-8/mwcnt derived electrospun composite carbon nanofabric as a high performance anode material for lithium ion battery. *ECS Transactions*, 72(1):69, 2016.

[120] S. M. P. Kalaiselvi, T. L. Tan, A. Talebitaher, P. Lee, S. P. Heussler, M. B. H. Breese, and R. S. Rawat. X-ray lithography of su8 photoresist using fast miniature plasma focus device and its characterization using ftir spectroscopy. *Physics Letters A*, 379(6):560–569, 2015.

[121] G. Kalita and M. Tanemura. Fundamentals of chemical vapor deposited graphene and emerging applications. In G. Z. Kyzas and A. Ch. Mitropoulos, editors, *Graphene Materials - Advanced Applications*. InTech, May 2017.

[122] S. R. Kamali, C.-N. Chen, D. C. Agrawal, and T.-H. Wei. Sulfur-doped carbon dots synthesis under microwave irradiation as turn-off fluorescent sensor for Cr(III). *Journal of Analytical Science and Technology*, 12(1):48, October 2021.

[123] R. Kamath and M. J. Madou. Selective detection of dopamine against ascorbic acid interference using 3d carbon interdigitated electrode arrays. *ECS Transactions*, 61(7):65–73, Mar 2014.

[124] R. R. Kamath and M. J. Madou. Three-dimensional carbon interdigitated electrode arrays for redox-amplification. *Analytical Chemistry*, 86(6):2963–2971, March 2014.

[125] C. W. Kang, Y.-J. Ko, S. M. Lee, H. J. Kim, J. Choi, and S. U. Son. Carbon black nanoparticle trapping: a strategy to realize the true energy storage potential of redox-active conjugated microporous polymers. *Journal of Materials Chemistry A*, 9:17978–17984, 2021.

[126] D. Kang, P. V. Pikhitsa, Y. W. Choi, C. Lee, S. S. Shin, L. Piao, B. Park, K.-Y. Suh, T.-I. Kim, and M. Choi. Ultrasensitive mechanical crack-based sensor inspired by the spider sensory system. *Nature*, 516(7530):222–226, 2014.

[127] I. Kang, Y. Y. Heung, J. H. Kim, J. W. Lee, R. Gollapudi, S. Subramaniam, S. Narasimhadevara, D. Hurd, G. R. Kirikera, V. Shanov, M. J. Schulz, D. Shi, J. Boerio, S. Mall, and M. Ruggles-Wren. Introduction to carbon nanotube and nanofiber smart materials. *Composites. Part B, Engineering*, 37(6):382–394, 2006. JCOM 731 "Nanoengineered composites and Ceramic Laminates" Special Issue.

[128] W. Kiciński, M. Szala, and M. Bystrzejewski. Sulfur-doped porous carbons: Synthesis and applications. *Carbon*, 68:1–32, 2014.

[129] I. B. Klimenko, T. S. Zhuravleva, A. A. Lactionov, and T. B. Komarova. Electrophysical properties of ex-rayon and ex-pan carbon fibers with different temperatures of heat-treatment. *Materials Chemistry and Physics*, 31(4):319–324, 1992.

[130] T. D. Y. Kozai, N. B. Langhals, P. R. Patel, X. Deng, H. Zhang, K. L. Smith, J. Lahann, N. A. Kotov, and D. R. Kipke. Ultrasmall implantable composite microelectrodes with bioactive surfaces for chronic neural interfaces. *Nature Materials*, 11(12):1065–1073, December 2012.

[131] W. Krätschmer, L. D. Lamb, K. Fostiropoulos, and D. R. Huffman. Solid c60: a new form of carbon. *Nature*, 347(6291):354–358, 1990.

[132] H. W. Kroto, J. R. Heath, S. C. O'Brien, R. F. Curl, and R. E. Smalley. C60: Buckminsterfullerene. *Nature*, 318(6042):162–163, 1985.

[133] V. Krstić, G. L. J. A. Rikken, P. Bernier, S. Roth, and M. Glerup. Nitrogen doping of metallic single-walled carbon nanotubes: n-type conduction and dipole scattering. *Europhysics Letters*, 77(3):37001, February 2007.

[134] K. Krukiewicz, D. Janas, C. Vallejo-Giraldo, and M. J. P. Biggs. Self-supporting carbon nanotube films as flexible neural interfaces. *Electrochimica Acta*, 295:253–261, February 2019.

[135] C. V. Kumar and A. Pattammattel. Chapter 1 – discovery of graphene and beyond. In C. V. Kumar and A. Pattammattel, editors, *Introduction to Graphene*, pages 1–15. Elsevier, 2017.

[136] M. Kumar and Y. Ando. Chemical vapor deposition of carbon nanotubes: a review on growth mechanism and mass production. *Journal of Nanoscience and Nanotechnology*, 10(6):3739–3758, 2010.

[137] A. Kwaśniewska, M. Świetlicki, A. Prószyński, and G. Gładyszewski. Physical properties of starch/powdered activated carbon composite films. *Polymers*, 13(24):4406, December 2021.

[138] K. F. Lang, H. Buffleb, and J. Kalowy. *Die Pyrolyse des Naphthalins. Chemische Berichte*, 90(12):2888–2893, December 1957.

[139] L. Larrondo and R. St. J. Manley. Electrostatic fiber spinning from polymer melts. I. Experimental observations on fiber formation and properties. *Journal of Polymer Science. Polymer Physics Edition*, 19(6):909–920, June 1981.

[140] J. W. Lee, Y. M. Choi, K. J. Kong, H. J. Chang, and B. H. Ryu. Single-walled carbon nanotube rope for gas sensor application. *MRS Proceedings*, 788:L4.4, 2003.

[141] S. J. Lee, W. Shi, P. Maciel, and S. W. Cha. Top-edge profile control for su-8 structural photoresist. In *Proceedings of the 15th Biennial University/Government/ Industry Microelectronics Symposium (Cat. No. 03CH37488)*, pages 389–390, 2003.

[142] H. Legall, H. Stiel, P. Nickles, A. Bjeoumikhov, N. Langhoff, M. Haschke, V. Arkadiev, and R. Wedell. Applications of highly oriented pyrolytic graphite (hopg) for x-ray diagnostics and spectroscopy. In *Laser-Generated, Synchrotron, and Other Laboratory X-Ray and EUV Sources, Optics, and Applications II*, volume 5918, 09 2005.

[143] L. Lei, F. Pan, A. Lindbråthen, X. Zhang, M. Hillestad, Y. Nie, L. Bai, X. He, and M. D. Guiver. Carbon hollow fiber membranes for a molecular sieve with precise-cutoff ultramicropores for superior hydrogen separation. *Nature Communications*, 12(1):268, January 2021.

[144] M. I. León, L. F. Castañeda, A. A. Márquez, F. C. Walsh, and J. L. Nava. Review—Carbon cloth as a versatile electrode: manufacture, properties, reaction environment, and applications. *Journal of the Electrochemical Society*, 169(5):053503, May 2022.

[145] B. Lersmacher, H. Lydtin, W. F. Knippenberg, and A. W. Moore. Thermodynamische Betrachtungen zur Kohlenstoffabscheidung bei der pyrolyse Gasförmiger Kohlenstoffverbindungen. *Carbon*, 5(3):205–217, June 1967.

[146] F. Li, A. M. Gañán-Calvo, and J. M. López-Herrera. Absolute-convective instability transition of low permittivity, low conductivity charged viscous liquid jets under axial electric fields. *Physics of Fluids*, 23(9):094108, September 2011.

[147] H. Li, F. Gao, P. Wang, L. Yin, N. Ji, L. Zhang, L. Zhao, G. Hou, B. Lu, Y. Chen, Y. Ma, and X. Feng. Biodegradable flexible electronic device with controlled drug release for cancer treatment. *ACS Applied Materials & Interfaces*, 13(18):21067–21075, May 2021.

[148] Y. Liao, R. Zhang, H. Wang, S. Ye, Y. Zhou, T. Ma, J. Zhu, L. D. Pfefferle, and J. Qian. Highly conductive carbon-based aqueous inks toward electroluminescent devices, printed capacitive sensors and flexible wearable electronics. *RSC Advances*, 9:15184–15189, 2019.

[149] L. Liu, Z. Shen, X. Zhang, and H. Ma. Highly conductive graphene/carbon black screen printing inks for flexible electronics. *Journal of Colloid and Interface Science*, 582:12–21, 2021.

[150] X. Liu and J. T. Guthrie. A review of flexographic printing plate development. *Surface Coatings International. Part B, Coatings Transactions*, 86(2):91–99, 2003.

[151] W. Lu, T. Wang, X. He, K. Sun, Z. Huang, G. Tan, E. G. Eddings, H. Adidharma, and M. Fan. A new method for preparing excellent electrical conductivity carbon nanofibers from coal extraction residual. *Cleaner Engineering and Technology*, 4:100109, 2021.

[152] P. Makushko, E. S. Oliveros Mata, G. S. Cañón Bermúdez, M. Hassan, S. Laureti, C. Rinaldi, F. Fagiani, G. Barucca, N. Schmidt, Y. Zabila, T. Kosub, R. Illing, O. Volkov, I. Vladymyrskyi, J. Fassbender, M. Albrecht, G. Varvaro, and D. Makarov. Flexible magnetoreceptor with tunable intrinsic logic for on-skin touchless human-machine interfaces. *Advanced Functional Materials*, 31(25):2101089, 2021.

[153] E. R. Mamleyev, S. Heissler, A. Nefedov, P. G. Weidler, N. Nordin, V. V. Kudryashov, K. Länge, N. MacKinnon, and S. Sharma. Laser-induced hierarchical carbon patterns on polyimide substrates for flexible urea sensors. *npj Flexible Electronics*, 3(1):2, January 2019.

[154] E. R. Mamleyev, N. Nordin, S. Heissler, K. Länge, N. MacKinnon, and S. Sharma. Flexible Carbon-based Urea Sensor by Laser Induced Carbonisation of Polyimide. In *2018 International Flexible Electronics Technology Conference (IFETC)*, pages 1–6, 2018.

[155] X. Mao, T. Hatton, and G. Rutledge. A review of electrospun carbon fibers as electrode materials for energy storage. *Current Organic Chemistry*, 17(13):1390–1401, June 2013.

[156] S. Martha, N. Dudney, J. Kiggans, and J. Nanda. Electrochemical stability of carbon fibers compared to metal foils as current collectors for lithium-ion batteries. *Journal of the Electrochemical Society*, 159, 08 2012.

[157] J. L. Martins, F. A. Reuse, and S. N. Khanna. Growth and formation of fullerene clusters. *Journal of Cluster Science*, 12(3):513–525, 2001.

[158] J. Millar and C. W. A. Pelling. Improved methods for construction of carbon fibre electrodes for extracellular spike recording. *Journal of Neuroscience Methods*, 110(1):1–8, 2001.

[159] S. A. Mirdehghan. Fibrous polymeric composites. In *Engineered Polymeric Fibrous Materials*, pages 1–58. Elsevier, 2021.

[160] R. Mishra, B. Pramanick, T. K. Maiti, and T. K. Bhattacharyya. Glassy carbon microneedles—new transdermal drug delivery device derived from a scalable C-MEMS process. *Microsystems & Nanoengineering*, 4(1):38, December 2018.

[161] J. Mitra, S. Jain, A. Sharma, and B. Basu. Patterned growth and differentiation of neural cells on polymer derived carbon substrates with micro/nano structures in vitro. *Carbon*, 65:140–155, 01 2013.

[162] R. M. Kakhki. A review to recent developments in modification of carbon fiber electrodes. *Arabian Journal of Chemistry*, 12(7):1783–1794, 2019.

[163] M. Mojica, J. A. Alonso, and F. Méndez. Synthesis of fullerenes: synthesis of fullerenes. *Journal of Physical Organic Chemistry*, 26(7):526–539, July 2013.

[164] J.-Y. Moon, S.-I. Kim, S.-K. Son, S.-G. Kang, J.-Y. Lim, D. K. Lee, B. Ahn, D. Whang, H. K. Yu, and J.-H. Lee. An eco-friendly, CMOS-compatible transfer process for large-scale CVD-graphene. *Advanced Materials Interfaces*, 6(13):1900084, July 2019.

[165] P. Mullai and V. Rajesh. Post treatment of antibiotic wastewater by adsorption on activated carbon. In *AIP Conference Proceedings*, page 020002, Jatinangor, Indonesia, 2018.

[166] H. Murata, Y. Nakajima, N. Saitoh, N. Yoshizawa, T. Suemasu, and K. Toko. High-electrical-conductivity multilayer graphene formed by layer exchange with controlled thickness and interlayer. *Scientific Reports*, 9(1):4068, March 2019.

[167] R. Natu, M. Islam, J. Gilmore, and R. Martinez-Duarte. Shrinkage of su-8 microstructures during carbonization. *Journal of Analytical and Applied Pyrolysis*, 131:17–27, 2018.

[168] L. Nayak, M. Rahaman, and R. Giri. *Surface Modification/Functionalization of Carbon Materials by Different Techniques: An Overview*, pages 65–98. Springer Singapore, Singapore, 2019.

[169] M. M. Nazemi, A. Khodabandeh, and A. Hadjizadeh. Near-field electrospinning: Crucial parameters, challenges, and applications. *ACS Applied Bio Materials*, 5(2):394–412, 02 2022.

[170] T. D. Nguyen and J. S. Lee. Electrospinning-based carbon nanofibers for energy and sensor applications. *Applied Sciences*, 12(12), 2022.

[171] S. Nimbalkar, E. Castagnola, A. Balasubramani, A. Scarpellini, S. Samejima, A. Khorasani, A. Boissenin, S. Thongpang, C. Moritz, and S. Kassegne. Ultra-capacitive carbon neural probe allows simultaneous long-term electrical stimulations and high-resolution neurotransmitter detection. *Scientific Reports*, 8(1):6958, 2018.

[172] J. J. Niu, J. N. Wang, Y. Jiang, L. F. Su, and J. Ma. An approach to carbon nanotubes with high surface area and large pore volume. *Microporous and Mesoporous Materials*, 100(1):1–5, 2007.

[173] K. S. Novoselov, A. K. Geim, S. V. Morozov, D. Jiang, Y. Zhang, S. V. Dubonos, I. V. Grigorieva, and A. A. Firsov. Electric field effect in atomically thin carbon films. *Science*, 306(5696):666–669, October 2004.

[174] A. Numan, Y. Zhan, M. Khalid, and M. Hatamvand. Chapter three – introduction to supercapattery. In N. Arshid, M. Khalid and A. N. Grace, editors, *Advances in Supercapacitor and Supercapattery*, pages 45–61. Elsevier, 2021.

[175] C. Oshima, E. Bannai, T. Tanaka, and S. Kawai. Carbon layer on lanthanum hexaboride (100) surface. *Japanese Journal of Applied Physics*, 16(6):965–969, jun 1977.

[176] C. Oshima and A. Nagashima. Ultra-thin epitaxial films of graphite and hexagonal boron nitride on solid surfaces. *Journal of Physics. Condensed Matter*, 9(1):1–20, jan 1997.

[177] K. Ouchi. Infra-red study of structural changes during the pyrolysis of a phenol-formaldehyde resin. *Carbon*, 4(1):59–66, 1966.

[178] K. Ouchi and H. Honda. Pyrolysis of coal. 1. thermal cracking of phenolformaldehyde resins taken as coal models. *Fuel*, 38(4):429–443, 1959.

[179] D. G. Papageorgiou, I. A. Kinloch, and R. J. Young. Mechanical properties of graphene and graphene-based nanocomposites. *Progress in Materials Science*, 90:75–127, October 2017.

[180] B. Park. *The Voltaic Cell: Its Construction and Its Capacity*. John Wiley and Sons, New York, 1893.

[181] D.-W. Park, A. A. Schendel, S. Mikael, S. K. Brodnick, T. J. Richner, J. P. Ness, M. R. Hayat, F. Atry, S. T. Frye, R. Pashaie, S. Thongpang, Z. Ma, and J. C. Williams. Graphene-based carbon-layered electrode array technology for neural imaging and optogenetic applications. *Nature Communications*, 5(1):5258, October 2014.

[182] M.-S. Park, M.-J. Jung, and Y.-S. Lee. Significant reduction in stabilization temperature and improved mechanical/electrical properties of pitch-based carbon fibers by electron beam irradiation. *Journal of Industrial and Engineering Chemistry*, 37:277–287, 2016.

[183] K. Parvez, Z.-S. Wu, R. Li, X. Liu, R. Graf, X. Feng, and K. Müllen. Exfoliation of graphite into graphene in aqueous solutions of inorganic salts. *Journal of the American Chemical Society*, 136(16):6083–6091, April 2014.

[184] P. R. Patel, P. Popov, C. M. Caldwell, E. J. Welle, D. Egert, J. R. Pettibone, D. H. Roossien, J. B. Becker, J. D. Berke, C. A. Chestek, and D. Cai. High density carbon fiber arrays for chronic electrophysiology, fast scan cyclic voltammetry, and correlative anatomy. *Journal of Neural Engineering*, 17(5):056029, October 2020.

[185] S. B. Patwardhan, S. Pandit, P. K. Gupta, N. K. Jha, J. Rawat, H. C. Joshi, K. Priya, M. Gupta, D. Lahiri, M. Nag, V. K. Thakur, and K. K. Kesari. Recent advances in the application of biochar in microbial electrochemical cells. *Fuel*, 311:122501, 2022.

[186] V. Penmatsa, H. Kawarada, and C. Wang. Fabrication of carbon nanostructures using photo-nanoimprint lithography and pyrolysis. *Journal of Micromechanics and Microengineering*, 22, 04 2012.

[187] P. K. Penumakala, J. Santo, and A. Thomas. A critical review on the fused deposition modeling of thermoplastic polymer composites. *Composites. Part B, Engineering*, 201:108336, 2020.

[188] L. A. Pesin. Review: structure and properties of glass-like carbon. *Journal of Materials Science*, 37:1–28, 2002.

[189] W. J. Peveler, M. Yazdani, and V. M. Rotello. Selectivity and specificity: pros and cons in sensing. *ACS Sensors*, 1(11):1282–1285, November 2016.

[190] A. C. Pina, A. Amaya, J. S. Marcuzzo, A. C. Rodrigues, M. R. Baldan, N. Tancredi, and A. Cuña. Supercapacitor electrode based on activated carbon wool felt. *C*, 4(2), 2018.

[191] V. Poursorkhabi, M. A. Abdelwahab, M. Misra, H. Khalil, B. Gharabaghi, and A. K. Mohanty. Processing, carbonization, and characterization of lignin based electrospun carbon fibers: A review. *Frontiers in Energy Research*, 8:208, September 2020.

[192] S. Prawer and I. Aharonovich, editors. *Quantum Information Processing With Diamond: Principles and Applications*. Woodhead Publishing series in electronic and optical materials, volume 63. Elsevier, Woodhead Publ, Amsterdam Heidelberg, 2014.

[193] X. Qian, J. Zhi, L. Chen, J. Zhong, X. Wang, Y. Zhang, and S. Song. Evolution of microstructure and electrical property in the conversion of high strength carbon fiber to high modulus and ultrahigh modulus carbon fiber. *Composites. Part A, Applied Science and Manufacturing*, 112:111–118, 2018.

[194] L. Qie, W. Chen, X. Xiong, C. Hu, F. Zou, P. Hu, and Y. Huang. Sulfur-doped carbon with enlarged interlayer distance as a high-performance anode material for sodium-ion batteries. *Advanced Science*, 2(12):advs.201500195, December 2015.

[195] J.-X. Qin, X.-G. Yang, C.-F. Lv, Y.-Z. Li, K.-K. Liu, J.-H. Zang, X. Yang, L. Dong, and C.-X. Shan. Nanodiamonds: Synthesis, properties, and applications in nanomedicine. *Materials & Design*, 210:110091, 2021.

[196] L. R. Radovic. *Chemistry & Physics of Carbon*, volume 28. Chemistry and Physics of Carbon. CRC Press, 2003.

[197] L. V. Radushkevich and V. M. Lukyanovich. The structure of carbon forming in thermal decomposition of carbon monoxide on an iron catalyst. *Russian Journal of Physical Chemistry (In Russian)*, 26:88–95, 1952.

[198] S. Sarkar, D. Banerjee, N. S. Das, and K. K. Chattopadhyay. A simple chemical synthesis of amorphous carbon nanotubes–mno2 flake hybrids for cold cathode application. *Applied Surface Science*, 347:824–831, 2015.

[199] R. Schirhagl, K. Chang, M. Loretz, and C. L. Degen. Nitrogen-vacancy centers in diamond: nanoscale sensors for physics and biology. *Annual Review of Physical Chemistry*, 65(1):83–105, April 2014.

[200] S. Shaheen Shah, S. M. Abu Nayem, N. Sultana, A. J. Saleh Ahammad, and Md. A. Aziz. Preparation of sulfur-doped carbon for supercapacitor applications: A review. *ChemSusChem*, 15(1), January 2022.

[201] C. S. Sharma, H. Katepalli, A. Sharma, and M. Madou. Fabrication and electrical conductivity of suspended carbon nanofiber arrays. *Carbon*, 49(5):1727–1732, 2011.

[202] S. Sharma. *Microstructural tuning of glassy carbon for electrical and electrochemical sensor applications*. PhD thesis, University of California, Irvine, CA, USA, December 2013.

[203] S. Sharma. Glassy carbon: A promising material for micro- and nanomanufacturing. *Materials*, 11(10):1857, 2018.

[204] S. Sharma and G. Agrawal. Biomedical applications of electrospun polymer and carbon fibers. In *Encyclopedia of Materials: Plastics and Polymers*, pages 681–696. Elsevier, 2022.

[205] S. Sharma, R. Kamath, and M. Madou. Porous glassy carbon formed by rapid pyrolysis of phenol-formaldehyde resins and its performance as electrode material for electrochemical double layer capacitors. *Journal of Analytical and Applied Pyrolysis*, 108:12–18, July 2014.

[206] S. Sharma and M. Madou. Micro and nano patterning of carbon electrodes for bioMEMS. *Bioinspired, Biomimetic and Nanobiomaterials*, 1:252–265, 2012.

[207] S. Sharma, A. M. Rostas, L. Bordonali, N. MacKinnon, S. Weber, and J. G. Korvink. Micro and nano patternable magnetic carbon. *Journal of Applied Physics*, 120(23):235107, 2016.

[208] S. Sharma, A. Sharma, Y.-K. Cho, and M. Madou. Increased graphitization in electrospun single suspended carbon nanowires integrated with carbon-MEMS and carbon-NEMS platforms. *ACS Applied Materials & Interfaces*, 4(1):34–39, January 2012.

[209] S. Sharma, A. Sharma, Y.-K. Cho, and M. Madou. Increased graphitization in electrospun single suspended carbon nanowires integrated with carbon-mems and carbon-nems platforms. *ACS Applied Materials & Interfaces*, 4:34–39, 2012.

[210] S. Sharma, C. N. Shyam Kumar, J. G. Korvink, and C. Kübel. Evolution of glassy carbon microstructure: In situ transmission electron microscopy of the pyrolysis process. *Scientific Reports*, 8(1):16282, 2018.

[211] S. Sharma, S. Zorzi, V. Cristiglio, R. Schweins, and C. Mondelli. Quantification of buckminsterfullerene (c60) in non-graphitizing carbon and a microstructural comparison of graphitizing and non-graphitizing carbon via small angle neutron scattering. *Carbon*, 189:362–368, 2022.

[212] O. A. Shenderova, V. V. Zhirnov, and D. W. Brenner. Carbon nanostructures. *Critical Reviews in Solid State and Materials Sciences*, 27(3–4):227–356, 2002.

[213] M. Singh and R. L. Vander Wal. Nanostructure quantification of carbon blacks. *C*, 5(1), 2019.

[214] W. Smith. The role of fuel cells in energy storage. *Journal of Power Sources*, 86(1):74–83, 2000.

[215] H. Staudinger. Über polymerisation. *Berichte der deutschen chemischen Gesellschaft (A and B Series)*, 53(6):1073–1085, 1920.

[216] G. Sun, X. Wang, and P. Chen. Microfiber devices based on carbon materials. *Materials Today*, 18(4):215–226, 2015.

[217] S. Suresh, A. Becker, and B. Glasmacher. Impact of apparatus orientation and gravity in electrospinning—a review of empirical evidence. *Polymers*, 12(11), 2020.

[218] N. Tajima, T. Kaneko, J. Nara, and T. Ohno. A first principles study on the cvd graphene growth on copper surfaces: a carbon atom incorporation to graphene edges. *Surface Science*, 653:123–129, 2016.

[219] Z. Tan, K. Ni, G. Chen, W. Zeng, Z. Tao, M. Ikram, Q. Zhang, H. Wang, L. Sun, X. Zhu, X. Wu, H. Ji, R. S. Ruoff, and Y. Zhu. Incorporating pyrrolic and pyridinic nitrogen into a porous carbon made from C_{60} molecules to obtain superior energy storage. *Advanced Materials*, 29(8):1603414, February 2017.

[220] J. Tang, Y. Wu, S. Ma, T. Yan, and Z. Pan. Flexible strain sensor based on CNT/TPU composite nanofiber yarn for smart sports bandage. *Composites. Part B, Engineering*, 232:109605, March 2022.

[221] M. Tawalbeh, S. Alarab, A. Al-Othman, and R. M. N. Javed. The operating parameters, structural composition, and fuel sustainability aspects of PEM fuel cells: A mini review. *Fuels*, 3(3):449–474, August 2022.

[222] G. I. Taylor. Disintegration of water drops in an electric field. *Proceedings of the Royal Society of London. Series A, Mathematical and Physical Sciences*, 280(1382):383–397, July 1964.

[223] G. I. Taylor. Electrically driven jets. *Proceedings of the Royal Society of London. Series A, Mathematical and Physical Sciences*, 313(1515):453–475, December 1969.

[224] A. Thess, R. Lee, P. Nikolaev, H. Dai, P. Petit, J. Robert, C. Xu, Y. H. Lee, S. G. Kim, A. G. Rinzler, D. T. Colbert, G. E. Scuseria, D. Tománek, J. E. Fischer, and R. E. Smalley. *Crystalline Ropes of Metallic Carbon Nanotubes*. Science, 273(5274):483–487, July 1996.

[225] A. Thiha, F. Ibrahim, S. Muniandy, I. J. Dinshaw, S. J. Teh, K. L. Thong, B. F. Leo, and M. Madou. All-carbon suspended nanowire sensors as a rapid highly-sensitive label-free chemiresistive biosensing platform. *Biosensors & Bioelectronics*, 107:145–152, 2018.

[226] N. Thongsai, K. Hrimchum, and D. Aussawasathien. Carbon fiber mat from palm-kernel-shell lignin/polyacrylonitrile as intrinsic-doping electrode in supercapacitor. *Sustainable Materials and Technologies*, 30:e00341, 2021.

[227] P. A. Thrower and L. A. Radovic. *Chemistry and Physics of Carbon: a Series of Advances*, volume 26. Marcel Dekker, New York, 1999. OCLC: 934009555.

[228] W. Tian, W. Li, W. Yu, and X. Liu. A review on lattice defects in graphene: types, generation, effects and regulation. *Micromachines*, 8(5):163, May 2017.

[229] G. G. Tibbetts. Vapor-grown carbon fibers: Status and prospects. *Carbon*, 27(5):745–747, 1989. Carbon Fibers and Composites.

[230] G. G. Tibbetts, G. L. Doll, D. W. Gorkiewicz, J. J. Moleski, T. A. Perry, C. J. Dasch, and M. J. Balogh. Physical properties of vapor-grown carbon fibers. *Carbon*, 31(7):1039–1047, 1993.

[231] R. Tortorich and J.-W. Choi. Inkjet printing of carbon nanotubes. *Nanomaterials*, 3(3):453–468, July 2013.

[232] M. M. J. Treacy, T. W. Ebbesen, and J. M. Gibson. Exceptionally high Young's modulus observed for individual carbon nanotubes. *Nature*, 381(6584):678–680, June 1996.

[233] K. Tse-Hao, Y. Ching-Chyuan, and C. Wen-Tong. The effect of stabilization on the properties of pan-based carbon films. *Carbon*, 31(4):583–590, 1993.

[234] S. Ullah, X. Yang, H. Q. Ta, M. Hasan, A. Bachmatiuk, K. Tokarska, B. Trzebicka, L. Fu, and M. H. Rummeli. Graphene transfer methods: A review. *Nano Research*, 14(11):3756–3772, November 2021.

[235] A. R. Urade, I. Lahiri, and K. S. Suresh. Graphene properties, synthesis and applications: A review. *JOM*, 75(3):614–630, 2023.

[236] V. Uskoković. A historical review of glassy carbon: synthesis, structure, properties and applications. *Carbon Trends*, 5:100116, 2021.

[237] L. Vaisman, H. D. Wagner, and G. Marom. The role of surfactants in dispersion of carbon nanotubes. *Advances in Colloid and Interface Science*, 128–130:37–46, 2006. In Honor of Professor Nissim Garti's 60th Birthday.

[238] D. Valdec, P. Miljkovic, and B. Auguštin. The influence of printing substrate properties on color characterization in flexography according to the iso specifications. *Technical Sciences*, 11(3):73–77, 09 2017.

[239] J. A. Viecelli, S. Bastea, J. N. Glosli, and F. H. Ree. Phase transformations of nanometer size carbon particles in shocked hydrocarbons and explosives. *Journal of Chemical Physics*, 115(6):2730–2736, 08 2001.

[240] M. Vomero, E. Castagnola, F. Ciarpella, E. Maggiolini, N. Goshi, E. Zucchini, S. Carli, L. Fadiga, S. Kassegne, and D. Ricci. Highly stable glassy carbon interfaces for long-term neural stimulation and low-noise recording of brain activity. *Scientific Reports*, 7(1):40332, January 2017.

[241] M. Vomero, C. Gueli, E. Zucchini, L. Fadiga, J. B. Erhardt, S. Sharma, and T. Stieglitz. Flexible bioelectronic devices based on micropatterned monolithic carbon fiber mats. *Advanced Materials Technologies*, 5(2):1900713, 2020.

[242] M. Vomero, A. Oliveira, D. Ashouri, M. Eickenscheidt, and T. Stieglitz. Graphitic carbon electrodes on flexible substrate for neural applications entirely fabricated using infrared nanosecond laser technology. *Scientific Reports*, 8(1):14749, October 2018.

[243] P. L. Walker. *Chemistry and physics of carbon: a series of advances*, volume 1. Marcel Dekker, Inc., New York, 1965.

[244] P. L. Walker. *Chemistry and physics of carbon: a series of advances*, volume 7. Dekker, New York, N.Y, 1971. OCLC: 85983795.

[245] P. L. Walker. *Chemistry and physics of carbon: a series of advances*, volume 11. M. Dekker, New York, 1973. OCLC: 1285458794.

[246] P. R. Wallace. The band theory of graphite. *Physical Review*, 71:622–634, May 1947.

[247] L.-S. Wang, J. Conceicao, C. Jin, and R. E. Smalley. Threshold photodetachment of cold c-60. *Chemical Physics Letters*, 182:5–11, 07 1991.

[248] L. Wang, M. Borghei, A. Ishfaq, P. Lahtinen, M. Ago, A. C. Papageorgiou, M. J. Lundahl, L.-S. Johansson, T. Kallio, and O. J. Rojas. Mesoporous carbon microfibers for electroactive materials derived from lignocellulose nanofibrils. *ACS Sustainable Chemistry & Engineering*, 8(23):8549–8561, 2020.

[249] Y. Wang, I. Ramos, R. Furlan, and J. J. Santiago-Aviles. Electronic transport properties of incipient graphitic domains formation in PAN derived carbon nanofibers. *IEEE Transactions on Nanotechnology*, 3(1):80–85, March 2004.

[250] Z. Wang, S. Wu, J. Wang, A. Yu, and G. Wei. Carbon nanofiber-based functional nanomaterials for sensor applications. *Nanomaterials*, 9(7), 2019.

[251] D. Wei, Y. Liu, Y. Wang, H. Zhang, L. Huang, and G. Yu. Synthesis of n-doped graphene by chemical vapor deposition and its electrical properties. *Nano Letters*, 9(5):1752–1758, 05 2009.

[252] G. R. Williams, B. T. Raimi-Abraham, and C. J. Luo. *Nanofibres in Drug Delivery*. UCL Press, September 2018.

[253] M. Winter and R. J. Brodd. What are batteries, fuel cells, and supercapacitors? *Chemical Reviews*, 104(10):4245–4270, 10 2004.

[254] M. Worgull. Chapter 7 – hot embossing technique. In M. Worgull, editor, *Hot Embossing*, Micro and Nano Technologies, pages 227–245. William Andrew Publishing, Boston, 2009.

[255] K. Wu, X. Tang, E. J. An, Y. H. Yun, S.-J. Kim, H. C. Moon, H. Kong, S. H. Kim, and Y. J. Jeong. Screen printing of graphene-based nanocomposite inks for flexible organic integrated circuits. *Organic Electronics*, 108:106603, 2022.

[256] Y. Wu, W. Liu, Y. Wang, X. Hu, Z. He, X. Chen, and Y. Zhao. Enhanced removal of antibiotic in wastewater using liquid nitrogen-treated carbon material: Material properties and removal mechanisms. *International Journal of Environmental Research and Public Health*, 15(12), 2018.

[257] F. Xu. Braided rope sensor based on carbon nanotube yarn. *Journal of Textile Science & Fashion Technology*, 3(3), July 2019.

[258] R. Yang, W. Zhang, N. Tiwari, H. Yan, T. Li, and H. Cheng. Multimodal sensors with decoupled sensing mechanisms. *Advanced Science*, 9(26):2202470, 2022.

[259] A. L. Yarin, S. Koombhongse, and D. H. Reneker. Bending instability in electrospinning of nanofibers. *Journal of Applied Physics*, 89(5):3018–3026, March 2001.

[260] G. R. Yazdi, T. Iakimov, and R. Yakimova. Epitaxial graphene on sic: A review of growth and characterization. *Crystals*, 6(5), 2016.

[261] M. Yi and Z. Shen. A review on mechanical exfoliation for the scalable production of graphene. *Journal of Materials Chemistry A*, 3:11700–11715, 2015.

[262] Y. Yin, C. Guo, H. Li, H. Yang, F. Xiong, and D. Chen. The progress of research into flexible sensors in the field of smart wearables. *Sensors*, 22(14), 2022.

[263] M.-F. Yu, B. S. Files, S. Arepalli, and R. S. Ruoff. Tensile loading of ropes of single wall carbon nanotubes and their mechanical properties. *Physical Review Letters*, 84(24):5552–5555, June 2000.

[264] P. Zaccagnini, C. Ballin, M. Fontana, M. Parmeggiani, S. Bianco, S. Stassi, A. Pedico, S. Ferrero, and A. Lamberti. Laser-induced graphenization of PDMS as flexible electrode for microsupercapacitors. *Advanced Materials Interfaces*, 8(23):2101046, December 2021.

[265] A. Zakhurdaeva, P.-I. Dietrich, H. Hölscher, C. Koos, J. G. Korvink, and S. Sharma. Custom-designed glassy carbon tips for atomic force microscopy. *Micromachines*, 8(9), 2017.

[266] H. Zhang, W. Han, K. Xu, Y. Zhang, Y. Lu, Z. Nie, Y. Du, J. Zhu, and W. Huang. Metallic sandwiched-aerogel hybrids enabling flexible and stretchable intelligent sensor. *Nano Letters*, 20(5):3449–3458, 05 2020.

[267] S. Zhang and N. Pan. Supercapacitors performance evaluation. *Advanced Energy Materials*, 5(6):1401401, March 2015.

[268] S. Zhang, A. Hao, N. Nguyen, A. Oluwalowo, Z. Liu, Y. Dessureault, J. G. Park, and R. Liang. Carbon nanotube/carbon composite fiber with improved strength and electrical conductivity via interface engineering. *Carbon*, 144:628–638, 2019.

[269] Z. Zhang, W. Yang, L. Cheng, W. Cao, M. Sain, J. Tan, A. Wang, and H. Jia. Carbon fibers with high electrical conductivity: laser irradiation of mesophase pitch filaments obtains high graphitization degree. *ACS Sustainable Chemistry & Engineering*, 8(48):17629–17638, 2020.

[270] Z.-L. Zhao, H.-P. Zhao, J.-S. Wang, Z. Zhang, and X.-Q. Feng. Mechanical properties of carbon nanotube ropes with hierarchical helical structures. *Journal of the Mechanics and Physics of Solids*, 71:64–83, 2014.

[271] J. Zhu, P. Xiao, H. Li, and S. A. C. Carabineiro. Graphitic carbon nitride: synthesis, properties, and applications in catalysis. *ACS Applied Materials & Interfaces*, 6(19):16449–16465, October 2014.

[272] X. Zhu, Q. Li, S. Qiu, X. Liu, L. Xiao, X. Ai, H. Yang, and Y. Cao. Hard carbon fibers pyrolyzed from wool as high-performance anode for sodium-ion batteries. *JOM*, 68, 08 2016.

[273] X. Zhu, B. Bruijnaers, T. V. Lourençon, and M. Balakshin. Structural analysis of lignin-based furan resin. *Materials*, 15(1), 2022.

[274] Y. Zhu, S. Murali, M. D. Stoller, K. J. Ganesh, W. Cai, P. J. Ferreira, A. Pirkle, R. M. Wallace, K. A. Cychosz, M. Thommes, D. Su, E. A. Stach, and R. S. Ruoff. Carbon-based supercapacitors produced by activation of graphene. *Science*, 332(6037):1537–1541, June 2011.

Index

3D-printing 66, 215

Activated carbon 53, 94
Allotrope 43
– Allotrope conversion 47
– Allotropes of carbon 43
– Classification 45
– Primary 45
– Secondary 45
Amorphous material 34
Analyte 20, 23, 213
Anisotropic material 35
Arc discharge 137

Battery 12, 13, 19, 83, 99
– Capacity 14
– Efficiency 14
– Li-ion battery 13
– Lifetime 14
– Ragone plot 20
– Rating 14
Bioelectronic medicine 25
Biofuel 75
Biogas 75
Biomedical device 25
Bottom-up manufacturing 9

Carbon black 122
– Manufacturing of 139
Carbon cloth 192
Carbon fiber 53
– Classification 181
– Electrical conductivity 177
– General purpose carbon fiber 184
– High-performance 184
– Mechanical properties 178
– Thermal conductivity 181
Carbon Fiber Reinforced Carbon, CFRC 18
Carbon Fiber Reinforced Plastic 157
Carbon Fiber Reinforced Polymer, CFRP 18
Carbon nanotube 53, 117
– Chirality 119
– nm notations 118
Carbon nitride 211
– Graphitic carbon nitride 213
Carbon yield 82, 86

Carbonization 78
Carbyne 41, 42
Cell culture 25
Charring 87
Chemical vapor deposition 125
– Boundary layer 133
– Graphene 129
– Mechanism 126
– of carbon filaments 130
– of diamond film 131
– Reactor 132
Coking 87
Composite 4
– Carbon fiber reinforced 158
Composition diagram 46
Crystalline material 34
Cytocompatibility 26

Deep carbon 81
Device 1, 2, 5
Diamond 38
Diamond-like carbon 131
Disordered material 34
Dry-jet wet spinning 167

Electrochemical cell 12
Electrolyzer 11, 18
Electrospinning 161
– Bending instabilities 164
– Collector 165
– Near-field 164
– Taylor cone 162
Etching 68

Flexible electronics 195
– Carbon-based 197
– Fabrication techniques 198
– Substrate 196
Fuel cell 11, 17–19, 99
– Air stoichiometry 19
– Efficiency 19
– Hydrogen 17
– Hydrogen evolution reaction 18
– Hydrogen oxidation reaction 17
– Oxygen evolution reaction 18
– Oxygen reduction reaction 17
Fullerene 40, 121

https://doi.org/10.1515/9783110620634-009

Glass-like carbon 53, 89, 93
Graphene 53, 112
– Crystal structure 114
– Defects 115
– Epitaxial growth 136
– Exfoliation 134
– Nomenclature 116
Graphite 39, 43, 53
– Hexagonal 39
– Rhombohedral 44
Graphitization 78
Graphitizing carbon 88

Highly oriented pyrolytic graphite 94, 126
Hot-embossing 63
Hybridization 37, 38, 45
– Rehybridization 41
– sp 37, 41
– sp^2 37
– sp^{2+n} 37
– sp^3 37
Hydrothermal synthesis 213

In-situ TEM 92
Intercalation 15
Isotopes of carbon 36
Isotropic material 35

Laser-patterned carbon 53, 202
– Carbonization mechanism 203
– Nomenclature 203
– Precursor 208
Lignin 174
Lithography 57
– 2.5-dimensional 58
– Electron-beam lithography 67
– Nano-imprint lithography 62
– Photolithography 58
– Two-photon lithography 66
– X-Ray LIGA 63
Lonsdaleite 43

Melt-spinning 166
Micro-Electro-Mechanical-Systems 24
Microfabrication 3, 96
Microstructure 6, 7
Microsystems engineering 3

N-containing carbon 211

Nano-Electro-Mechanical-Systems 24
Nanoengineering 3
Nanofabrication 3
Nanomanufacturing 3
Nanomaterial 1–3, 34, 108
Nanoscience 3, 8
Nanostructure 3
Nanotechnology 3, 8
Non-graphitizing carbon 88
– Microstructural model 89

Pattern transfer 152
Phase diagram 50
Pi-orbital axis vector POAV 41
Pitch 171
Polyacrylonitrile 168
Polycrystalline material 34
Polymer-derived carbon 80, 86
Polymorph 43
Porous carbon 94
Powers of ten 4
Printing 140
– Carbon ink 141
– Flexography 150
– Gravure 148
– Inkjet 144
– Screen 146
Pyramidalization 41
– Pyramidalization angle 40
Pyrolysis 71, 72
– Characterization 78
– of cellulose 87
– of furfuryl alcohol 73
– of gaseous hydrocarbon 126
– of naphthalene 73
– of phenol-formaldehyde resin 73
– of Urban Solid Waste 76, 217
Pyrolytic graphite 94

Quantum dots 213

Rayon 173

S-containing carbon 214
Scaffold 26
Semimetal 4
Sensor 20–24
– Biosensor 21, 25, 100
– Chemosensor 21

– Colorimetric 21
– Cross-sensitivity 23
– Electrochemical 21
– Limit of detection 23
– Linearity 24
– Multi-modal sensor 21, 23
– Optical 21
– Recovery time 24
– Resolution (of a sensor) 24
– Response time 24
– Selectivity 23
– Sensitivity 23
– Specificity 23
– Thermal 21
– Topography 101
Solubility of carbon 127
Solvothermal synthesis 213
Soot 123

Spray coating 151
Sputtering 68
Stone–Wales defect 115
Supercapacitor 14–16, 19, 99
– Electrochemical Double Layer Capacitor, EDLC 14
– Equivalent series resistance 16
– Pseudocapacitor 15
– Specific capacitance 16
Synchrotron 64
Syngas 75

Top-down manufacturing 9

Vapor deposited diamond 131
Viscoelasticity 175

Waste-derived carbon 217
Wet spinning 167

www.ingramcontent.com/pod-product-compliance
Lightning Source LLC
Chambersburg PA
CBHW061403210326

41598CB00035B/6079